Also in the Variorum Collected Studies Series:

JERRY STANNARD
Herbs and Herbalism in the Middle Ages and Renaissance

WILLIAM A. WALLACE
Galileo, the Jesuits, and the Medieval Aristotle

BRUCE S. EASTWOOD
Astronomy and Optics from Pliny to Descartes: Texts, Diagrams and Conceptual Studies

MARGARET DEACON
Scientists and the Sea, 1650–1900: A Study of Marine Science

URSULA LAMB
Cosmographers and Pilots of the Spanish Maritime Empire

ALLAN CHAPMAN
Astronomical Instruments and Their Users: Tycho Brahe to William Lassell

CRAIG G. FRASER
Calculus and Analytical Mechanics in the Age of Enlightenment

SILVIO A. BEDINI
Patrons, Artisans and Instruments of Science,1600–1750

CURTIS WILSON
Astronomy from Kepler to Newton

W.G.L. RANDLES
Geography, Cartography and Nautical Science in the Renaissance: The Impact of the Great Discoveries

A. RUPERT HALL
Science and Society: Historical Essays on the Relations of Science, Technology and Medicine

A.J. TURNER
Of Time and Measurement: Studies in the History of Horology and Fine Technology

A. RUPERT HALL
Newton, his Friends and his Foes

JOHN GASCOIGNE
Science, Politics and Universities in Europe, 1600–1800

VARIORUM COLLECTED STUDIES SERIES

Renaissance Astrolabes and their Makers

Professor Gerard L'E. Turner

Gerard L'E. Turner

Renaissance Astrolabes and their Makers

Ashgate
VARIORUM

Published in the Variorum Collected Studies Series by

Ashgate Publishing Limited
Gower House, Croft Road,
Aldershot, Hampshire
GU11 3HR
Great Britain

Ashgate Publishing Company
Suite 420
101 Cherry Street
Burlington, VT 05401–4405
USA

Ashgate website: http://www.ashgate.com

ISBN 0–86078–903–9

British Library Cataloguing-in-Publication Data
Turner, Gerard L'E. (Gerard L'Estrange)
Renaissance astrolabes and their makers. - (Variorum collected studies series)
1. Astrolabes - History - To 1500 2. Astrolabes - History - 16th century 3. Astrolabes - Design and construction
I. Title
681.7'5'09031

US Library of Congress Cataloging-in-Publication Data
Turner, Gerard L'Estrange.
Renaissance astrolabes and their makers / Gerard L'E. Turner.
p. cm. – (Variorum collected studies series ; 766)
Includes bibliographical references and index (alk. paper).
1. Astrolabes. 2. Astronomy, Renaissance–Instruments. 3. Instrument manufacture–Europe–History–Modern period, 1500–Biography.
I. Title. II. Collected studies ; CS766.
QB85.T85 2003
522'.4–dc21 2002043963

The paper used in this publication meets the minimum requirements of the American National Standard for Information Sciences – Permanence of Paper for Printed Library Materials, ANSI Z39.48–1984. ∞ ™

Printed and bound in Great Britain by TJ International Ltd, Padstow, Cornwall

VARIORUM COLLECTED STUDIES SERIES CS766

CONTENTS

This volume contains xii + 294 pages

PREFACE

The astrolabe was the pre-eminent astronomical instrument of the Middle Ages and the Renaissance. It is a model of the universe that one can hold in one's hands, representing the heavens with respect to the horizon of a given locality. The instrument is in the form of a brass disc, with the rim divided into 360°. Held vertically, one takes an altitude reading of the sun or of a star with a sighting device on the back. On the front is a stereographic projection of the globe appropriate to the observer's latitude, and the pointer for the sun or star is set according to the measured altitude. Thus the configuration of the heavens at the time of the sighting is fixed for various calculations in astronomy or astrology, especially time-telling by day or night.

The origins of the astrolabe are lost in time, although they will lie in the Classical period. The earliest to remain is from the ninth century, and is Islamic. That there was a Byzantine tradition is attested by a fine astrolabe from Asia Minor dated 1062, which shows no Islamic influence. In 1962 there was published a Western astrolabe, known as the 'Carolingian', that has been thought to be datable to AD 980, but the present author considers it to be 1100 or later. The West European astrolabe evolved from this time until the early seventeenth century, after which it rapidly died out through the much superior instruments based on the telescope and precision machine dividing of the degree scales. The Islamic astrolabe continued to be produced in its traditional form until AD 1900. Because the astrolabe has existed for so long, it is a unique source for the transmission of culture, and the artefact itself is the only means of access to the development of the ideas, usage, and craft skills. Astrolabes are often very beautiful, in addition to being 'ideas made brass'.

In the study of scientific instruments a very different approach is required when dealing with instruments before and after 1700. During more modern times there are printed books, the proceedings of learned societies, descriptive pamphlets, advertisements, and then photographs and vastly more printed material. For the earlier period there is very little other than a limited amount of printed matter. And with the period before 1600, there is really nothing to speak of except the occasional manuscript. But there *are* the instruments themselves, and it is these we must learn to read, for they are full of all types of information, mathematical, astronomical, metallurgical, as well as providing knowledge on craftsmanship, and economic and social developments.

Since written information about makers of scientific instruments is so rare, particularly in the early period, our knowledge of the workshops producing such artefacts, and their locations, is bound to depend largely on examination of the instruments themselves. Such examination, in particular the study of the letters and numbers engraved or stamped on these brass objects, can be the most important aid to attributing unsigned instruments to known makers, through comparison with signed ones. This makes it possible to establish a larger corpus of instruments for a given period and location. With a few signed examples it is possible to define the characteristics of manufacture, and then to apply these to unsigned instruments. This technique has worked well with English Elizabethan scientific artefacts, where about 30 signed instruments have been used to identify nearly 90 unsigned ones as belonging to the period 1550–1600.

In Florence, four signed quadrants and a nocturnal have made possible the identification of 26 instruments, comprising 8 astrolabes, 5 plain theodolites, 7 quadrants, and 5 others as products of the workshop of Giovan Battista Giusti, working between 1556 and 1575. The finest item is the very large astrolabe– nearly a metre in diameter – that is often called Galileo's astrolabe. As Galileo will have been about 5 years old when the instrument was constructed, he cannot have had any part in its making.

In February 1992, at the Istituto e Museo di Storia della Scienza in Florence, an astrolabe with six latitude plates and a map plate was examined that Dr Elly Dekker and I were able to attribute conclusively to Gerard Mercator. We postulated that the astrolabe itself and five of the plates are by the hand of Gerard, while the map plate, although definitely from his workshop, may have been engraved by his son, Rumold. The sixth latitude plate was made in Florence by a local craftsman for the exact latitude of the city, and the craftsman has been identified as Giusti, thus helping to narrow the dating spectrum to around 1570.

In the case of the Mercator workshops at Louvain and Duisburg, a rather different method of comparison is needed. There were thought to be no surviving instruments signed by Gerard Mercator or either of his sons. What is known, however, is that he did make astrolabes and other instruments, some for the Holy Roman Emperor, Charles V. His globes and maps are, of course, available for study. These provide their evidence through printing, so the engraved copperplates for the gores and maps provide the link with the engravings on brass instruments, but in mirror image. The printing process removes the sharpness of the edges and the depth of cuts with the burin, but the style of letters and numbers, and the skill, are clear. Although the Mercator astrolabe in Florence can be clearly identified in a late-eighteenth-century inventory, it cannot be identified in earlier inventories because the entries are not sufficiently exact and detailed.

The attribution of this unsigned instrument rests upon detailed comparisons between the calligraphy on Mercator's maps and globes, and the lettering and numbering on the astrolabe and its plates. In addition, the positional data of stars plotted on the rete of the astrolabe have been used to check whether this information is consistent with the conclusion that Mercator was the maker. The results have revealed that the astrolabe's star positions correspond closely to those used on Mercator's celestial globes.

Another question that had to be answered is whether this unsigned astrolabe in Florence could have been the product of the famous Arsenius workshop in Louvain, from which signed examples are of outstanding workmanship, comparable to that of the astrolabe under study. Five features were selected for comparison between Arsenius astrolabes and that attributed to Mercator: the selection of stars and their names, the pattern of the rete, certain constructional techniques, the calligraphy, and the design of the throne. On the basis of these features, it was possible confirm the attribution. Shortly after the identification of the astrolabe in Florence, two more astrolabes made by Mercator were found. Two papers on Mercator's astrolabes are reproduced in the present volume. The great Flemish cartographer has also been found to be linked to the origin of the scientific instrument trade in London through the use on an astrolabe made by Thomas Gemini for King Edward VI, of astrological information presented in the same form as on a disc published by Mercator in 1551.

An intriguing problem concerned a Germanic astrolabe dated 1462. When the late Derek Price published in 1959 his paper on this small astrolabe, thought to be by Regiomontanus, he called it 'The first instrument of the Renaissance'. Other people thought it was not genuine, as it appeared (as some claimed) to have been copied from the mid-16th century astrolabes of Georg Hartmann, of Nuremberg. After a journey through France, Germany, and Austria, in the early 1990s, Professor David King of Frankfurt and myself found and examined ten more astrolabes like the Regiomontanus example. One was like a brother, and the others were cousins at least. Those that bore a date placed the group in the period between 1457 and 1482. Again through a detailed examination of many elements of their construction, it has been shown that this group points to a German tradition of the mid-fifteenth century, and suggests a craft centre that may be Vienna. Therefore, contrary to the Regiomontanus astrolabe being false, and based on Hartmann, it has been shown that Hartmann was continuing a pattern of astrolabe that had originated in Germany some 80 years before he began to trade. Thus a detailed study of the Regiomontanus astrolabe has resulted in a newly-identified corpus of historical material.

In a number of fields, such as archaeology, numismatics, and the decorative arts, objects have been examined systematically, and displayed to show variations with time. But within the history of science, surprisingly little attention has been paid to instruments, and those studies that have been made are quite recent. Universities have been taking an increasing interest in scientific history since the middle of the twentieth century, but scholars have argued along philosophical and theoretical lines, rather than giving thought to the data-creating tools of science. Instruments have much to tell us about economic and social history, as well as how science was actually done. The study of who made what, where, when, and why, in the precision technology field, provides remarkable new insights, but it is only possible if collections of apparatus are preserved, maintained, and catalogued with expert knowledge. Such work is infinitely rewarding, as the instrument historian shares the excitement of the archaeologist in letting the artefacts tell their secrets to the receptive mind.

G.L'E. TURNER

July 2002

ACKNOWLEDGEMENTS

Five of the papers reproduced in this selection were joint publications, and I thank all my collaborators. Professor David King and I established the genuineness of an astrolabe dated 1462 that had been queried unjustifiably (IV). My identification in 1993 of a large astrolabe in the Museo di Storia della Scienza, Florence, as by the hand of Gerard Mercator (VII) owed much to the astronomical knowledge of Dr Elly Dekker. Her expertise was again called on for my study of the Elizabethan silver globe by Charles Whitwell (XI). I explored the links between the London instrument maker, Thomas Gemini, and Mercator with the co-operation of Dr K. Van Cleempoel. Finally, Dr A. Morrison-Low provided information for my critique of the dating of certain German instruments by Ernst Zinner (XII). The many people who provided help in various ways are acknowledged in each paper published here. As always, my wife, Helen Turner, and my family, have given unstinting support.

People who study instruments rely on the co-operation of museums where they work, and I thank the Director, Professor Paolo Galluzzi, Dr Mara Miniati, and other staff at the Museo di Storia della Scienza, Florence, where much of my research has been done. I am also grateful for funding support from the Renaissance Trust, and the Leverhulme Trust.

Permission to reprint the articles was generously granted by the following journals and presses: British Museum Press, London (I); Leo S. Olschki Editore (II, IV, V, VI); Harrassowitz, Wiesbaden (III); Taylor & Francis Ltd, London (VII, VIII) [*Annals of Science* http://www.tandf.co.uk]; The Society of Antiquaries of London (IX, XI); Christie Manson & Woods Ltd (X); The Scientific Instrument Society (XII).

PUBLISHER'S NOTE

The articles in this volume, as in all others in the Variorum Collected Studies Series, have not been given a new, continuous pagination. In order to avoid confusion, and to facilitate their use where these same studies have been referred to elsewhere, the original pagination has been maintained wherever possible.

Each article has been given a Roman number in order of appearance, as listed in the Contents. This number is repeated on each page and is quoted in the index entries.

I

Later Medieval and Renaissance Instruments

Astronomy is the systematic study of the heavens, and is the foundation of all the sciences. It was necessary for time-telling, for keeping a record of the seasons, and for guidance in travel by sea and land. In view of the universally visible marvels of the heavenly bodies in the night sky, it is not surprising that all early human societies believed that the stars influenced people's fate. Therefore, alongside what we today would see as the scientific study of the heavens, and closely related to it, there existed prognostication, or astrology, the study of the same phenomena for the purpose of casting horoscopes and predicting future events. The first practitioners of astronomy and astrology used simple instruments and methods of computation. The earliest observatories were monumental in scale, using natural features allied with stone obelisks as foresights and backsights. As the requirement for greater precision grew, instruments of a more sophisticated kind gradually came to be designed and used.

The true watershed for serious astronomical observation came, however, at the end of the fifteenth century when Christopher Columbus set sail from Spain to the West in 1492. The discovery of the Americas created the economic requirement for accurate astronomical observation, particularly related to some means of finding the longitude at sea when out of sight of land. For over two centuries this constituted an intractable problem, the final solution of which had to wait till the design of the marine chronometer by John Harrison (1693–1776) in the mid eighteenth century. But from the end of the fifteenth century, it became of great practical importance to make accurate astronomical observations, and for this purpose instruments were devised for use in the observatory, by navigators and by land surveyors.

Though scientific instruments of engraved brass were made from the thirteenth century in Europe, it was not until the second half of the fifteenth century that the trade began to develop into a serious and lucrative one. A major event that contributed to the change in the approach to astronomical observation was the invention of the printing process in the 1450s. The skills in calligraphy and the engraving that this revolutionary process generated, associated with the economic pressures that resulted from the age of discovery, formed a base for the precision instrument-making trade in Europe. The pioneers were Gemma Frisius (1508–55) and Gerard Mercator (1512–94), of Louvain in the Low Countries, whose influence in globe, map and instrument making spread rapidly across the continent of Europe and to England.

To summarise, there occurred, in the period under discussion, a transformation in world view that is embodied in the whole meaning of the word 'Renaissance'. This profoundly affected the whole concept of cosmology, and so the way in which the heavens were observed and studied. From the point of view of instrumentation, this change produced an overriding requirement for greater accuracy, so that more, and above all, more precise instruments were needed.

Astronomical instruments

The armillary sphere and celestial globe

The armillary sphere was the earliest of demonstration and teaching instruments associated with astronomy. The concept of a spherical earth and sky is by no means self-evident, but it was generally accepted in the Greek world by the time of Aristotle (382–322 BC). The Ptolemaic world system placed the earth in the centre of a series of nested spheres representing the planets and the starry sky. It was this system that the armillary sphere reproduced. Its basic form was a series of metal or wooden rings or circles representing the equator, ecliptic, tropics, polar circles, solsticial colure, and other meridian circles, generally with the polar axis extended to form a handle. Its purpose was to represent the position and motion of those circles, and to explain how they were related to one another. In its later, more elaborate form, the armillary was supported in a stand carrying a horizon ring, into which the meridian ring slotted, and against which it was adjustable. Early versions, showing the handle, are illustrated in thirteenth and fourteenth-century manuscripts, while elaborate and elegant examples on stands, from the sixteenth century, survive. It is interesting to note that the armillary was used as a device or badge by Manoel I, who came to the throne of Portugal in 1495. It was during his reign that astronomical navigation was first seriously promoted in a country noted for its seafaring, with the result that the sea route to India was discovered, and the Portuguese empire founded. The instrument was both a potent symbol and a practical teaching device, by means of which students could be shown the workings of the heavenly bodies.

The armillary is, of course, a close relative of the celestial globe, of which the design and mounting were first described by Ptolemy in the *Almagest*. But though numerous celestial globes of Islamic origin are known, it was not until the end of the fifteenth century that globes were produced as part of an established craft in Europe. The celestial globe made by Hans Dorn in 1480 marks the start of this development. It coincided with the great voyages of exploration, and the increased concern both with deep-sea navigation and with map-making, and its most famous exponent was the great

77 European astrolabe, *c.* 1200. Brass, diam. 11.3 cm. This is one of the earliest known European astrolabes, with a rete inspired by Hispano-Moorish designs. There are 13 star pointers, none of them named. On the back of the instrument are scales for 360°, the zodiac and the calendar. Crudely drawn is a shadow square and an unequal hour diagram. The three plates are marked out for latitudes from 23° to 48°. IC no. 161. (British Museum, MLA 1961, 12–1, 1)

map-maker, Gerard Mercator. His celestial globe of 1551, following the production of a terrestrial globe ten years earlier, was the largest printed globe made until that time, and embodied the latest astronomical knowledge.

Astrolabe

In many ways the planispheric astrolabe is the archetypal scientific instrument because of its antiquity and remarkable sophistication. It is a flat, circular brass instrument, which embodies a stereographic projection of the globe of the earth and of the hemisphere of the heavens – a sort of flattened armillary sphere (Figs 77, 78). The origin of the stereographic projection is nearly always the South Pole, and the plane of the projection the equator. Although it can be used for time-telling either by the stars or by the sun, the astrolabe has an additional function as an analogue computer, important for solving problems the mathematics of which would be too daunting to solve in other ways.

The European university arose from the cathedral schools during the twelfth century, and by 1350 their number had reached thirty. Science was taught in the arts faculties as part of the *quadrivium*: arithmetic, geometry, astronomy, music. The astrolabe had, at that time, an obvious role in teaching astronomy, which encompassed astrology, since in the medieval cosmological view, the planets ruled the life of mankind, and so prognostications were vitally important. By 1500 the number of universities had increased to seventy, which explains the fifteenth-century increase in astronomical texts and in the making and use of astrolabes and associated instruments. A number of examples have

78 English astrolabe, *c.* 1295. Brass, diam. 46 cm. The rete has 40 star pointers, of which 33 are named in Arabic and 7 in Latin. The mater has a stereographic projection for 42° (ROMA), while three separate plates are marked out for latitudes 48°30', 51°, 52° (LVNDONIARVM), 53°, 54°, 55°. The back bears comprehensive calendrical information, including a table of epacts in the medieval form beginning with 0. This astrolabe belonged to Sir Hans Sloane and formed part of the foundation collection of the British Museum. IC no. 290. (British Museum, MLA Sloane 53)

survived of fairly simple and small astrolabes that were made at that period in Germany, and it has been postulated that these were made for the use of scholars at the growing number of universities founded at that time in the Germanic states. These were used largely for teaching. During the sixteenth century some very large astrolabes (50 to 80 cm in diameter) were constructed with as great an accuracy as possible for calculating purposes. In Western Europe the planispheric astrolabe ceased to be made very suddenly after 1600, having been made redundant by new instruments and modes of calculation.

The astrolabe was introduced to Europe through Spain by the Islamic peoples who occupied the north coast of Africa and part of Spain in the tenth century. With the Christian reconquest of Spain, knowledge of the astrolabe penetrated medieval Europe during the eleventh century, and craft centres were established by the thirteenth in France, Germany, the Low Countries and England. The *Treatise on the Astrolabe*, written in 1391 by Geoffrey Chaucer, the great English literary figure, is not only an excellent introduction to the use of the instrument, but it is important as the first technical treatise written in the English native language.

Radius astronomicus or cross-staff

The purpose of the instruments used by astronomers in the Middle Ages was to measure for the stars the following celestial co-ordinates: altitudes and azimuths; declinations and right ascensions; celestial longitudes and latitudes. The *radius astronomicus* or cross-staff enjoyed a life of over 300 years, and was a late medieval invention. It was published in 1328 by the southern French astronomer and mathematician Rabbi Levi ben Gerson.

The cross-staff consists of a long wooden staff (whose cross-section is about 3 × 3 cm) with a perpendicular vane that moves over it. The staff is graduated trigonometrically so that angles can be measured by holding the staff to the eye and moving the vane until its ends are level with the points that are to be measured. Its main function was to measure directly angular distances between stars or planets, or the diameter of the sun or moon. Because of its simplicity, however, it was extremely versatile, and could also be used to measure the length of a comet's tail, and both celestial altitudes and terrestrial longitudes using the method of lunar distances. Its practical functions extended to surveying, where it was used to measure heights, lengths and distances, and to navigation. It soon became the sailor's preferred instrument for taking the altitude of the sun or stars.

The cross-staff was light, easy to construct and easy to dismantle and carry. To provide accuracy for the astronomer it could be as long as 450 cm, but was generally 90–180 cm in length for use by surveyors, and even smaller to take on board ship. During the sixteenth century, the surveyor's and seaman's staff was improved by such scholars as William Bourne (fl.1556–88) and John Dee (1527–1608), while Dee and Thomas Digges (*c.* 1546–95) also worked on the astronomer's staff. The improvements pioneered in England were gradually adopted by Continental makers.

Nocturnal

The nocturnal is a simple instrument for giving a rough indication of the time, perhaps to a quarter of an hour, during the night. Its use depends on being able to see the Pole Star and the Great Bear (Ursa major) or Plough, and on the fact that the stars appear to rotate about the Pole once in 24 hours (less 4 minutes each day). In appearance, the nocturnal looks like a small table-tennis bat, with a calendar scale engraved on it. Rotating over this is a disc marked with two periods of 12 hours. The hour positions are usually notched for counting by feel when used in the dark. Above this disc is a long rotating pointer, and the central rivet has a hole. To tell the time, the mark for 12 midnight is set to the date, the instrument is held upright, the Pole Star sighted through the rivet, and the pointer turned to be in line with the Guards of the Bear, that is the two prominent stars α and β in the constellation, which always line up with the Pole Star. The time is then shown by the pointer cutting across the hour disc.

The nocturnal was described in sixteenth-century texts, and examples exist dating from about 1500. It found a use on sea and on land into the seventeenth century, when the pocket watch made it redundant.

The reform of observatory instruments

Scientific knowledge progresses through increasing accuracy of measurement. The making of accurate angular measurements is a fundamental task of the astronomer, and from ancient times until the nineteenth century cataloguing of celestial positions was

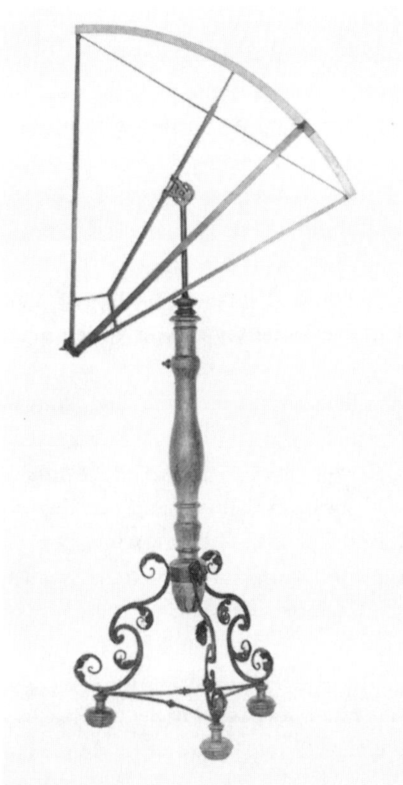

79 Astronomical sextant made in 1600 at Prague by the Swiss instrument-maker Jost Bürgi (1572–1632). The whole instrument is made of iron, except for the brass limb (radius 112.2 cm) which carries the degree scale. It was used to measure the angular distance of any two objects in the sky in any position, provided they were not more than 60° apart. It was based on a design of Tycho Brahe, who died in Prague. The sextant was used during the years 1602–4 by Kepler and again on 20 January 1628 at Prague to observe the lunar eclipse. (National Technical Museum, Prague, 17–195)

his principal responsibility. Angular measure involves dividing a circle into 360° (or a quadrant into 90°), an inconvenient figure that would not have been chosen unless it had been accepted over many millennia. The techniques necessary to perform such divisions with exactitude occupied the best craftsmen for centuries. The quality of the instruments, and the skills in their proper use, enabled cosmology to rise above mere speculation.

There has, however, to be good reason for designing improvements, and one such was the problem of deep-sea navigation across the Atlantic to the New World. Navigation was now reliant on stellar astronomy through the use of printed tables. But the tables of the co-ordinates of some 1,000 stars remained, in effect, those produced by Ptolemy *c.* AD 150 (see p. 87), which had merely been adjusted for precession through the centuries. The Dane, Tycho Brahe (1546–1601), while a student at Leipzig, realised that the then current astronomical tables were defective. Tycho observed that a conjunction of Saturn and Jupiter in August 1563 varied by a month from the date expected from the tables. It was this observation (measured using a simple cross-staff) that Tycho regarded as the turning point in his career, and he abandoned the study of law for astronomy. Eventually, under the patronage of the Danish king, Tycho began to construct a large observatory in 1576 on the island of Hven in the Danish Sound, and conducted most of his systematic plotting of stellar positions during the 1580s using instruments to his own design made by the finest craftsmen in Germany (see pp. 207–10). Although all the instruments used on the island of Hven are lost, there remain two sextants constructed for Tycho now preserved in the National Technical Museum in Prague. Both were made in the year 1600, one by Erasmus Habermel, the other by Jost Bürgi (Fig. 79). These sextants were so ingenious they served as a model for subsequent decades.

Tycho Brahe was the first to break through the accuracy barrier in the measurement of angles. With his instruments and with his skill, he was able to record co-ordinates to an accuracy of better than 1' of arc. This surpassed the measurements made by the best Asian astronomers, notably Ulugh Beg (1394–1449) in Samarqand (see p. 164), whose accuracy was between 10' and 5' of arc. The stellar co-ordinates measured by Tycho enabled Johannes Kepler (1571–1630) to formulate his laws of planetary motion, which vindicated the Copernican view that the earth moved round the sun, and laid the groundwork for the celestial mechanics of Isaac Newton (1642–1727).

Navigational instruments

Astronomical navigation

While a ship is sailing along a coast, the determination of its position is no great problem. In the open sea, out of sight of land, a different sort of navigation has to be applied. The latitude (the angular distance from the Equator, 0°, to the Pole, 90°) can be measured by a quadrant or cross-staff by taking the altitude above the horizon of the Pole Star. The altitude of the sun at noon, when it crosses the meridian, can be taken by a quadrant or sea astrolabe, cross-staff or back-staff. Knowing the day of the month, the altitude, and hence the latitude, can be found from tables.

The longitude is more difficult. This is the angle around the globe, east to west. Longitude can be measured by a clock, because if you know the time at your port of departure, and can find out the local time by the sun, then the difference in time gives you the longitude, because 1 hour equals 15° of longitude. But clockwork mechanisms were not good enough to withstand the ship's motion and changes in temperature until after the middle of the eighteenth century when John Harrison perfected the marine chronometer. Before this, seamen had to rely on the compass, the log-line, and the traverse board to keep a rough tally of the progress of the journey.

When the Portuguese, during the early fifteenth century, began to explore the coast of West Africa, it was initially by coastal navigation. But with attempts to reach India, instrumental navigation became necessary. The Portuguese Prince Henry the Navigator (1394–1460) founded a training school at Sagres for officers. Here were brought mathematicians and astronomers and cartographers to found the nautical science of navigation, to devise instruments, and to draw up charts.

During the middle of the fifteenth century, Portuguese sailors used the plain altitude quadrant for angle measurements. The astronomical quadrant is known from classical times; for ship use they are small, about 30 cm in radius, and were engraved with a scale from 0° to 90°. With a pair of pinhole sights correctly aligned, the plumbline marked the angle. The Portuguese navigators knew that they could return to Lisbon by taking the winds out into the Atlantic until the latitude of Lisbon (38° 42') was reached, as shown by the altitude of the Pole Star. Then it remained to run east 'down the latitude'. Despite the invention of other instruments, the quadrant found favour among some sailors until the seventeenth century.

The sixteenth century dawned with maps, charts and terrestrial globes that depicted land masses unrecognisable in comparison with modern versions. These poor charts made navigation exceedingly hazardous, like riding a horse in a thick fog: ships were wrecked, men were killed, voyages were longer, and trade was diminished. Settlements made in America one year could not even be located the next because of the gross inaccuracies in plotting positions. The problem was that of finding the longitude sufficiently accurately, and the solution lay in astronomy with improved instrumentation. The technology of clocks was too primitive to be of assistance; accurate measurements of the moon in relation to stars were attempted in major observatories, but the resulting tables to carry on board were not sufficiently good, and the navigator and surveyor did not have satisfactory instruments for practical use. A great deal of effort was made, but a satisfactory advance came only with the new national observatories in Paris (1671) and Greenwich (1675).

An example of the rigorous search for more accurate navigation techniques is provided by Edward Wright. David Waters, in his masterly history of navigation, wrote of him: '. . . his book set the seal on the supremacy of the English in the theory and practice of the art of navigation at the end of the sixteenth century'. Wright was a Cambridge University mathematician, who became a fellow of his college, Caius. He was brought into the service of Queen Elizabeth as a result of the Armada, and he travelled on naval ships to gain practical experience. The result was *Certaine Errors in Navigation, arising either of the ordinarie erroneous making or using of the sea Chart, Compasses, Crosse staffe, and Table of declination of the Sunne, and fixed Starres detected and corrected*, published in 1599. Here Wright provided a most thorough mathematical treatment of errors in measurements and in the practices of seamen.

Mariner's astrolabe

For measuring the altitude of the sun near the meridian a new instrument was devised late in the fifteenth century. The mariner's astrolabe is a great simplification of the much earlier astronomer's planispheric astrolabe. It carries a degree scale around the rim, and it has an alidade or sighting rule, and there the similarity ends. The mariner's astrolabe is made of bronze or brass and is very heavy, especially at the bottom. The weight helps to keep it steady, and there are considerable portions cut away to reduce wind resistance, which would otherwise make its use harder or impossible. The mariner's astrolabe does no more than the quadrant: the sights are aligned and the angle read from the degree scale. The earliest recorded use of a mariner's astrolabe is during a voyage in 1481 by Diogo d'Azambuja down the coast of Africa. They were used by Portuguese, Spanish and Dutch navigators through the sixteenth century, their end coming during the 1640s.

Back-staff

For the seaman, the typical measurement made with the cross-staff was to find the ship's latitude by measuring the altitude of the Pole Star above the horizon. The sun's altitude could also be found, but this required the observer to face the sun, with the result that its light was in his eyes. To avoid this the back-staff was devised, and the cross-staff became known as the fore-staff.

The back-staff was invented in about 1594 by the English sea-captain John Davis. It was intended to be an improvement on the quadrant, mariner's astrolabe and cross-staff for finding the meridian altitude of the sun. The back-staff was also known as the Davis quadrant, and as the English quadrant by Continental seamen. The name quadrant arises because 90° can be measured although there is no 90° arc. The instrument looks like a large triangle with a 30° arc at one end, and a small 60° arc at the other. One-sided vanes with pinholes move over these arcs, and at the end opposite the large arc is a push-on vane with a slit through which the horizon can be viewed. By adjusting the vanes the sun's angle with the horizon can be found. The back-staff remained in use until John Hadley invented the octant in 1731.

Surveying

During the first half of the sixteenth century the scene was being set for the development of surveying into a profession, requiring increasing skill and accuracy, and also the use of instruments devised to make angular measurements, again basically addressing the same problems as those faced by astronomers and sailors. Surveying was to become a science alongside, and to a certain extent as a result of, the development of deep-sea navigation and the growing military skills of fortification and siege associated with the use of artillery. The accurate determination of position, both at sea and on land, became of increasing economic importance as ships left the security of coastal waters, and as new tracts of unexplored territory were discovered. As a result of these needs the art of globe and map-making grew rapidly in importance at a time when the necessary skills of engraving and printing were available, and most notably in the Low Countries. The engraver of copper-plate illustrations could also divide the circle on brass instruments for the use of the astronomer, seaman and surveyor. Another important requirement, both for instrument makers and for those who practised surveying and navigation, was mathematical knowledge. As with printing and engraving, mathematics, in particular in the practical applications, was developed on the Continent of Europe by such notable scholars as Martin Waldseemüller (1470-*c.*1518) in Lorraine, and Gemma Frisius at Louvain. During the mid sixteenth century, John Dee, a founder Fellow of Trinity College, Cambridge, spent three years studying in Paris, Brussels and Louvain, and on his return home he pioneered the provision of textbooks and translations of Continental practice in English. At much the same time, in 1543, the highly influential work by Robert Recorde, *The Ground of Artes Teaching the Worke and practise of Arithmetike*, was published, which, running through twenty-eight editions in the next 150 years, introduced pen reckoning and the use of Arabic numerals to a wide public. Before the new, angle-measuring techniques could be satisfactorily used, a prior requirement was a knowledge of arithmetic: Arabic had to replace Roman numerals. In the Exchequer Records Arabic numerals were first used at the end of the sixteenth century, but Roman did not disappear for another 60 years. Accounts were done by using ruled boards or cloths, and casting-counters: hence the phrase 'to cast one's accounts'. Recorde may be regarded as the founder of English mathematics, since his textbooks opened the way for self-education for the new class of technicians. He was also the founder of the English school of mathematical practitioners, among whom were later numbered many of the great London instrument makers.

What, then, were the techniques of the sixteenth-century surveyor? The basic requirements, then as now, comprise: measuring angles and the length of a baseline. After sighting prominent features with the angle-measuring instrument first at one and then at the other end of the baseline, and knowing its length, it is a simple matter to calculate the distance of a feature by the theorem of similar triangles. It is also possible to draw a plan or map to scale. Other requirements are to measure heights and depths and distance, to ascertain the orientation, and to achieve a level. When it came to opening up new lands in the Americas, the surveyor was obliged to fix his location by the stars, and thus his skills were close to those of the navigator.

Navigating and surveying are intensely practical activities, so that the development of instruments for their use was always conditioned by factors such as compactness, and speed and ease of use. The theory was established by the late sixteenth century and changed little over the next 300 years. It was the instruments that evolved, becoming more accurate and efficient, and better adapted to differing conditions.

The makers

Printers and copper-plate engravers were craftsmen from continental Europe who were to influence English practice. Robert Recorde's *The Ground of Artes* was printed by Reynor Wolfe, who left his native Drenthe in the north of the Netherlands to settle in London in 1533. Even 50 years later it was difficult to find an English printer who could set up a mathematical work correctly. Throughout the sixteenth century the English were indebted to the Continent, and especially to the Netherlands, north and south, for the skills of printing, of engraving book illustrations, maps and instruments, and for surveying techniques. All these activities are melded together in the work of Christopher Saxton (d.1596), who was obliged to farm out his engraving needs to several men to hasten the production of maps. His *An Atlas of England and Wales*, published at London in 1579, contains thirty-four maps, twenty-three of them bearing the engraver's name. There are seven different signatures, four of them by Flemings, and three by Englishmen. The English are: Augustine Ryther (four maps), Francis Scatter (two), and Nicholas Reynolds (one). Ryther was to become a leading instrument maker. The migration of skilled men increased during mid-century when the grip of Spanish religious persecution was tightening on the Netherlands, finally to result in the Revolt which broke out in 1564.

The instruments

The main materials used for making mathematical instruments were brass and wood, usually boxwood. Prior to the sixteenth century, brass that was needed for such uses as candlesticks, or memorials in churches, had been imported from the Continent, and was expensive. Now with the military threat from Catholic Europe and growing demand for brass and bronze, the Royal Charters for the Company of Mineral and Battery Works and the Company of Mines Royal for the production of brass and brass plate amongst other things, were granted in May 1568, thus allowing English manufacture for the first time. Humfrey Cole, of whom more will be said later, was closely involved in the setting up of the Mineral and Battery Works, which necessitated bringing in German craftsmen to train Englishmen in metalworking skills, and in prospecting for ores. Cole was engraver of dies for the Royal Mint, as well as the leading instrument-maker of the period.

It was from the Louvain area that came the man who can be regarded as the first to establish the scientific instrument-making trade in England. Thomas Gemini (*c.*1510–

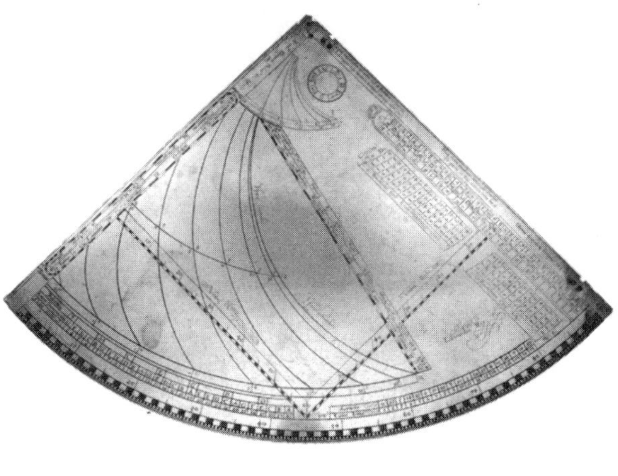

80 Horary quadrant made in London by Thomas Gemini; dated 1551. Radius 27 cm. Engraved 'Edwardus Rex' (for King Edward VI, who reigned from January 1547 to July 1553). The curved lines, calculated for the latitude of London, 51°34', are for time-telling in equal hours by the elevation of the sun in conjunction with a pair of sights and a plumb-line, now lost. Across the centre is a shadow square, for measuring elevations; comprehensive calendrical tables are distributed around the edges. (British Museum, MLA 1958,8–21,1)

62) came from a village near Liège, and as an engraver made his reputation with his plates for his own printing of the *Anatomy* of Andreas Vesalius, issued in 1545. This earned him an annuity of £10 from Henry VIII. At Blackfriars, he carried on the business of map-engraver and mathematical instrument-maker (Fig. 80). An astrolabe by him is in a Belgian museum bearing the arms of the Duke of Northumberland and of King Edward VI; it is dated 1552. In 1555 Gemini printed the *Prognostication* of Leonard Digges, and in the following year, Digges's *A Booke Named Tectonicon.* It was said there that the instruments could be obtained from Gemini. Another astrolabe was made for Queen Elizabeth I. This is dated 1559, and is engraved with the queen's name and the royal arms.

There is no doubt that Humfrey Cole (*c.*1530–91) was London's foremost mathematical instrument-maker of the sixteenth century. Cole was from the north of England, and was employed at the Royal Mint. He undertook to supply all the instruments described in the 1571 edition of Digges's *Pantometria.* Cole's masterpiece must surely be the large, two-foot diameter astrolabe, dated 1575, in the possession of the University of St Andrews, Scotland. This has several resemblances to the Gemini astrolabe made for Elizabeth I, both instruments having on the back a horizontal projection of the

81 Case of architectural drawing instruments in gilt brass, made by Bartholomew Newsam of London in about 1570. Ht 19.3 cm. All four sides are engraved with classical figures, representing Peace, War, Poverty and Abundance. The instruments plug into shaped holes in a wooden block. There are twenty items (a few have been lost), which include 3:1 proportional dividers, adjustable dividers, compasses, beam compasses, folding rule, folding set-square, scissors, knives, pens and scribers. (British Museum, MLA 1912,2–8,1)

sphere derived from the planisphere of Gemma Frisius. Cole's production is both varied and extensive, judging by the twenty-six known instruments, and two engravings. It is clear, also, that he had an influence on subsequent makers, as would be expected. The existing instruments by Cole include: astronomical compendia, sundials, nocturnals, sectors, theodolites, armillary sphere and an astrolabe.

A key factor in the establishment of the English instrument-making trade was the way in which it could become grafted on to the existing guild structure. The City of London Guilds were medieval and, by the sixteenth century, what are known as the Twelve Great Livery Companies had emerged as leaders. To learn a craft and to practise it meant that an apprenticeship had to be served, and the arrangement properly recorded and approved by a City Company. Practising a new craft, mathematical instrument makers had to find a company as best they might; one way was to join a father's company, whichever it might be, under the patrimony arrangement. New companies were formed, of course: the Spectacle Makers in 1629, the Clockmakers in 1631. But the mathematical instrument-makers were captured to a great and surprising extent by one of the Twelve Great Livery Companies, the Grocers. Once a master-apprentice succession was established, the instrument-makers remained in the Company, and so a school was built up.

By 1600, the number of men trained as engravers had increased, and workshops had grown in number and size. Instruments became more complex and varied, with competition in producing new designs to catch attention and to do down a rival (Col. Pl. XVII). Earlier, craftsmen had worked in collaboration with the scholar-inventor; now they were to become capable of independent invention. Instruments for astronomy, time-telling, navigation, and surveying were created for economic reasons. There was, too, a market for wealthy men interested in scientific matters, and elaborately embellished instruments were provided for their delectation (Fig. 81). But this group is more obviously catered for by the products of the leading Continental makers rather than the English, whose instruments are austerely functional.

Bibliography

Bennett, J.A. 1987. *The Divided Circle: A History of Instruments for Astronomy, Navigation and Surveying.* Oxford: Phaidon Christie's.

Brown, J. 1979. *Mathematical Instrument-Makers in the Grocers' Company 1688–1830.* London: Science Museum.

Brown, J. 1979. Guild organization and the instrument-making trade, 1550–1830. *Annals of Science* 36, 1–34.

Brown, Lloyd A. 1949. *The Story of Maps.* Reprinted 1977. New York: Dover Publications.

Chapman, A. 1984. Tycho Brahe in China: the Jesuit mission to Peking and the iconography of European instrument-making processes. *Annals of Science* 41, 417–43.

Chapman, A. 1990. *Dividing the Circle: The Development of Critical Angular Measurement in*

Astronomy 1500–1850. 2nd edn, 1995. Chichester: John Wiley & Sons.

Cotter, Charles H. 1968. *A History of Nautical Astronomy*. London: Hollis & Carter.

Dekker, Elly 1993. Epact tables on instruments: their definition and use. *Annals of Science* 50, 303–24.

Dekker, Elly 1995. An unrecorded medieval astrolabe quadrant from *c.* 1300. *Annals of Science* 52, 1–47.

Gibbs, S., with Saliba, G. 1984. *Planispheric Astrolabes from the National Museum of American History*. Washington, D.C.

Gunther, R.T. 1932. *The Astrolabes of the World*. Reprinted 1976. London: Holland Press.

Horský, Z. and Škopová, O. 1968. *Astronomy Gnomonics: A Catalogue of Instruments of the 15th to the 19th Centuries in the Collections of the National Technical Museum, Prague*. Prague.

Karrow, R.W., Jr. 1993. *Mapmakers of the Sixteenth Century and their Maps*. Chicago: Newberry Library.

King, D. A. and Turner, G. L'E. 1994. The astrolabe presented by Regiomontanus to Cardinal Bessarion in 1462. *Nuncius: Annali di Storia della Scienza* 9, 165–206.

Krämer, K.E. 1980. *Mercator: Eine Biographie*. Duisburg.

Mackensen, L. von 1979. *Die erste Sternwarte Europas mit Ihren Instrumenten und Uhren: 400 Jahre Jost Bürgi in Kassel*. Munich: Callwey.

Maddison, F.R. 1969. *Medieval Scientific Instruments and the Development of Navigational Instruments in the XVth and XVIth Centuries*. Coimbra.

Mörzer Bruyns, W.F.J. 1994. *The Cross-Staff: History and Development of a Navigational Instrument*. Zutphen: Walburg Pers.

Pledge, H.T. 1939. *Science since 1500*. Reprinted 1966. London: Science Museum.

Poule, E. 1983. *Les Instruments astronomiques du Moyen Age*. Rev. edn. Paris.

Price, D.J. 1957. Precision instruments to 1500. In *A History of Technology*, ed. Charles Singer *et al*, vol. III, pp. 582–619. Oxford: Clarendon Press.

Ræder, H., Strömgren, E. and B. (eds) 1946. *Tycho Brahe's Description of his Instruments and Scientific Work as given in Astronomiae instauratae mechanica (Wandesburgi 1598)*. Copenhagen: Munksgaard.

Richeson, A.W. 1966. *English Land Measuring to 1800: Instruments and Practice*. Cambridge, Mass.

Roche, J.J. 1981. The radius astronomicus in England. *Annals of Science* 38, 1–32.

Šima, Z. 1993. Prague sextants of Tycho Brahe. *Annals of Science* 50, 445–53.

Stimson, A. 1988. *The Mariner's Astrolabe: A Survey of Known, Surviving Sea Astrolabes*. Utrecht: HES.

Taylor, E.G.R. 1954. *The Mathematical Practitioners of Tudor and Stuart England*. Cambridge University Press.

Taylor, E.G.R. 1971. *The Haven-Finding Art: A History of Navigation from Odysseus to Captain Cook*. Rev. edn. London: Hollis & Carter.

Turner, A.J. 1987. *Early Scientific Instruments: Europe 1400–1800*. London: Sotheby's.

Turner, G.L'E. 1991. *Gli Strumenti: Storia della Scienza*, vol.I. Gen. ed. Paolo Galluzzi. Turin: Einaudi.

Turner, G.L'E. 1991. *Scientific Instruments and Experimental Philosophy 1550–1850*. Aldershot: Variorum.

Tyacke, S. (ed.) 1983. *English Map-Making 1500–1650: Historical Essays*. London: British Library.

Ward, F.A.B. 1981. *A Catalogue of European Scientific Instruments in the Department of Medieval and Later Antiquities of the British Museum*. London: British Museum Publications.

Waters, David W. 1958. *The Art of Navigation in England in Elizabethan and Early Stuart Times*. London: Hollis & Carter.

Whitfield, Peter 1995. *The Mapping of the Heavens*. London: British Library.

Zinner, E. 1956. *Deutsche und niederländische astronomische Instrumente des 11. bis 18. Jahrhunderts*. 2nd edn 1967; reprinted 1979. Munich: Beck.

II

THE CRAFTSMANSHIP OF THE 'CAROLINGIAN' ASTROLABE, IC 3042

The so-called 'Carolingian' astrolabe shows what can happen when firm conclusions are earnestly desired from the contemplation of one, isolated, example, long before detailed (or any) studies have been made on some 100 instruments amongst which it can reasonably be numbered. This astrolabe was owned and published by Marcel Destombes in 1962. Throughout his account he assumes as the major premise that the instrument was made in about 980, because: «After the year 986, the maker could no longer write Francia for Barcelona, but perhaps Burcinona».[1] His argument then becomes Aristotelian. I prefer to take the Baconian route, looking at

* I thank Professor Wesley M. Stevens of the University of Winnipeg, Canada, and Mme J. Mouliérac, Musée de l'Institut du Monde Arabe, Paris, for permission to examine the astrolabe. I am grateful to Dr E. Dekker for helpful discussions.

[1] M. DESTOMBES, *Un astrolabe carolingien et l'origine de nos chiffres arabes* [1962], in *Selected Contributions to the History of Cartography and Scientific Instruments*, ed. by G. Schiler, P. van der Krogt, S. de Clerq, Utrecht, HES Publishers, 1987, p. 6.

the object to find out all its small particulars, and so building up towards a conclusion.

First of all, I shall give a brief description of the astrolabe, which I saw for the first time in 1968 at the owner's apartment in company with the late Derek Price. I remember he thought it a fake, and it is so marked in his *Checklist* of 1973.[2] It is a conventional pattern of astrolabe, made of a copper-based alloy, and consisting of a mater, two plates, a rete, rule, alidade, pin and horse. It is extremely simple, a fitting object for the cell of a friar or a bachelor of arts. The diameter across the rim is 154 mm, or about 6 inches. Although seven surfaces are properly marked out with mathematical-astronomical lines, most are not labelled. The crude markings one sees are not those of the maker. One test shows this. The mater and plates have been inscribed with the latitude in alphanumeric numerals, thus: LV (36°), LT (39°), MAL (41.5°), ME (45°), MZL (47.5°). When the two projections that have been marked with half degrees are examined, one sees that there is no half present, and they are divided for 42° and 48°. We have here the customary Ptolemaic climates from III at 36° to VII at 48°. Thus the marking of Roma MAL (41.5°) on the plate for Climate V (42°) should probably be regarded as nothing more than a memorandum. Roma is a city; Francia points to a region or state. But what is intended must remain a matter of speculation. It is surely a *non sequitur* to link Roma with another region named Francia just because it lies on the same parallel.

The mater is engraved with a stereographic projection for Climate III, with almucantars at every 6 degrees and no azimuths. The same goes for plate 45°, Climate VI, and 47.5°, Climate VII. Both sides of plate 1, Climates IV and V, have azimuths at every 10 degrees. The back of the instrument has scales of degrees, zodiac, and calendar (eccentric). A quarter of a shadow square is at the bottom right of an otherwise blank central portion. All sections are crudely lettered, excepting three-quarters of the degree scale which is left blank. The same crude lettering labels the almucantars and hour lines in the mater, while the other plates have only their latitude indicated in letters. In addition, plate 41.5° is inscribed: ROMA ET FRANCIA. The rete is fitted out for 20 stars, but none is named. The ecliptic circle has the names for the signs of the zodiac lettered in round Gothic (evolved from 13th century), unlike the lettering elsewhere.

---

[2] S.L. Gibbs, J.A. Henderson, D.J. de Solla Price, *A Computerized Checklist of Astrolabes*, New Haven, Department of History of Science, Yale University, 1973.

Destombes thought he could identify some other numbers or letters. On plate 45° he saw a gothic 4; while on plate 47.5° a P was found, said to stand for Paris.[3] These are random scratches, of which there are many, and the top of the P is a fissure in the metal.

THE CONSTRUCTION

The material is a copper alloy similar to dirty brass. The mater, plates, alidade, rule, and horse, are all formed from beaten sheet with a thickness between 0.8 and 1.15 mm. One plate is convex on the 45° side, a sure indication of hammering. The rim is soldered to a disc most probably from the same sheet as the plates. The rim itself is spliced and soldered in two places, at the bottom and to the left (corresponding to 170° and 270°); the splice extends for 13 mm (Fig. 1).

This method of splicing can be seen on fourteenth to mid-sixteenth-century astrolabes, and probably many more if carefully examined. The simple, three-lobed throne is formed from a continuation of the rim and the back plate. The sights are soldered to the alidade, and by their thickness could have come from the same sheet as the rest. They are curious in being peg sights, a stalactite, as it were, descending to the centre of a wide circular aperture (Fig. 2). It is of importance to attempt to find sights of a similar pattern. It is unsound to *assume*, although it frequently happens, that these vulnerable parts are, in general, replacements.

Destombes has described the instrument as made of *laiton* (brass), but he claims that both the alidade and the rule are in *cuivre* (copper), and so are «later additions because they are not the same colour as the rest». These parts do, indeed, appear slightly more reddish in tone than the main parts of the instrument, and so a higher copper content could be presumed. However, X-ray fluorescence and other kinds of analysis have shown that the copper content is less for the alidade and rule than for the rest. The comments on colour are valueless, because a 67/32 brass should appear more yellow than a 81/16 alloy.[4] A reddish tone may be explained as a surface effect (cuprus oxide). Finally, it must be remarked that the layout lines, and the beating, trimming, and filing, on the alidade and rule

---

[3] M. DESTOMBES, *op. cit.*, p. 3.

[4] Copper/zinc ratio taken from B. GRATUZE, J.-N. BARRANDON, *Nouvelles analyses de l'astrolabe latin AI. 86-31*, *infra*, pp. 433-438.

all point to a unity with the astrolabe. It may be that a metallurgical examination could help to explain any apparent anomalies. Such an examination can reveal the history, from the act of casting to the present day, of a piece of metal, and not merely the proportions of the constituent elements. It seems that the chemical analysis of this instrument has been interpreted without any controls by comparisons with other astrolabes, and without consideration of the way in which an astrolabe maker would found or acquire his brass.

To subject one instrument to forms of chemical analysis may perhaps be of benefit in the very long term, but unless a clearly defined programme is made on a very large number of well-substantiated and dated instruments no conclusions may legitimately be drawn. As Professor A.M. Pollard has written

> all that can be done is to compare [the instrument's] composition with those of known genuine pieces [...]. It is also possible that perfectly genuine instruments will have analyses outside the expected range for a particular country and date.[5]

THE ENGRAVED LINES

A pair of compasses has been used to draw out the lines. The centres of circles are revealed by pronounced dots that would have located the leg of the compass firmly. Many very short arcs show how positions were stepped by the compass. The back of the rete shows the centre dot on the pointer for Lyra used to scribe the ecliptic circle. The N-S and E-W lines are clear, and show that the lining was not well controlled in places, and that in piercing the plate the metal was not removed right up to the lines. The same is true for the front of the rete, so the laying out was done on both sides. Around the border between Gemini and Cancer the layout marks for the subsequently engraved degrees are clear. A layout letter can be seen at the end of Scorpio (Fig. 3), so there were probably scratch marks to identify each zodiac sign. The latitude plates show layout lines for the hours after 6 because the man who held the compass swung the arm to the left over the tropic having started from the outer edge. For hours before 6, the start is on the tropic and the finish is over the edge. In

---

[5] A.M. POLLARD, *Authenticity of Brass Objects by Major Element Analysis*, unpublished proceedings of the 1984 Archaeometry Symposium.

the mater, a false line was drawn for 2 o'clock, and the correct line is cut deeply near by. There are some other slips and misses.

The front of the rim and the outer edge of the back are divided into single degrees, fully partitioned every five. The degree marks are modestly well placed, which is not true for the days of the months. The worst of the marking is on the ecliptic circle in the rete. The lines indicating the single degrees slant in various ways, but most are radial to the centre of the ecliptic circle, and not to the centre of the astrolabe. The dividing lines between the signs of the zodiac are, however, aligned with the centre of the instrument. Some of the degree divisions are so variable that they are twice the width of others (see centre of Scorpio, Fig. 3). The number of divisions in the signs is anomalous, as Libra to Pisces have 31, except for Capricorn which has 32, and Aries to Virgo have 16 each, presumably two unit divisions because of the cramped spacing in that part of the ecliptic (Fig. 4). Adding up the divisions round the ecliptic one finds a total of 379 unit divisions. I presume the divider thought of 30 or 15 divisions and cut 30 or 15 *marks*, which would, of course, explain what we find. The degree scale on the front of the rim is off the N-S axis by nearly one degree West, and it does not align with the scale on the back, which is accurate.

This astrolabe shows an intelligence behind it, but one lacking in craft skills. The brass could be acquired fairly readily. The mathematical lines are drawn in the same manner as would be done on vellum or paper. The impression given by the astrolabe is minimalist. The shadow square could not be more exiguous, azimuths are omitted on three latitude projections, degree numbers and star names are completely omitted, and there is no decoration. The partial labelling that one sees on the astrolabe is not, in my opinion, done by the man who produced the instrument. It is too crude, the inscribing tool does not seem to be a burin, and is inexpertly handled.

In a study on Gerard Mercator published by Dr Dekker and myself, we showed the similarities and subtle differences in the 'tulip' rete patterns of six Flemish or Flemish inspired astrolabes made over a period of 80 years.[6] Here is a line of study that might extend the time period and point to the location of workshops. The 'Carolingian' astrolabe has, therefore, been treated in a similar way by comparing it with five others (Table I and Fig. 5).

---

[6] G.L'E. TURNER, E. DEKKER, *An Astrolabe Attributed to Gerard Mercator c. 1570*, in «Annals of Science», L, 1995, pp. 403-443.

Table I - Origin and date as in the literature

	Checklist number	Origin as claimed	Date as claimed	Aries March	Diameter in millimetres
(1)	IC 1026	Syro-Egyptian	9th C		112
(2)	IC 2	Byzantine	1062		380
(3)	IC 161	European	*c.* 1195	15	113
(4)	IC 300	English	*c.* 1300-1350	14	86
(5)	IC 4556	Medieval	*c.* 1300	13.5	258
(6)	IC 3042	'Carolingian'	10th C	16 [15?]	154

(1) 17 stars, Kufic and Armenian script
(2) 14 stars, names in Greek; Brescia
(3) 12 stars, no names; The British Museum
(4) 21 stars, names in Gothic script; Oxford
(5) 32 stars, names in Gothic script; for location see note 8
(6) 20 stars, no names; Paris

All six have their retes fretted leaving the Ecliptic circle above a 90-degree arc that follows the Equator, a 'Cup and Ball' shape. Their places of origin as stated in the literature are very vague. For example: Syro-Egyptian, Byzantine, European, English. The dating is even more unreliable, going from late ninth century through AD 1062, to 1300-1350. I have arranged these six instruments in the order of the shape of their star pointers. The earliest have been referred to as 'dagger' pointers. Triangle upon square is fully descriptive, and serves for two retes. Then comes an example with just two square bases, the triangles much elongated with decorative lobes at the base. Next is a rete with three very shallow rectangular bases,[7] and thin pointers rising from pierced single, double, or triple lobed bases. Three of these retes have straight East-West bars, spanning the

[7] Such shallow rectangular bases to star pointers may be seen on Arsenius astrolabes, see G. L'. E. TURNER, E. DEKKER, *op. cit.*, p. 433.

complete range in date. One has a single change, another two, while the 'Carolingian' boasts three counter-changes in its bar. While remembering that it is a crude piece of work, the 'Carolingian' star pointers resemble more closely those on the so-called 14th-century English astrolabe. The lobes are not pierced, but lobes are present, and the points are long. Pierced lobes with long pointers are on an astrolabe of the late thirteenth century (Fig. 5, no. 5).[8]

It is my view that rete patterns are likely to be a matter of fashion, and through inertia have a life of one to two hundred years depending which part of the Medieval or Modern periods we are concerned with. Similarities might point to a route of imitation, but care must be taken not tacitly to assume a Darwinian evolution of rete pattern. A craftsman could offer more than one pattern to suit the desires of his customers. When greater accuracy is both desired and attainable, as was the case in the sixteenth century, more stars are introduced, and the astrolabe becomes larger both to accommodate them and to be more accurate in measuring the parameters.

THE VERNAL EQUINOX

The Vernal Equinox as marked on the 'Carolingian' astrolabe falls on 16 March: *but* the line marking the division between Pisces and Aries bisects the arc occupied by the days in March, so one might as well take the date as 15 March (Figure 6). But I have already shown how sloppy and unreliable are most of the divisions on this instrument. Now what about the dating of the other instruments I list and all those many hundreds of others that are not actually dated? They are most frequently allocated dates on the basis of the point of Aries. This has been done over the past century by different people, and mostly we do not know how they arrived at their dates. However, Henri Michel, in his *Traité de l'Astrolabe*, published a graph of date and year from 1360 to 1500 that was based on modern calculations.[9] Michel took as his origin 1600 and worked backwards, which is unhistorical. But for the ancient craftsman no such 'exactness' was possible. The information needed to fix the Vernal Equi-

[8] Christie's South Kensington, sale of scientific instruments, 29 September 1994, lot 136, A Rare Medieval Brass Astrolabe.

[9] H. MICHEL, *Traité de l'Astrolabe,* Paris, Cauthiers-Villars, 1947, p. 139.

nox was taken from manuscript tables relating dates in the calendar ring with positions in the Zodiac. The precise relationship depends on the relative position of the often eccentric calendar and zodiacal circles. Such tables are not easily adapted for a later epoch, and instruments may be expected to have been based on grossly out of date information.

R.T. Gunther claimed that: «the epoch of construction is accurately indicated by the day of the month of the vernal equinox», and he counted one day in 130 years, with his origin 21 March 1923.[10] In his *Astrolabes of the World*,[11] Gunther frequently notes the date of the Equinox, as do other authors who also believe in the validity of dating from the Vernal Equinox. The following table connecting dates actually on instruments with the Vernal Equinox marked on them, is compiled from Gunther, from other astrolabes examined, from astrolabe quadrants and manuscripts.

Table II - Dated Instruments and the Vernal Equinox

IC	Date	Equinox	IC	Date	Equinox
1099	1080	16	4501	1482	12
MS	1295	13.5	642	1486	11
5509	1300*	15	248	1491	9.5
MS	1324	13	250	1491	11
MS	1325	16	258	1525	10.5
291	1326	13	221	1542	10.5
292	1342	13	176	1543	10.5
2111	1415†	14	224	1560	10
549	1457	10.3	230	1565	11
640	1462	10.5	271	1568	10.5
592	1466	10.5	306	1577	10
530	1481	13	208	1580	10

* Astrolabe quadrant, datable between 1291 and 1310.
† Astrolabe quadrant, dated 1415, with star position for the epoch of 1088.[12]

[10] R.T. GUNTHER, *Early Science in Oxford*, 14 vols., Oxford, Oxford University Press, 1921-1945, vol. II, 1923, p. 187.

[11] R.T. GUNTHER, *The Astrolabes of the World*, Oxford, 1932, reprinted London, Holland Press, 1976.

[12] E. DEKKER, *An Unrecorded Medieval Astrolabe Quadrant from c. 1300*, in «Annals of Science», LII, 1995, 1, pp. 1-47.

The complexities buried within this rule of thumb, which in some hands has become a means for precise dating, cannot be gone into here. Suffice to say that we see that 10.5 March occurred on instruments for some 120 years, and that 14 March can continue to within 40 years of 10.5 March. If dated instruments show such variety in timing the Vernal Equinox, then so will those that are undated, making *circa* dating on this basis impossible. In fact, it seems that an 'equinoctial date' may be accommodated within a span of 200 years or more. This means that all *circa* dates for astrolabes are virtually meaningless when based on the equinox.

The Vernal Equinox on the 'Carolingian' astrolabe is at 16 March (but see above), which following Henri Michel could indicate a date of manufacture of around AD 950. However, one must always remember that it was made in a careless way. The date of the Autumnal Equinox is 17 September, which means the two dates are 185 days apart, whereas one would expect 187 days. The number of days between the equinoxes depends on the position of the solar apogee, and when one examines dated astrolabes between 1326 and 1588 one finds a difference between the day of the September date and the day in March from 4.5 to 2 days. More commonly the difference is 2.5 to 3.5, so with the 'Carolingian' astrolabe September 17 could be expected to correspond with 14 March, and a 'date' of around 1200, with the possibility of a further extension to 1400. We can find incompatibility between the pair of equinoctial dates in the case of two German astrolabes both dated 1491. Here the Autumnal Equinox falls on 14 September for both, and the Vernal Equinox falls on 9.5 and 11 March, the latter being not unreasonable for the year.

The Stars

There is no particular problem in deducing which stars the twenty pointers represent. They are all well-established astrolabe stars. The topic has been fully covered by Paul Kunitzsch and Elly Dekker.[13] I shall make a brief reference to their findings because they support my thesis on the quality of the instrument. Twelve stars may be considered to be in or near their correct position; eight are not well placed, with two very seriously out of place by 20 degrees. In summary, Kunitzsch and Dekker say: «Cer-

[13] P. Kunitzsch, E. Dekker, *The Stars on the Rete of the «Carolingian Astrolabe»*, to be published.

tainly, when judged by eleventh-century standards, the 'Carolingian' astrolabe is not a very accurate piece of work». And in a private communication, Dr Dekker points out:

> Makers did not use up-to-date astronomical information. [...] The reason is simple: the computational techniques were lacking. One must not forget that only by 1175 had the Almagest been translated from the Arabic and that it took a long time before this highly advanced learning was understood. Only after 1325 did the Alphonsine tables become available, but only the mathematically skilled knew how to use these tables. I doubt that makers could do so. Therefore, rather than trying to carry out the most complex calculations imaginable, makers would use the data from manuscripts available to them, which could easily be 100 years old.

Indeed, it can be worse than that. An astrolabe quadrant (IC 5507) at the National Museum of American History, Washington, DC, which can be dated *c.* 1325, has star positions from a star list for the epoch 1067, about 260 years earlier. Another astrolabe quadrant (IC 2111) at Angers, France, which bears the date 1415, has star positions for the epoch 1088.[14]

LETTERING

Ancient letter forms are not my province, but even so I have a few observations to make. I regard the lettering on the rete, that is the signs of the zodiac, as done by a different hand from the lettering elsewhere. The letters are all capitals in a rounded form known as Gothic that may be seen in the thirteenth to fifteenth centuries.[15] One can pick out: A with a bar on top, and E, G, M.

Elsewhere on the astrolabe the lettering is roman, but with the alphanumeric numeral system. The letters appear to have been formed with a stylus and not with a burin. The scribe was unable to manage curves: his G, O, R, and S are poor. The rectangular C occurs as an alternative to the rounded C on a memorial stone in Augsburg, late-ninth century (after 887), and it occurs in other countries, sometimes alongside the curved C. On medieval English seals the rectangular C occurs in the XIth and XIIth centuries. It is also to be found on a cross in Bologna,

---

[14] E. DEKKER, *op. cit.*, p. 35.

[15] N. GRAY, *A History of Lettering: Creative Experiment and Letter Identity*, Oxford, Phaidon Press, 1986, pp. 111-121, 227-228, 248; showing examples of round Gothic letters, inter *alia c.* 1275 (England), 1320 (Siena), 1413 (Florence).

dated 1159.[16] The epigraphy of early mathematical instruments has not been studied, and clearly it is high time that this be rectified.

The lettering performed with a stylus has nothing to do with the making of this instrument. With the instrument cut loose from the 10th century, and from Barcelona, we are free to consider the Byzantine regions. Naturally unknown to Destombes, and still known to but a few, is the Byzantine portable sundial with calendrical gearing, dated to *c.*500 and now in the Science Museum, London, that was, in 1985, published by Dr Field and Mr Wright in *Annals of Science*.[17] The numerals are Greek, and match well with those on the 'Carolingian' astrolabe. As the authors say at the end of their *Summary*:

> The emergence of Byzantine mathematical gearing indicates that the Hellenistic tradition attested by the Antikythera machine (first century BC) continued to be active in the Byzantine period, and suggests that it may have influenced the Islamic tradition. Once again an artefact has shown up the inadequacy of the evidence derived from literary sources.

CONCLUSIONS

This astrolabe is a one-off piece by a scholar; no craftsman could have made it. Having scribed the projections, the scholar would not necessarily have needed markings. Later on another person marked it for his own use. So this rough lettering provides no secure indication to the place or time of manufacture. But this lettering may point to a date after which such scribing would have been most unlikely, let us say after 1300 in Western Europe.

In museums there are a few examples of scholar-craftsmen's work, so we should expect to find them on occasion. There is also an early example of an astrolabe where a later user marked the plates according to his own purpose (IC 303). The instrument was made much earlier than the date of *c.*1390 hitherto allotted. The projections are lightly scribed, and the lines

---

[16] B. BISCHOFF, *Die Karolingischen Inschriftensteine aus der Krypt-Grabung 1961/1962*, in *Die Ausbrabungen in St.Ulrich und Afra in Augsburg 1961-1968*, hrsg. von J. Werner, Munich, C.H. Beck, 1977, pp. 263-267, tables 82, 83; H. S. KINGSFORD, *The Epigraphy of Medieval English Seals*, in «Archaeologia», LXXIX, 1929, pp. 149-178; B. BREVEGLIERI, *Scritture lapidarie romaniche e gotiche a Bologna*, Bologna, Istituto per la storia di Bologna, 1986, pp. 12-14.

[17] J.V. FIELD, M.T. WRIGHT, *Gears from the Byzantines: A Portable Sundial with Calendrical Gearing*, in «Annals of Science», XLII, 1985, pp. 87-138.

on one plate and on the back of the astrolabe have been erased and new lines cut, apparently with a burin. These revised parts are marked with numbers in arabic numerals of a period that could be *c.*1390. The original maker was rather lacking in skill, and it has been up-graded in a later age.[18] In this respect it bears some similarity to the 'Carolingian' astrolabe.

This astrolabe may have been made at some time between 1100 and 1300, but it cannot tell us anything certain, nothing about craft centres, and very little about the transmission of ideas. Perhaps when the study of astrolabes and associated instruments has truly begun, then the place of this so-called 'Carolingian' astrolabe will be clearer.

---

[18] IC 303, Merton College. There are other astrolabes with rubbed-down plates that have been re-engraved, for example that cited in note 8.

Fig. (a) Front of astrolabe (IC 3042). Courtesy of the Museé de l'Institut du Monde Arabe, Paris (AI. 86-31).

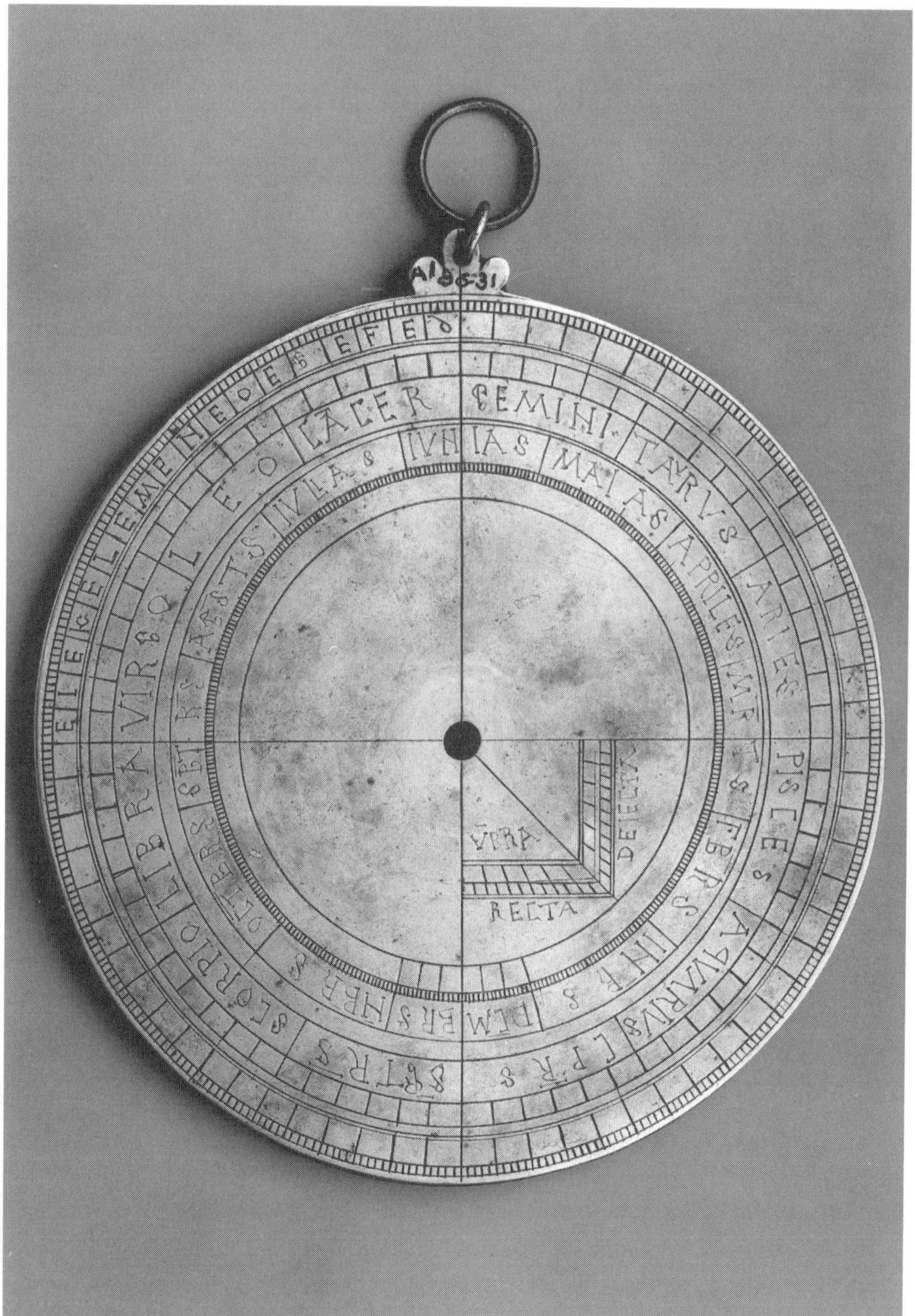

Fig. (b) Back of astrolabe, without the rule.

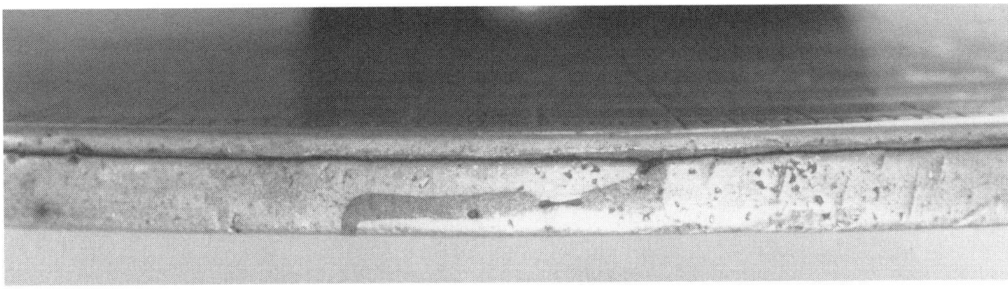

Fig. 1 – A soldered splice in the rim attached to the back plate.

Fig. 2 – A sight vane on the alidade.

Fig. 3 – The scratched letter is to the right of the final O in Scorpio. Notice the variation in width of the degree divisions.

Fig. 4 – Single degrees under Aquarius and double degrees under Pisces. Notice variations in spacing and angle of the marks, which are radial to the centre of the ecliptic circle, and not aligned with the centre of the astrolabe.

Fig. 5 – Rete patterns for six astrolabes. See Table I for the key.

III

A Critique of the Use of the First Point of Aries in Dating Astrolabes

It is a truism that all historians, antiquarians, archaeologists, geologists, indeed all who are trying to make sense of past events, people, and artefacts, have to develop a systematic dating scheme. All schemes have to be under constant refinement, and new techniques developed. Unless there is a reliable ordering in time, it is not possible to elucidate the factors that bring about change.

In the study of scientific instruments it is the Medieval and Renaissance periods that give trouble in dating. Signatures and dates on the instruments are commonly lacking, and this is especially so up to around 1500. The astrolabe is the instrument with the longest span of existence. Those that have survived of Eastern origin date from 900, while those of Western origin date from around 1100. In this essay I am concerned with Western astrolabes only, because I have become involved with this group. The Eastern astrolabes are much more likely to be both signed and dated. However, dating through the use of the First Point of Aries is common to both groups, but applied much more frequently to the Western group. The International Checklist of Astrolabes (1973), initiated by Derek Price, has a table of the Western group in date order, beginning with *c.* 1195 (were there to be a revision, then the table would begin in the 10th century). Between *c.* 1195 and *c.* 1500 inclusive, 135 astrolabes are listed, of which only 20 bear a date. Some of the conjectured dates will have come through the maker's signature, but probably most – certainly the earliest 60 or more – will have been dated by employing the time and the date of the Sun's entry into the First Point of Aries.

One could have expected other ways of dating to have been brought to bear, but unfortunately this field has received scanty attention. Fine art and applied art are much further advanced in identification, attribution, and dating. In the case of unsigned astrolabes, quadrants, and other types of instrument where it appertains, the day in March in the Julian calendar when the Sun enters the Sign of Aries, namely the Vernal Equinox, has been seen as a straightforward means to establish a date of construction. So it is time to take a closer look at the idea behind this way of dating, and to see how it has been applied.

A dating scheme proposed

It has become an established practice to attempt such a method of dating. There are writers on astrolabes who have put forward this practice strongly. R. T. Gunther, in the second volume of his series *Early Science in Oxford* (1923) set the tone:

> The epoch of construction is accurately indicated by the day of the month of the vernal equinox, when the sun enters the first point of Aries. In 1923 it entered at 3.29 p.m. on March 21st. Chaucer, in his *Treatise on the Astrolabe*, addressed to his son Lowys, and dated 1391, makes the sun enter Aries on the 12th of March. [There follows a list of 14 astrolabes and the March dates.] The sun enters Aries one day later after a lapse of 130 years.[1]

Gunther assumed the civil day, that is the day numbered from midnight to midnight. The alternative is the astronomical day, which is numbered from noon to noon. Thus the first half of the 21st of the month in the civil calendar is the second half of the 20th of the month according to the astronomical calendar. Clearly, the marked date of the Equinox depends on which calendar was used by the craftsman.

Gunther's view was modified by Henri Michel in his influential *Traité de l'Astrolabe.* Having quoted the above passage from Gunther, he explained the procedure for calculation, which was one day's retardation every 128 years. This derives from the length of time the Julian calendar overestimates the solar year per year, which is 11′ 14″. It is clear from his arguments and from his calculations to four decimal places that Michel was hoping to date to the year. To find the First Point of Aries in the Julian calendar, Michel calculated back from AD 1600 (a leap year) when the Sun entered Aries on 9.93 March, using the formula: $9.93 - (N \times 0.2422) + N/4$, where N stands for 1600 minus the year in question. He gives a worked example to find the Equinox for 1454.[2]

$$N = 1600 - 1454 = 146$$
$$146 \times 0.2422 = 35.3612, \text{ rounded to } 35.36$$
$$N/4 = 36.5, \text{ rounded up to } 37$$
$$9.93 \text{ March} - 35.36 + 37 = 11.57 \text{ March}$$

Gunther and Michel have calculated the Equinoctial date backwards in time, beginning with accurate modern figures; Gunther from 1923, and Michel from 1600 (see Table 1). This shows that the methods of Gunther and Michel are effectively the same, and that the choice of the civil or astronomical calendar is significant. One might think that the only valid approach to interpreting the Equinoctial dates is to take figures from ancient texts. Even here there is a danger because of the continuing use of out-of-date texts by instrument makers, particularly before the sixteenth century. Professor Poulle has considered medieval astrolabe treatises, where there are also problems in dating. He refers to Michel, and he is sceptical of what he describes as the *mirage scientifique* based on modern astronomical knowledge and overlooking the historical perspective.[3]

1 R. T. Gunther, *Early Science in Oxford*, ii, 'Astronomy' (Oxford, 1923), p. 187. The year chosen is the third in the leap year cycle.

2 Henri Michel, *Traité de l'Astrolabe* (Paris, 1947; reprinted Paris, 1976), p. 139.

3 E. Poulle, "Peut-on dater les astrolabes médiévaux?", *Revue d'Histoires des Sciences et de leurs applications*, 9 (1956), 301–22 (p. 321).

Equinox March	Year: Michel astronomical	civil	Year: Gunther civil
$9\frac{1}{2}$			1728
10	1591		1663
$10\frac{1}{2}$	1527	1591	1598
11	1463	1527	1533
$11\frac{1}{2}$	1399	1463	1468
12	1335	1399	1403
$12\frac{1}{2}$	1271	1335	1338
13	1207	1271	1273
$13\frac{1}{2}$	1143	1207	1208
14	1079	1143	1143
$14\frac{1}{2}$	1015	1079	1078
15	951	1015	1013

Table 1: The year derived from the Vernal Equinox according to Michel and Gunther

The problem of Leap-Year

Michel then pointed out that the moment of the Equinox did not follow a smooth progression, rather a saw-tooth pattern. So he published a calculated graph for the years 1360 to 1500, showing that there is nearly a day's difference in the timing of the Equinox according to the 4-year leap year cycle. Taking 1400 as an example from his graph: leap year gives a civil date of 12 March, one year on 12.25 (i. e., 12th day, 6 hours), two years on 12.5, three years on 12.75, and back to the next leap year at 11.95 March. If the astronomical year were chosen, then these dates would be half a day less. He then observed: 'La date du point vernal ne nous permet donc aucunement la détermination de l'année de construction'.[4] After this emphatic statement, Michel proceeds to try to reduce the uncertainty 'à un intervalle beaucoup plus court' through knowing which year in the cycle the maker chose. Of course this is impossible – even assuming it occurred to a maker – with the exception of Arsenius, who chose the second year after the leap year as his marker.[5] Arsenius in Louvain, the preeminent craft centre in Europe, was exceptional in that he was making the largest and most finely crafted astrolabes from about 1550 to 1575.

The *Traité de l'Astrolabe* by Michel has become a lauded text. Certainly it describes the mathematics of how astrolabes are constructed and how they function. Michel went exhaustively into the problems of dating by a variety of methods, but did not take account of the ability of the craftsman to deliver the accuracy he wished for. If he had, he would not have bothered about his leap-year problem. Taking his example given above, if 12 March is the date for 1400, then if such dates existed on an instrument one would construe them as: for 12.25 March, 1433; 12.5 March, 1465; 12.75 March, 1498; instead of just three years further on. This makes it clear that Michel thought the

4 Michel, *loc. cit.*

5 Gemma Frisius, *De astrolabo catholico et usu ejusdem* (Antwerp, 1556).

craftsman calculated the actual date of the Vernal Equinox in the year he was making the astrolabe. With such an idea (invalid as the dates cannot be distinguished) he hoped to provide a way of fixing this year. But such an instrument, costly in time, materials, and expertise, would have to be useful for at least 50 years. One way, of course, was to settle upon the epoch, and not the date. Taking just one example, a clear-cut one: Mercator plotted the star positions on his celestial globes of 1551 and 1579 for the epoch of 1550. His three recently identified astrolabes, one *c.* 1545, two *c.* 1570, are also designed for the epoch of 1550.[6]

The dating scheme in practice

Gunther, Michel, and most other writers commonly cite the date of the Vernal Equinox. D. J. Price listed it whenever he could.[7] F. A. B. Ward in his catalogue of the instruments in the British Museum recorded the dates of both the Vernal and Autumnal Equinoxes.[8] Sharon Gibbs is particular about dating, and explains the system used for the Smithsonian catalogue:

> Signed but undated astrolabes have been assigned to the middle of the maker's productive period. In the absence of any inscriptions, the astrolabe's date has been estimated from the celestial longitudes of the stars in its rete or from the date of the vernal equinox indicated on its calendar scale.[9]

The dating from rete stars is also problematic, but will not be considered here. A table of dated instruments has been compiled, with the dates of the Vernal Equinox as read from the instrument, and the year derived from the First Point of Aries on the basis employed by Gunther, where one day's change occurs in 130 years. The difference between Gunther's derived year and the year as engraved, is listed in Table 2. The derived year can be some 280 years too early, or over 150 years too late. This Table of dated instruments also shows that the same March date may be found over a period of 140 years.

One would, of course, not expect accuracy on a medieval instrument, the whole concept is unthinkable. Even so, one would expect to find some regularity in the marked, approximate date for the Vernal Equinox. In practice one does not. This

6 G. L'E. Turner and Elly Dekker, "An Astrolabe attributed to Gerard Mercator *c.* 1570", ***Annals of Science***, 50 (1993), 403–43 (pp. 430–1); G. L'E. Turner, "The Three Astrolabes of Gerard Mercator", *Annals of Science*, 51 (1994), 329–53 (p. 350).

7 D. J. Price, "An International Checklist of Astrolabes", *Archives Internationales d'Histoire des Sciences*, 8 (1955), 243–63, 363–81. Extended edition, S. L. Gibbs, J. A. Henderson, and D. J. de Solla Price, *A Computerized Checklist of Astrolabes* (New Haven: Department of History of Science, Yale University, 1973). Column 7 gives the date of the Equinox.

8 F. A. B. Ward, *A Catalogue of European Scientific Instruments in the Department of Medieval and Later Antiquities of the British Museum* (London, 1981). Ward wrote (p. 141) that: "Since the equinox date changes by three-quarters of a day every leap year and by a quarter of a day in other years, the method is not a very precise one". Even so, he thought it "a useful guide to the date of construction of the instrument, especially over the period between AD 325 and 1582".

9 Sharon Gibbs with George Saliba, *Planispheric Astrolabes from the National Museum of American History* (Washington, D.C.: Smithsonian Institution Press, 1984), p. 61. The basis for the link between date and year is not given.

is clearly shown by listing the day of the Equinox as marked on dated instruments (Table 2).

IC	Date	♈	Gunther's Date	Error
5509	1300*	15	1013	–287
291	1326	13	1273	–53
292	1342	$12\frac{1}{2}$	1338	–4
2111	1415†	14	1143	–272
640	1462	$10\frac{1}{4}$	1631	+169
248	1490	10	1663	+173
250	1491	11	1533	+42
251	1507	10	1663	+156
252	1521	10	1663	+142
175	1525	$10\frac{1}{2}$	1598	+73
258	1525	$10\frac{1}{2}$	1598	+73
261	1537	$10\frac{1}{2}$	1598	+61
262	1537	10	1663	+126
221	1542	$10\frac{1}{2}$	1598	+56
204	1548	$10\frac{1}{2}$	1598	+50
224	1560	10	1663	+103
229	1565	11	1533	–32
271	1568	$10\frac{1}{2}$	1598	+30
306	1574	11	1533	–41
208	1580	10	1663	+83
165	1588	10	1663	+75
282	1594	11	1533	–61
312	1631	10	1663	+32

Table 2: The actual date compared with the derived date
IC: International Checklist Number.[10]
Date: as engraved on the instrument.
♈: the day in March of the Vernal Equinox as read on the instrument.
Gunther's date: as derived from the Vernal Equinox.
Error: – Gunther's date too early; + Gunther's date too late.
*: astrolabe quadrant datable between 1290 and 1310. See Dekker.[11]
†: astrolabe quadrant dated 1415. See Dekker.[12]

---

10 See note 7.

11 Elly Dekker, "An Unrecorded Medieval Astrolabe Quadrant from *c.* 1300", *Annals of Science*, 52 (1995), 1–47.

12 Dekker, *op. cit.*, p. 36. The maker used a very outdated star list.

The craftsman's accuracy

There is a prevailing factor: the accuracy with which the degree and calendar scales were constructed. Dr Chapman has shown that there are systematic errors in the layout of such scales, while on many medieval astrolabes one may see variations in the width of a degree of one to two, by simple inspection. As Chapman says:

> A cursory examination of the thirteenth- and fourteenth-century instruments in the Museum's collection revealed graduations of such manifest irregularity as not to warrant critical study, from which one might conclude that Chaucer's astrolabe was probably not accurate much beyond half a degree at the most, for any single observation.[13]

This means that, although one may try to read the equinoctial point as accurately as possible, the result one gets may not be what the craftsman actually intended. He could easily be one day out, which, according to Gunther, is 130 years. This does not mean that the instrument was not useful, far from it. It could still tell the approximate time, and be a portable almanac to serve the purpose of the Middle Ages.

The astrolabe is a hand-held instrument, the ring being slipped over the thumb and held vertical. The sighting of a star is then made by eye, viewing through a pair of pin-holes in vanes at each end of the alidade. The craftsmen would know of this potential source of error and therefore not be too concerned for any advanced accuracy in locating the First Point of Aries. Michel's proposals are additionally vitiated by user error.

A symposium on what was called "The Oldest Latin Astrolabe" (also known as the Carolingian Astrolabe), was held during the XIX International Congress of the History of Science of 1993. My contribution was "The Craftsmanship of the 'Carolingian' Astrolabe, IC 3042". This was the only paper that was concerned with the engraved lines. One of my observations is as follows:

> The degree marks are modestly well placed, which is not true for the days of the months. The worst of the marking is on the ecliptic circle in the rete. The lines indicating the single degrees slant in various ways, but most are radial to the centre of the ecliptic circle, and not to the centre of the astrolabe. The dividing lines between the signs of the zodiac are, however, aligned with the centre of the instrument. Some of the degree divisions are so variable that they are twice the width of others. The number of divisions in the signs is anomalous, as Libra to Pisces have 31, except for Capricorn which has 32, and Aries to Virgo have 16 each, presumably two unit divisions ...[14]

The Vernal Equinox is marked as falling on 16 March, but because of the errors in division one can as well take the date as 15 March. For Gunther, these two March dates (civil day) point to the years 883 and 1013, and for Michel (civil day) 887 and 1015. Taking Michel's calculation for the astronomical day one derives 823 and 951. For all manner of reasons one can discard the use of the astronomical day here. In my opinion, this particular astrolabe was made by a scholar and not a craftsman,

13 Allan Chapman, 'A Study of the Accuracy of Scale Graduations on a Group of European Astrolabes', *Annals of Science*, 40 (1983), 473–88 (p. 475). The collection is that of the Museum of the History of Science, Oxford.

14 G. L'E. Turner, "The Craftsmanship of the 'Carolingian' Astrolabe, IC 3042", *Physis: Rivista internazionale di Storia della Scienza*, new series, 32 (1995), 421–32.

so it can be regarded as untypical. But it has been the date of 16 March that has been accepted. With later astrolabes, clearly generally better made, it is possible to find single degree divisions whose widths vary by 2 : 1. The possibility for error is considerable. A latitude plate may have the stereographic projection cut for 49°, but with the plate marked 48°. The star positions on the rete, or on the universal projection for instruments of the sixteenth century, may be out of position by as much as 10°. If the astrolabe is to be used to tell the time, the nearest hour may be good enough. For astrological or medical use who is going to notice any error?

Conclusions

The schemes for estimating the date of early instruments from the marked day in March of the Vernal Equinox do not warrant the status they have been accorded because there are too many uncertainties. Michel was accurate in his calculations, and was aware of a few of the problems in achieving the exactness he sought. He pointed out the uncertainty that results from not knowing the year in the leap-year cycle that the maker had chosen, but failed to appreciate the futility of his calculations in an historical context. Constructional and dividing difficulties until the mid-16th century introduce an error easily amounting to one day, and therefore about 130 years. This means that those medieval instruments that are undated, and that have been given *circa* dates based on the First Point of Aries, need re-examination. The equinoctial trap must be avoided, and full attention paid to all the constructional details typical of the many European craft centres.

IV

THE ASTROLABE PRESENTED BY REGIOMONTANUS TO CARDINAL BESSARION IN 1462

DAVID A. KING

GERARD L'E. TURNER

1. On astrolabes and the state of the art

The astrolabe was the pre-eminent astronomical instrument of the Middle Ages and the Renaissance. It is a model of the universe that one can hold in one's hands, representing the heavens with respect to the horizon of a given locality.[1] One takes an altitude

[1] The best introductions to the astrolabe are North, «Astrolabe», and Hartner, «Astrolabe». See also Michel, *Astrolabe*, Trento; *Astrolabio*, and London NMM *Astrolabe*.

measurement of the sun, or a star, with the sighting device on the back and sets the position of the sun, or the pointer for that star on the rete (the celestial part) on top of the appropriate altitude circle on the plate (the terrestrial part). The combination portrays the configuration of the heavens at the time the sighting was made. All manner of astronomical problems, particularly timekeeping, can be solved by means of the astrolabe.

Today, over one thousand Islamic and European astrolabes are preserved, the majority not properly documented.[2] Astrolabes are historical sources. Like manuscripts they demand attention to their basic design (equivalent to the text and its literary and geographic milieu) as well as to details such as inscriptions and special markings (equivalent to colophon, copyist, script, and marginalia). We still lack a reliable corpus of data on the surviving instruments of the European Middle Ages and early Renaissance, but at least the earliest ones (up to *c.* 1550) will find a place in a new critical catalogue currently being prepared at Frankfurt.[3]

When the first (Price, 1959) and second (G.L'E. Turner, 1989) descriptions of the Bessarion astrolabe were published,[4] no other fifteenth-century German astrolabes had ever been studied in detail. Indeed, very few were known to exist, and some that were known were not recognized as German.[5] Now Bessarion's astrolabe can be compared with several other contemporaneous pieces.[6]

---

[2] The basic work on individual astrolabes remains the monumental study listed in the Bibliography as GUNTHER, *Astrolabes*. However, it treats only one quarter of the astrolabes now known, many only summarily. The errors in Gunther's work have not been corrected in the main, and the section on early German astrolabes is particularly weak. Specifically, many of the most historically important astrolabes – including virtually all of those mentioned in this study – were not available to Gunther. Some reliable catalogues of important collections or exhibitions have appeared in recent years: Milan MAA *Catalogue* and PA *Catalogue* (Tullio Tomba), Rockford TM *Catalogue* (A. J. Turner), Washington NMAH *Catalogue* (Sharon Gibbs and George Saliba). A German translation of a catalogue of the important Nuremberg collection (by D. A. King) is in Nuremberg 1992-1993 *Catalogue*, I, pp. 101-114, and II, pp. 568-602, 640-643. Some of the largest and most important collections, including the Museum of the History of Science, Oxford, the Science Museum, London, the National Maritime Museum, Greenwich, the Deutsches Museum, Munich, and the Museo di Storia della Scienza, Florence, are without catalogues (or without an adequate catalogue in the case of the last).

[3] See KING, «Report», on this new catalogue, and idem, «Astronomical Instruments».

[4] PRICE, «First Renaissance Instrument»: G. L'E. TURNER, *The Regiomontanus Astrolabe*.

[5] In this study, «German» refers to the German-speaking world in order to include both Vienna and Nuremberg under one rubric. Even though the Bessarion astrolabe is based on an Italian rete design, it is still a «German» astrolabe.

[6] In what follows we refer to these instruments by the numbers used in PRICE, *Checklist*, with any additions (4001 and above) designated by D. A. King for his forthcoming *Catalogue*. These numbers are written thus: IC 123.

2. Regiomontanus and Cardinal Bessarion

The academic exchanges between the German astronomer, Regiomontanus, and Bessarion, the Greek bishop turned Catholic cardinal, are well documented and need only be summarized here.[7] Paul L. Rose, in his book *The Italian Renaissance of Mathematics*, wrote:

> Regiomontanus was the first European mathematician to undertake a complete restoration of mathematics and astronomy and the large measure of success which he achieved must in part be attributed to his connections with Italian humanists and classicists, most of all his close friendship with Cardinal Bessarion. Through the cardinal, Regiomontanus was brought to Italy from Vienna, set to work on his important critique of the *Almagest*, introduced to the leading humanist circles where he was taught Greek, and given a free rein to peruse both literary and mathematical manuscripts. Though he began his career north of the Alps, and spent only a few years in Italy, Regiomontanus' work was intimately related to the flowering of an Italian humanist culture sympathetic to mathematics.

Johannes Müller was born on 6 June 1436 at Königsberg in Franconia. He latinized his name as «Joannes de Regio monte», from which comes Regiomontanus. From 1447 to 1450 he studied astronomy and logic at the University of Leipzig, and was attracted to the University of Vienna in 1450 by the reputation of the humanist and astronomer, Georg Peuerbach (1423-1461). The latter had arrived at the university in 1446, and from 1454 he lectured on the Latin poets and also continued the Viennese astronomical tradition founded by John of Gmunden (d. 1442). Peuerbach is known to have visited Italy, but the most decisive turn of events occurred in May, 1460, when Bessarion arrived in Vienna. Bessarion (1403-1472), originally one of the Greek Orthodox representatives at the Council of Florence, and renowned for his scholarship, had remained in the West, and had been made a cardinal and appointed papal legate to the Holy Roman Empire. The Cardinal had planned to prepare a new translation of Ptolemy's *Almagest*, but once in Vienna he handed the project over to Peuerbach. Since Peuerbach did not command enough Greek for a new translation, he started to prepare an epitome from an earlier Latin translation, which was

---

[7] See, for example, Zinner, *Regiomontanus*, especially chapter 5; Rose, *Italian Mathematics*, chapter 4, Regiomontanus in Italy, pp. 90-117; and the article on Regiomontanus by Edward Rosen in *Dict. Sci. Biog.*, XI, 348-352.

completed by Regiomontanus after the untimely death of Peuerbach in April 1461. By providing easier access to Ptolemy's complex masterpiece, the Peuerbach-Regiomontanus *Epitome* contributed to current scientific research by giving astronomers an understanding of Ptolemy that they had not before had access to. Finally printed in Venice in 1496, it was read by Copernicus, and helped him to turn his attention to problems with the Ptolemaic system of planetary motions. Copernicus was to overthrow this system with his own heliocentric system in his book of 1543: *De Revolutionibus*.

In the September of 1461, Bessarion returned to Rome, taking Regiomontanus with him. Regiomontanus stayed in Italy until 1467, when he moved to Hungary to take up a professorship of mathematics at the newly-established University of Pressburg (modern Bratislava).[8] In 1471 he moved to Nuremberg, because of the availability of instruments and ease of communication with learned men. The city was on the great trade routes of central Europe.

Regiomontanus set up an observatory and installed a printing press in his own house to publish scientific texts, which were normally avoided by printers because of diagrams and symbols they did not understand. Johann Gutenberg (1397-1468) invented printing by moveable type at Mainz by 1448. Regiomontanus, the greatest astronomer of the fifteenth century, was the first publisher of astronomical and mathematical literature with texts relatively free of scribal and typographical error. The point was that sound texts in large numbers could only be obtained quickly and cheaply by printing, but to ensure accuracy an expert astronomer would have to be involved in the printing process. He also used the humanist «antiqua» type instead of the German «gothic» in his printing. His first book was Peuerbach's *Theoricae novae planetarum* (*c.* 1474), and the second his own *Ephemerides* (1474) for the years 1475 to 1506, a copy of which was taken by Columbus on his fourth voyage of 1504.

In 1475, Regiomontanus responded to a papal summons to participate in the plans for calendar reform by returning to Rome. It was there that he died, most likely a victim of the plague, in July 1476. His earlier benefactor, Bessarion, had already died in 1472.

Among the astronomical and mathematical works of Regiomontanus that he dedicated to the Cardinal in Rome are the *Epytoma Almagesti* started by Peuerbach and completed by himself

---

[8] ZINNER, *Regiomontanus*.

(1462), *De compositione et usu cuiusdam meteoroscopi* (1462), and *De triangulis* (1463). It is hard to imagine that the «Ioannis» in the astrolabe's inscription refers to any other than Regiomontanus. In about the year 1465, Regiomontanus wrote a treatise for Cardinal Bessarion on an armillary sphere, which he called *Meteoroscopium armillare* (printed in Nuremberg in 1514).[9] This shows that the astrolabe was not the only instrument that the two of them might have discussed.

Regiomontanus dedicated several of his known works to Bessarion. What more appropriate gift for the Cardinal than a model of the universe that he could hold in his hand? In fact, the instrument is fully in keeping with contemporary German instrument making, and incorporates a novel feature, the moveable grid on the back, about which Regiomontanus had learned from his Viennese teachers. One other contemporary German astrolabe incorporates the same feature (Section 5 below: 2a, IC 256). Other astrolabes of this school incorporate a lunar volvelle on the back (Section 5 below: 1a, IC 549; 1b, IC 452), another feature that was novel at that time.

In brief, the Bessarion astrolabe is important for the reasons that Price and Turner indicated, namely, because it is unique amongst European instruments from the thirteenth to the sixteenth century in possibly bearing the name of a distinguished astronomer. The instrument fits squarely in the scene of fifteenth-century German astrolabe making, being one of eleven known instruments from the same school. Although there is no proof that Regiomontanus himself made the instruments, there is written evidence from 1527 that he devised several astrolabes and that they all had a certain feature, the choice of latitudes for the plates, and that feature is shared by the instrument he dedicated to Bessarion.[10] But there are good reasons – the elegance of the piece and the quality of its construction, the accuracy of the astronomical markings, the use of antiqua (specifically roman) type face throughout, the full use of Latin star

---

[9] ZINNER, *Instrumente*, pp. 481-482.

[10] See Section 5, note 40. The only published queries about the authenticity were written by «trade-journal» writers who knew nothing about German instrument making in the fifteenth century, and who were not familiar with any other astrolabes of the Regiomontanus type. See ROY (Paris), and ANONYMOUS (London). In passing, we may mention that the doubts raised even by Derek Price on the so-called «Peuerbach» and «Regiomontanus» brass diptych sundials, of which 7 are known, are ill-founded. Some are dated, and the range is from 1451 to 1463. See PRICE, «Review», p. 132. In WARD, «Sundials», some evidence for their authenticity is provided, but more is available. This becomes apparent when the instruments are studied in detail and compared one with another.

names, and the standardized «universal» span of latitudes for the plates – to retain the description of this astrolabe by a leading historian. Price called it simply: «the first scientific instrument of the Renaissance»; certainly no other instrument qualifies for this noble title.

3. Cardinal Bessarion's astrolabe: description and provenance

i. *Description*

The whole instrument is made in brass. The dimensions (all in millimetres) are:

Mater, outer diameter 115; thickness at rim 3.8
Plates, diameter 108.2; thickness 0.3.
Rete, diameter 108.3; thickness 1.0
Disc on the reverse; diameter 91.5; thickness 0.6
Rule, length 115; thickness 1.0
Alidade, length 116; thickness 1.7
Pin, length 12.5; diameter of shaft 4.0 (may not be original)
Wedge, now lost, replacement fitted

The mater is constructed from a plate, about 0.5mm thick, formed from a casting filed flat. One region shows fissures formed in casting. This plate is held by four rivets to a rim (width 12mm) on the back. From the front a faint inscribed circle can be seen just 12mm from the outer edge, which shows the inner limit of the rim on the back. This rim, which is 1mm thick, is so made that one side of the throne is contiguous with it. The other side of the throne is integral with a narrow (3.5mm) rim on the front of the mater. The two portions of the rim are tightly joined, presumably by solder. The thin line of the join can be seen right through the instrument. At the throne the join is more noticeable because the thicker, front part has cracked across having been stressed against the thin part behind, which has not cracked.

On the reverse, the rim is only some 0.8mm proud of the central disc. The engraving of the rim into its degree divisions has been done after assembly because one can easily see how the engraver's burin slipped forward on the cutting stroke and came to a stop in the disc leaving a nick. On the front there are very few nicks, presumably because the rim is some 2mm proud of the plate, leaving greater depth for the forward cut to be arrested. Those nicks that do occur on this side are much further towards the centre than on the reverse side.

The rim on the face side is engraved in one-degree intervals and numbered 0° to 360°. On the reverse the rim is engraved in degrees, numbered 90°-0°-90°-0°-90°. Inside the degree scale is the Zodiac, each sign divided to 30°, and inside this is a calendar scale, the months divided into their days. Finally, the innermost scale is in two parts. The upper semicircle is divided in degrees, 0°-90°-0°. The lower forms a shadow square centred on the centre of the astrolabe, both quadrants divided 0-12-0. This is used in height/distance measurements, as well as to find shadow lengths.

Into the mater fit three plates, each with a locating lug at the top to prevent rotation. Each plate is engraved with circles of altitude (almucantars) and azimuth at 5°-degree intervals. Punched dots have been employed to lay out the co-ordinate lines. The latitude for which the plate is engraved is marked on each side.

1.	LATIT̄	39 GŘ	/	LAT̄	54
2.	LATIT̄	48 GŘ	/	LATIT̄	45 GŘ
3.	LAT̄ ROMAE	42 Ǧ	/	LATIT̄	51 GŘ

The three plates cover Europe from 39° to 54° in 3° intervals.

Other engraved circles indicate the tropics, equator, and solar depression at twilight. There are also lines of unequal hours, numbered 1 to 12. The plate engraved for 48° is also engraved with the boundaries of the astrological houses.

The position of the Vernal Equinox – first point of Aries – is indicated on the rim on the reverse, and it is opposite a position 10.5 days into March.

Both the Zodiac and the calendar scales have been labelled in an unusual way, but only to save space. On most contemporary German astrolabes one finds that each sign and each month occupy an annular space above the degree and day divisions. But on this astrolabe the names of the signs are abbreviated to their first two letters, and these are placed where normally 30° would be engraved. The calendrical scale is marked in an analogous manner, where the first two letters occupy the space normally filled by 28, 30, or 31. In the case of February, the final 3-day space is too small to accommodate the letters FE, so they have been placed in the earlier 5-day space.

The rete is delicately made from cast brass. It has 30 star

pointers, 18 inside the ecliptic circle, and 12 outside. Each end of the Capricorn band terminates in carefully worked decoration, while adjacent to the first point of Capricorn is a trilobed motif. The pointer for *Cor Leonis* is decorated by an eight-pointed star at its root. The degrees on the ecliptic circle have been marked out in 5° intervals on the otherwise plain back of the rete. These are layout lines for transference to the front where the edge is bevelled. The one-degree intervals are then added, presumably by eye. The alignment is by no means perfect. The lettering on the rete has been executed with great care.

It is astonishing that the above description of the rete can serve in every detail for the rete on a different astrolabe (see Section 5, 1b; IC 452). The floral chasing is to the *identical pattern*, and the root of *Cor Leonis* is the same. The layout lines on the back and the letter forms on the front follow exactly as here. The minor difference lies in the naming of some stars, which are all in Latin on Bessarion's astrolabe, and partly derived from Arabic on the second instrument. There can be no question that both of the astrolabes were made by the same hand.

In the recess on the reverse of the mater fits a rotating disc, engraved with an angel above and the dedication below the *organum Ptolemei*, which runs across the middle. This is discussed fully below. The back of this disc is plain, and shows casting flaws and file marks. There are two small circles with a centre prick that have been made using scribing compasses. A scribed arc passes through 10 prick points, an end pair being divided by four prick points into five divisions. Perhaps this was an aid to dividing the ecliptic circle on the rete. The dedication reads:

SVB DIVI BESSARIONIS DE CARDINE DICTI
PRAESIDIO ROMAE SVRGO IOANNIS OPVS:- 1462

[Under the protection of the divine Bessarion called Cardinal
I arise in Rome as the work of Johannes: 1462]

The rule is divided in celestial declinations, and the arms are engraved: MERIDIES SEPTEMTRIO.[11] The plain, cast alidade has

---

[11] Sometimes the comment is made that the M in Septemtrio is incorrect, the use of N being classical Latin (see note 24 below). One can regularly find either find either spelling on astrolabes, and any Latin dictionary will give, for example, «septemtriones (septentriones)» for the seven stars in the plough, or Great Bear.

lost the vanes from its hinged sights. The pin may not be original; the wedge has been lost.

The throne has a shackle and ring. It is decorated on the front with a motif comprising three roses. On the reverse there is one rose above, and below, a pair of six-pointed stars.

Around the outer edge of the mater are twenty-four divisions representing the hours in a day/night period. Each section has been engraved with the hours, so the numbering runs from 1 to 12 twice. The style of the numbers is similar to those on the other parts of the instrument, but using later (modern) forms of 4, 5, and 7.

Two metallurgical examinations have been made on the brass of this astrolabe. By modern standards the brass used to construct this astrolabe is a very dirty *alpha* brass alloy. The results, and the problems of metal analysis for historical dating purposes, are considered below.

ii. *The Names of the 30 Star Pointers on the Astrolabe*

As engraved	*Full name*	*Modern name*	*Popular name*
VM AND	Umbilicus Andromedae	β Andromedae	*Mirach*
CAVDA CETI	Cauda Ceti	β Ceti	*Diphda*
VENT CET	Venter Ceti	ζ Ceti	*Bata kaitos*
NAR CET	Naris Ceti	α Ceti	*Menkar*
CAP GOR	Caput Gorgonis	β Persei	*Algol*
OCVL TAV	Oculus Tauri	α Tauri	*Aldebaran*
HIR AG	Hircus Aurigae	α Aurigae	*Capella*
PES DEX OR	Pes dexter Orionis	β Orionis	*Rigel*
CAN MA	Canis Major	α Canis Majoris	*Sirius*
CAN MI	Canis Minor	α Canis Minoris	*Procyon*
YDRA	Hydra	α Hydrae	*Alphard*
COR LE	Cor Leonis	α Leonis	*Regulus*
VAS	Vas Aquarium*	α Crateris	
CAV LE	Cauda Leonis	β Leonis	*Denebola*
CORVVS	Corvus	γ Corvi	*Algorab*
SPICA	Spica Virginis	α Virginis	*Spica*
CAV VR	Cauda Ursae	η Ursae Majoris	*Alkaid*
BOET	Boetes	α Boötis	*Arcturus/Alramech*
CORONA	Corona	α Coronae Borealis	*Alphecca*

As engraved	*Full name*	*Modern name*	*Popular name*
MAN SER	*Manus Serpentarii*	δ Ophiuchi	*Yed*
HER	Hercules	α Herculis	*Rasalgethi*
CAP SER	Caput Serpentarii	α Ophiuchi	*Rasalhague*
CA DR	Caput Draconis	γ Draconis	*Eltanin*
VVL CA	Vultur Cadens	α Lyrae	*Vega*
VVL VOL	Vultur Volans	α Aquilae	*Altair*
DELPH	Cauda Delphini	ε Delphini	*Deneb Adolfin*
GAL	Gallina	α Cygni	*Deneb*
MV EQ	Musita Equi	ε Pegasi	*Enif*
CRVS PEG	Crus Pegasi	β Pegasi	*Scheat*
HV EQ	Humerus Equi	α Pegasi	*Markab*

* Variously named Vas, Patera, Crater = bowl for libations.

The identifications of the engraved star names with the prominent stars have been made on the basis of the usage of the time. As may be expected, the star pointers being thin metal can be fairly easily bent away from their intended positions. If measurements are made of right ascension and declination of the positions of the tips of the star pointers, one finds, however, very little discrepancy after allowing for precession. This shows the quality of the craftsmanship that went into making this astrolabe and also the care with which it has been kept for 527 years. It might be thought that a confusion might arise when a star is named, for example, merely as Hercules, as there are three stars that could have been chosen. But the right ascension shows that α *Herculis*, a variable brightness star, is the correct one. The engraver has named β *Orionis* as *Pes dexter Orionis* instead of the customary *Pes sinister Orionis*, and the co-ordinates confirm that β *Orionis* is the star intended. The confusion may have arisen through the name given to α *Orionis, Dexter humerus Orionis* (Betelgeuse), which does not feature on this astrolabe.

iii. *The Provenance*

A short mention of Bessarion's astrolabe was published in *The Times* of London on 9 February, 1959, headed: «15th century Astrolabe at Maritime Museum». The report began:

Commander M. H. Hardcastle has lent a small brass astrolabe, made in 1462 by Johannes Muller of Konigsberg, to the National Maritime Museum, Greenwich, which is now putting it on view. Muller, more generally known by his Latin name, Regiomontanus, stands at the beginning of the renaissance of science, and this astrolabe was the first scientific instrument made by him. He made it for Cardinal Bessarion, who had taken the young scholar to Rome and given him access to his manuscripts of classical Greek astronomy. Regiomontanus afterwards returned to Nuremberg and there built an observatory, a printing press for scientific books, and a workshop for scientific instruments. It was these activities that, in the words of the musueum's announcement, «marked the re-birth of astronomy and indeed of all modern science».

The National Maritime Museum's inventory (No. A.56/58-207) adds little, except that the first point of Aries was datable as 10.75 March, and the latitudes on the three plates were given.

Clearly, such an interesting astrolabe would attract the attention of the late Professor Derek Price, whose very detailed monograph on a fourteenth-century manuscript dealing with an equatorium had been published in 1955 by the Cambridge University Press.[12] In the first issue of the new Italian journal, *Physis*, which appeared in 1959, Price published a short paper briefly illustrating and describing the Regiomontanus astrolabe.[13] Here Price gave no more than the bare bones of a description, though he did add a paragraph on provenance, which he presumably obtained from the owner. The paragraph reads as follows:

> The provenance of this historic specimen has not been without interest. It remained in Rome apparently until March 1848 when it was presented by some person unknown to Dr. Somerville, husband of Mary Somerville (*Connection of the Physical Sciences*, London, 1834) who was then wintering in Italy with his family, as was their custom at this time. It was then given by the Somervilles to their close friend, Sir John F. W. Herschel, the astronomer, and it has continued in the possession of that family ever since, being now the property of Commander M. H. Hardcastle, R. N. He has been kind enough to allow the present author to have it on indefinite loan for study and exhibition in Europe and America.

In passing, one must remark that this passage is not entirely accurate. The Somervilles lived a peripatetic life in Italy from 1840. Apart from visits to England in 1844 and 1848, they saw out their lives abroad. Both William Somerville (1771-1860) and his wife, Mary Somerville (1780-1872) died in Italy.[14] And the instrument did leave the Herschel family.

Sir John Herschel was a renowned astronomer and mathematician, and was a friend of Dr. Somerville for a long time: he was a supporter for Somerville's election to Fellowship of the

---

12 PRICE, *Equatorie*.

13 PRICE, «The First Renaissance Instrument».

14 PATTERSON, *Life*; PATTERSON, *Mary Somerville*.

Royal Society in 1817. Mrs. Somerville dedicated *Physical Geography*, published in 1848, to her scientific advisor for over twenty years, Sir John Herschel. If in 1848 the Somervilles were journeying to England, it would not be surprising that they should take a present of an astrolabe for an old friend. But where did the astrolabe come from? Possibly the Rome observatory. The old, conservative Pope Gregory XVI died in 1846, and was succeded by the younger Pius IX. Since he encouraged modernization, it may be that in clearing out the mess of ages in some institution, the ancient brass astronomical instrument was thought by some Roman astronomer to be a worthy gift to his world-famous English colleague.

Sir John Herschel died on 11 May, 1871. The Will of Herschel records that Edward Hardcastle, a Manchester merchant, and Herschel's widow were the two executors. The Will says nothing more specific than «my scientific books and instruments». It would seem that the astrolabe passed to the Hardcastle family through this connection. Commander Hardcastle lent the astrolabe at the end of 1956 to Derek Price who was about to embark on a lecture tour of the United States. In October 1958 it was placed on extended loan with the National Maritime Museum, and was withdrawn twenty-seven years later. On 28 September, 1989, it was sold at Christie's South Kensington, lot 170.

4. Remarks on features of the Bessarion Astrolabe

i. *The engraved letters and numbers*

All the letters are majuscules (upper-case, commonly called capitals), and they are engraved with skilled control. Their form comes from the Italian scribal hand, and it was this that was the model for much of the new printing industry. Typefounders followed the desires of their extremely conservative customers and produced type faces to look as nearly like manuscript as possible. Thus for German readers the type was «gothic», while for Italian and French readers the type was «antiqua».[15] It is usual to divide «antiqua» and

---

[15] Steinberg, *Printing*.

talk of roman and italic.[16] Thus it was natural that the first book to be printed in Italy, Cicero's *De oratore*, imitated the round hand of the Renaissance scribes. It was thought that for the spread of humanism the use of «littera antiqua» was the most appropriate. This influence was soon seen in the instrument trade. The forms on the Bessarion astrolabe are similar to those on, to take one example, the Viennese compendium dated 1491 in the British Museum (No. 94 6-15 1) bearing the name of the Benedictine monk, Hans Dorn (before 1440-1509). Even so, it should be noted that the form of the 4 is still gothic.

It is obvious that the design of the Regiomontanus astrolabe and the several others like it became the inspiration for the somewhat similar astrolabes produced in the next century by Georg Hartmann (1489-1564), who was vicar of the Church of St. Sebaldus in Nuremberg from 1518 to 1544. Here he wrote on sundials, and began to manufacture dials, astrolabes, and other mathematical instruments. His astrolabes copied those of the previous century, down to the shape of the rete and of the throne, repeating the rose motif. There is one signal difference: letters and numbers are punched and not engraved. This allowed for faster production in batches.[17]

The astrolabe is engraved with arabic numbers that, apart from 4, 5, and 7, are in their modern form. Around the outer edge there are numbers all in the modern form. In the old style, 4 is a loop with the two ends below; 5 is a vertical stroke, like 1, with a small curve to the left that is open at the top; the 7 is an inverted V. The modern forms of these numbers were in use in Italy by about 1400.[18] The usage varied among trades, the lapidaries being amongst the first at the beginning of the fifteenth century. Medallists were employing some of the new forms by 1417, and printers, a trade that started mid-century, were using all modern by about 1467, and painters likewise in about 1472. To the north the change-over was slower; German printed books were all modern by 1488, but in the Low

---

16 MORISON, *Printing*. Pointing out that the terminology is inexact, Morison writes: «the Germans have the best of it; for our «roman», their equivalent is «antiqua», for «italic», «kursiv» (p. 17).

17 Batch production was employed by Hartmann of Nuremberg. G. L'E. Turner has found an identifying number or letter scratched on all the parts of several of his astrolabes. This implies a parts shaper, a mathematical line engraver, and a letter/number puncher.

18 See, for example, HILL, *Arabic Numerals*; MENNINGER, *Zahlwort und Ziffer*, pp. 213-265 («Die Wanderung der indischen Zahlzeichen nach Westen»); FLEGG, *Numbers*.

Countries the modern form had not penetrated all trades until after 1500. In England one has to look to 1550 before the modern forms of all the numbers are in general use.

It seems, therefore, that the use of the older forms of 4, 5, and 7 on the astrolabe could be accounted for by the designer coming from Vienna, or the craftsman who made the engravings being of the older tradition. The numerals used to mark the hours around the edge of the instrument could have been added shortly after it was completed, or sometime within perhaps fifty years of completion. The manuscripts of Regiomontanus show that he gradually changed from the old to the new forms of 4, 5, and 7. Even so, his printed *Kalender* of 1474 employed the old forms, most probably because his readers were used to them.

ii. *The* Organum Ptolemei *on the reverse*

In a manuscript of a series of treatises by earlier Viennese astronomers copied by Regiomontanus in his own hand *c.* 1456, the grid known as the *organum Ptolemei* and its use for timekeeping by the sun at any terrestrial latitude are described.[19] Regiomontanus made no claim to have invented this «instrumentum universale ad inveniendum horas in quocunque climate fueris fabricare», and he was well aware that the underlying theory was due to Ptolemy. In addition, the instrument is described in several late-fifteenth-century German texts (in Latin), and features in writings of Apian in the early sixteenth century. In Paris there was published in 1550 a more complete grid based on the same projection and also serving the stars: this is now known as the Rojas projection.[20]

The only studies of this mathematical grid in the modern literature are by E. Zinner (on the *organum Ptolemei*), published in 1956, and F. R. Maddison (on the de Rojas projection), published in 1966.[21] The former dealt with the early history of the projection in Germany, which is overlooked in the latter. Maddison concluded his account of the de Rojas projection by recording the existence of a few instruments of earlier date than the publication of de Rojas's book in 1550, «which nevertheless use the same projection described

---

19 Zinner, *Instrumente*, p. 132.

20 Juan de Rojas, *Commentariorum in astrolabium...* (Paris, 1550; 2nd edn 1551). See also Gunther, *Astrolabes*, pp. 353-354; London NMM *Astrolabe*, pp. 32, 40.

21 Zinner, *Instrumente*, pp. 131-134; Maddison, «Rojas Projection».

by him». One of these was the Bessarion astrolabe (at that time displayed at the National Maritime Museum, Greenwich); another, in Cracow, attributed to Hans Dorn (before 1440-1509); another at Florence, also attributed to Dorn, although the lettering appears dissimilar to that on the only known signed Hans Dorn instrument (dated 1491) at the British Museum.

We can now point to a couple of mid-fifteenth-century Regiomontanus type astrolabes with the *organum Ptolemei* on the back, one in a private collection at Paris (see Section 5, 2a; IC 256) and the other Bessarion's. The fact that de Rojas first described this projection in 1550 and that it is commonly called the «de Rojas projection», is one more example of an unfortunate misnomer, and underlines how little we knew about contemporaneous German astronomy until the work of Ernst Zinner.

The *organum Ptolemei* was developed as a device for reckoning time from solar altitude *accurately in equinoctial hours*. It was an alternative to the universal horary quadrant popular on astrolabes both before and after the fifteenth century. This was an inferior device for reckoning time from solar altitude, giving approximate values in seasonal hours. Both devices were popular because they worked for all latitudes, but the quadrant was easier to construct and was far more widely used.

On most examples of the grid, including those on the Bessarion astrolabe and the other mid-fifteenth-century German example, and the astrolabe attributed to Hans Dorn, the grid rotates within an altitude-cum-latitude scale. As such it functions as an observational instrument; the alidade is aligned towards the sun, and the grid is aligned parallel to the celestial equator for a specific latitude – one then simply reads off the time from the grid. In this fashion the instrument is described by Regiomontanus in his treatise. If the grid is fixed with respect to the surrounding scale, then it functions as a calculating device. One must first measure the altitude of the sun by means of the alidade, note it, and then feed the value into the grid and read off the time.

The disc on Bessarion's astrolabe could not be engraved on both sides because of abrasion when rotated. The other similar astrolabe also has a rotating disc that is blank on the inside. Consequently, the grid, angel, and dedication scroll all had to be fitted on to the one side.

iii. *The Dedication*

The attribution of the astrolabe to Regiomontanus rests on the inscription in Latin on the back plate of the instrument. This gives the date, 1462, the name Ioannes (Johannes) as that of the maker, the place of dedication as Rome, and refers to Cardinal Bessarion as the maker's patron.

The study of inscriptions on medieval and early Renaissance scientific instruments is in its infancy; there are very few experts on medieval scientific Latin, and none who have turned their attention to inscriptions on scientific instruments. The dedication on the Bessarion astrolabe is one of several on medieval and Renaissance instruments in which the instrument «speaks» in the first person.[22] The verse form of the dedication is described as «elegiac couplets» by Dr. Scholderer in a Note on Price's paper of 1959.[23] Writing for librarians and bibliophiles, he said:

> All the inscriptions on the instrument are in majuscules of much the same style as those which Regiomontanus cut for his smaller type when he came to print at Nuremberg in 1474, and indeed the splendid craftsmanship of the piece is obviously akin to the skill and accuracy so impressively displayed in the pages of his printed ephemerides.

Some objections have been raised about the quality of the Latin. After the sale, the *Antiques Trade Gazette* quoted Dr. J. V. Field, at that time employed at the Science Museum, London, as having found mistakes in the Latin. Her view was reported thus:

> ... she is unhappy about the association with Regiomontanus because there are mistakes in the Latin on the rete and on the label and one does not expect mistakes from Regiomontanus. «As a humanist he was a leading proponent of the use of correct classical Latin».[24]

This last sentence is true, but irrelevant since there are no demonstrable mistakes; more to the point is the fact that humanists enjoyed learned puns. An even more ill-considered opinion was printed in a Parisian paper,[25] where the dedication is described as:

---

[22] See KING, «Medieval Picard Astrolabe», p. 60-61, especially note 24.

[23] V. SCHOLDERER, «A Note on the Astrolabe of Regiomontanus», *The Library*, 5th series, 15 (1960), p. 209.

[24] Anonymous (London). For rete, read rule. See note 11 above.

[25] ROY (Paris).

«en latin de cuisine». Professor Noel Swerdlow of the University of Chicago, who has worked a great deal on the writings of Regiomontanus, sees no problem with the Latin dedication.[26] Dr. Nancy K. Zumwalt, of the University of California, has pointed out that the inscription is neat and clever:[27]

> The writer of this dedicatory inscription was learned in classical Latin. The inscription shows skilful command of periodic style. It also contains two witty astronomical puns.

These puns can only be understood by someone familiar with astronomy. The reference to Bessarion as «called *de cardine*» could relate to the *cardines*, a concept of prime importance in astronomy and astrology, and would then constitute a pun on the title Cardinal, the original form of which was, indeed, *de cardine*, and possibly a further pun based on the use of *cardo* to mean the East (Bessarion came to Rome from the East).[28] Zumwalt saw a pun in the word *surgo*, «I arise», or «I come forth», the association being with an astronomical rising (usually described by the verbs: *ascendere*, to mount, ascend, go up; *orire*, to arise, spring from). In any case, the choice of words was to some extent forced by the verse form of the inscription.

iv. *The Angel*

Who is going to dare to explain why any craftsman decided to choose a particular motif with which to ornament his handiwork? On the astrolabe dedicated to Cardinal Bessarion there are floral motifs on the rete, roses and six-pointed stars in rosettes forming the throne, a *trompe l'oeil* scroll that carries the dedication, and an angel. This figure was part of the original construction as becomes obvious through the hands, which were cut before the straight line border that they overlap.

---

26 Private communication to D. A. King.

27 In a report made available to Christie's of London.

28 The «cardinal» points of the compass are North, South, East, West. The «cardinal» signs in astrology are those that mark the seasons: Aries, Cancer, Libra, Capricorn. Similarly, the astrologer's four main Houses of Heaven that relate to one's surroundings are called the «angles» or «cardines». See EADE, *Astrology*. Both cardines and cardinal derive from the Latin *cardo*, a hinge. This embraces the idea of swinging, a point around which anything turns, a pole, the ends of the Earth's axis; *Cassell's New Latin-English... Dictionary*, 5th edn (London, 1968).

Those who are unthinking claim the rose as English Tudor, and the star as that of David. It is quite otherwise. In ancient Rome, roses were obligatory at ceremonies. In heraldry the rose forms the cadency mark of a seventh son. In early and medieval Christianity it symbolised the Virgin Mary and Paradise. The rosette was used in ancient Egyptian ornament, and in Assyria, the Middle East, and in Greek and Romanesque art. «The rosette is one of the most widely disseminated of all decorative motifs».[29] The interior of a rosette may have a rose or floral form, with symmetries of 4, 6, 8, 10, or 12-fold; 5-fold symmetry is not obligatory. This is why the rosettes in the throne with 6-fold symmetry need not be associated with David: it is just a pattern. The rose motif is found on the other Regiomontanus-type astrolabes (see below), and on those made by Georg Hartmann.

The angel is a typical example of those found in many parts of Europe in the mid-fifteenth century. Not that angels are not found at other times; it is the very long wings, the plain tunic with simple collar, the hair and the clear «young» face that are characteristic of the time. In fact, the brass angel could easily have flown out of the superb *A Choir of Angels* painted on wood by Simon Marmion from Flanders, who was active in 1449, and who died in 1489. Any visitor to Tuscany cannot avoid all the angels in the many lunettes and panels done in glazed white terracotta on a blue ground. Of those by Luca della Robbia (1400-1482), one may point to the four angels in the lunette of the *Resurrezione* in the Duomo (1443) at Florence, and also to the *Madonna della Cintola* at Arezzo (*c.* 1490) by Andrea della Robbia (1435-1525).[30]

There is evidence that some other astronomical instruments of the fifteenth century feature angels. Johannes of Gmunden (*c.* 1380-1442), the predecessor of Peuerbach in Vienna, described, before 1435, a bronze plate with a representation of the seven celestial spheres with the Trinity, the Virgin Mary, and various angels. An unsigned surviving instrument from the fifteenth century features precisely these.[31] Then there is the celestial globe supported by two angels that was placed by King Matthew of Hungary in his

[29] OSBORNE, *Decorative Arts*, p. 673.

[30] Simon, Marmion, National Gallery, London. Duomo plate 20, Arezzo plate 58 in *Della Robbia: A Family of Artists*, Milan: Scala, 1992. A Bohemian chasuble embroidered in silk and gold thread, dated 1487, shows the Madonna with four flyng angels of the same style; Moravian Gallery, Brno, Czech Republic, inv. no. 27411.

[31] ZINNER, *Regiomontanus*, p. 323.

library at Buda after he became King of Bohemia on 3 May 1469. The instrument was fixed to show the heavens at the time of his coronation.[32] Insofar that angels are heavenly beings, the presence of one of them on an astronomical instrument dedicated to a cardinal could be seen as entirely appropriate. One need not labour the point: always popular in Christian imagery, they proliferated during the fourteenth and fifteenth centuries, usually sexless in appearance reflecting the Hellenic ideal of the beautiful youth.[33]

v. *The Metal*

A metallurgical examination has been made on the brass of this astrolabe by light and scanning electron microscopy, and X-ray dispersive analysis.[34] The brass was found to be of very poor quality. The mater appears to be a single phase *alpha* brass, which contains many small grain boundary cracks that appear to have been produced at the time of manufacture. These cracks are attributable to embrittlement of the grain boundaries of the metal during cooling or subsequent mechanical handling. The cracks contain many inclusion particles based on zinc sulphide, together with lesser amounts of iron and some antimony. The disc on the reverse (with engraved angel) is also *alpha* brass, but with a higher zinc content than the mater. The mater and the disc are made from different batches of *alpha* brass. The disc has undergone mechanical deformation, and it may be assumed that the original metal was mechanically worked into a thin plate from which the disc was cut.

By modern standards the brass used to construct this astrolabe is a very dirty alloy, and the zinc sulphide would cause the observed embrittlement. Metallic zinc was not known in the West until the sixteenth century, and for the production of brass, zinc ores (for example, carbonate or sulphide) were used. Such a technique would not have been employed in western Europe in modern times (after 1800).

A completely separate analysis was made using a different technique at a different laboratory. The results in both cases are consistent in their conclusions.

The second analysis was part of a pilot experiment made to

---

32 ZINNER, *Regiomontanus*, p. 160.

33 OSBORNE, *Decorative Arts*, pp. 45-46.

34 Matcon Report No. 1603, August 1989.

compare Bessarion's astrolabe with a few other German astrolabes. These were by Hartmann of Nuremberg, made some 60 years later. The work was done by Dr. R. Brownsword, Department of Applied Physical Sciences, Coventry University, Coventry.[35] It would, of course, be worthwhile to perform a similar analysis on all ten of the known Regiomontanus-type astrolabes. Comparison with contemporaneous brass of known origin might reveal differences in the composition of pieces from, say, Nuremberg as opposed to Vienna. Dr. Brownsword's technique does not involve transporting an instrument to his laboratory, but requires a very small sample of about 20 milligrams, which is not consumed by the X-ray fluorescence analysis. In this programme of chemical analysis there are nine elements that constitute the key: the big four, all under the metal-founder's control: copper, zinc, tin, lead; the small five, not under the founder's control: antimony, arsenic, nickel, iron, silver. The profile of these nine metallic elements has been proved to be useful in dating and geographically locating certain classes of brass specimens.[36]

Table 1 - Chemical analysis of Bessarion's astrolabe dated 1462

Metal	mater	rete	45/48° plate	disc	alidade	rule
Cu	73.64	68.71	72.44	73.16	78.15	76.65
Zn	23.22	28.24	25.58	25.74	19.45	21.44
Sn	0.98	0.91	0.26	0.15	0.59	0.15
Pb	1.04	0.48	0.39	0.17	0.63	1.38
Sb	0.13	0.14	0.20	0.13	0.21	0.08
As	0.37	0.51	0.39	0.10	0.26	0.14
Ni	0.28	0.29	0.33	0.08	0.49	0.29
Fe	0.25	0.65	0.35	0.46	0.15	0.00
Ag	0.07	0.07	0.07	0.01	0.07	0.02

Percentage of nine elements present in six different parts.

[35] Dr. Brownsword's report is in the hands of the authors. The alloy of a Hartmann astrolabe dated 1524 is generally similar to the alloy of the Bessarion, but the Hartmann is higher in lead content.

[36] BROWNSWORD, «Metalwork».

The nature of the material used by craftsmen is one source of historical information about scientific instruments that has not hitherto been tapped. A thorough and soundly based programme of research and analysis is desirable, in particular, on mathematical instruments before about 1600. There are many astrolabes, quadrants, armillary spheres, etc., where, although they are genuine enough, it is as yet impossible firmly to fix a date for them more closely than within one hundred or even two hundred years, and next to impossible to allocate them to a region of manufacture. Without a programme that builds up a data bank of a very considerable number of analyses, such problems can hardly be solved. The ratio of the percentage of copper to zinc in the alloy forming the brass in a single sample has been erroneously regarded as an adequate basis for establishing, or undermining, authenticity; indeed, as a *surer* basis than connoisseurship.

Professor A. M. Pollard, then of the Research Laboratory for Archaeology and the History of Art, Oxford, chemically analyzed by X-ray probe many scientific instruments at the National Maritime Museum, Greenwich, and elsewhere, while his colleague, Miss C. Mortimer, examined some instruments in the Museum of the History of Science, Oxford.[37] Amongst Pollard's findings was that chemical analysis cannot be used to date a piece of brass:

> ... all that can be done is to compare its composition with those of known genuine pieces... It is also possible that perfectly genuine instruments will have analyses outside the expected range for a particular country and date...

Mortimer's conclusion reads:

> Although some compositional patterns are emerging in certain workshops and countries, it still seems unlikely that judgements on authenticity can be made purely on the basis of major element chemical analysis.

Another approach is more invasive (it involves polishing and etching a small area) but more telling: this is metallographic examination using a special optical microscope. This kind of examination reveals the microstructure of the metal, and shows whether it is cast, beaten, rolled, annealed, etc. One such

---

37 POLLARD, «Authenticity of Brass Objects»; MORTIMER, «X-Ray Fluorescence».

examination has been made by Dr. Robert B. Gordon on a Hartmann astrolabe dated 1537.[38]

5. Remarks on «Regiomontanus-Type» Astrolabes

Bessarion's astrolabe has, for over three decades, presented a problem in that it seemed to stand alone. It is only since 1989 that a significant group of German astrolabes of the fifteenth century has been discovered, and systematic study of them begun. So far, ten closely related instruments form a context in which the astrolabe of Bessarion holds a key position. These eleven astrolabes will be referred to as «Regiomontanus-type».[39]

There is evidence that Regiomontanus did devise astrolabes. Georg Hartmann, the leading instrument maker in Nuremberg in the first half of the sixteenth century, mentions, in his treatise on astrolabe construction written in his own hand in 1527, a feature of all the astrolabes of Regiomontanus that he had seen, namely that the plates served the latitudes 39°, 42°, 45°, 48°, 51°, and 54°.[40] This is a feature of the Bessarion astrolabe, and one other, amongst the Regiomontanus-type instruments. There is other documentary evidence that Regiomontanus himself devised astrolabes, namely:

a) About 1470 he designed a large astrolabe with three plates and with «Italian lettering», i.e. antiqua, now lost, for King Matthew Corvinus of Hungary. In 1512 this was passed on to Kurfürst Friedrich of Saxony.[41]

b) In the autumn of 1478, the second year after Regiomontanus' death, King Matthew of Hungary sent Hans Dorn to Nuremberg to purchase his books and instruments. Dorn, although he remained

---

[38] Gordon, «Metallography»; Gordon, «Technology».

[39] It is a sad reflection on the state of the field that these astrolabes were not properly documented decades ago. A similar situation prevails for fifteenth-century French and also Italian astrolabes, let alone for earlier astrolabes from these three, and other, regions. Detailed descriptions of all of the astrolabes mentioned in this study are in King, *Catalogue*.

[40] See Zinner, *Regiomontanus*, p. 222, and Zinner, *Instrumente*, p. 480. The relevant passage in MS Vienna Staatsbibliothek 12768, fol. 2*r*, translates roughly as follows: «Johannes de Monte Regio made [hat gemacht] astrolabes in brass. He usually [gewonlich?; reading uncertain] took three plates and put the pole elevation on all sides. On the first he made 39° pole elevation, on the other 42°, the third 45°, the fourth 48°, the fifth 51°, and the sixth 54°. The[se] pole elevations he used [illegible] on all astrolabes made by him that I have seen».

[41] Zinner, *Regiomontanus*, p. 160, quoting MS Jena Universitätsbibliothek Msk. Appendix 22; and Zinner, *Instrumente*, p. 480.

several months in Nuremberg, failed to achieve his mission.[42] In April 1479, Bernhard Walther purchased some of Regiomontanus' brass instruments. A catalogue of 1512 includes mention of two astrolabes and three incomplete ones.[43] In 1480, according to Walther, the executors of Regiomontanus' legacy had several instruments made by Regiomontanus deliberately destroyed and their metal sold to the locksmiths. It is not clear which of the instruments now in the Germanisches Nationalmuseum are to be associated with these two complete and three incomplete astrolabes.[44] A prospectus of books that Regiomontanus hoped to print in Nuremberg, published in about 1473 or 1474,[45] ends thus:

> In the workshop an astrarium is constantly under construction, an achievement certainly worthy to be viewed as a wonder; some astronomical instruments for observations of the heavens are also being made, and likewise other instruments for common daily use, to recite the names of which would be tedious.

An astrolabe often associated with Regiomontanus in the modern literature is most probably only a piece that was owned by him if not merely associated with him by later tradition.[46] It is an astrolabe plate with gothic inscriptions serving a latitude of about 46 1/2°, suitable for some locality in the Italian Alps, and it is dated 1468. It is difficult to imagine Regiomontanus designing such a «medieval» instrument six years after his stay in Italy. Also, if the piece was in Nuremberg in the early sixteenth century Hartmann would surely have seen it, and had he thought that Regiomontanus had made it he would not have written his sweeping remark about all of Regiomontanus' astrolabes which he had seen.

Earlier (fourteenth century?) German astrolabes are a varied

---

[42] ZINNER, *Regiomontanus*, p. 245, and ZINNER, *Instrumente*, p. 293.

[43] ZINNER, *Regiomontanus*, pp. 221, 246.

[44] For a discussion of the provenance of these instruments, see Nuremberg 1992-1993, *Catalogue*, part 1.

[45] Information kindly supplied by Professor Noel Swerdlow, University of Chicago.

[46] IC 550: Nuremberg, Germanisches Nationalmuseum, inv. no. WI 5. See GUNTHER, *Astrolabes*, pp. 280-281 (no. 137), with a lithograph of the front captioned «Astrolabe of Regiomontanus» (in the text this piece is confused with a twelfth-century Syrian astrolabe with Zodiac figures depicted on the rete); ZINNER, *Regiomontanus*, pp. 217, 222 (on Regiomontanus' scientific estate), and figs. 72-73 on plate 40 (minuscule reproductions of the front and back); ZINNER, *Instrumente*, p. 480 (brief description, in which the latitude is miscalculated to 50°); Nuremberg 1976-1977 *Catalogue*, pp. 90-91 (no. 47); Linz 1990 *Catalogue*, p. 33 (no. 30: «Astrolabium des Regiomontanus»), with an illustration of the back only; Nuremberg 1992-1993 *Catalogue*, item 1.76 (first detailed description by D. A. King).

group, including a quatrefoil astrolabe owned by Ludolf de Scicte of Einbeck,[47] and an astrolabe displaying French influence.[48] An illustration of an astrolabe in a fourteenth-century treatise, perhaps by Heinrich of Langenstein, shows that rete types were used in that century bearing no obvious relation to any surviving German instruments.[49] An interesting example of transmission southwards across the Alps is afforded by an astrolabe signed by Bernadinus Aurifex in 1565,[50] which has the same unusual rete design as the one featured in the fourteenth-century German manuscript.

German astrolabes of the fifteenth century can be divided into four main groups: instruments of the Regiomontanus-type (eleven pieces), with retes based on an Italian model and precursors of the better-known Hartmann astrolabes (Type A, ten pieces), or more complicated, and precursors of the Hans Dorn-type (Type B, one piece); simplified astrolabes associated with Hans Bamman of Schmalkalden (six pieces), dated between 1484 and 1493;[51] some instruments usually associated with Hans Dorn (four pieces);[52] the

---

[47] IC 2027: Cracow, Jagellonian University Museum, inv. no. 4043. See Zinner, *Regiomontanus*, plate 42, figs. 78, 79; Zinner, *Instrumente*, pp. 535-536.

[48] IC 610: Cologne, Städtisches Museum, inv. no. KSM-L 181. On this piece, see Dieckhoff, «Arsenius-Astrolabium», pp. 23-24.

[49] See Zinner, *Instrumente*, p. 142, for details but no illustration.

[50] IC 2058: Fano, Museo Civico, inv. no. 103. See Panicali, *Orologi*, pp. 38, 39, 46 for the front, back, and signature.

[51] On some of these, see Gunther, *Astrolabes*, pp. 423-424 (IC 248 and IC 249), and Zinner, *Instrumente*, p. 244.

[52] On Dorn and the instruments associated with him (except for IC 492), see Zinner, *Regiomontanus*, pp. 46, 245, and plate 41 (IC 642); Zinner, *Instrumente*, pp. 292-297. Also A. J. Turner, *Instruments*, pp. 35-39, figs. 25, 26 (IC 642); Washington NGA 1992 *Catalogue*, pp. 221-224 (by F. R. Maddison). The *only one* of his instruments that is signed is a compendium in the British Museum, London (IC 250); the attribution of the others to him still needs to be confirmed. Astrolabic instruments associated with Dorn, which are quite different in conception (except for the rete design) and execution to the «Regiomontanus-type», are:

IC 4519: dated 1480; celestial globe, with sundial on horizontal frame and *organum Ptolemei* on astrolabic astrological plate attached to the top: Cracow, Jagellonian University Museum, 4039-37/V; Ameisenova, *Bylika Globe* (detailed description by art historian).

IC 492: dated 1483; astrolabe, fitted with *organum Ptolemei* and eclipse calculator (in fact part of an albion): Florence, Museo di Storia della Scienza, inv. no. 1096; Florence, 1954 *Catalogue*, pp. 57-59 (no. 1096) (identified as German); Florence, 1991 *Catalogue*, p. 8, I. 14; Goldstein, «Transversal Scale», pp. 103-104, figs. 1, 2 (front and detail); G. L'E. Turner, *Strumenti*, p. 73 (front).

IC 642: dated 1486; astrolabe: Cracow, Jagellonian University Museum, 4040-38/V;

IC 250: dated 1491; signed «HANNS DORN»; compendium with astrological astrolabic plate: London British Museum, inv. no. 94 6-15 1; London BM *Catalogue*, p. 74, no. 207, plate 28 (back); G. L'E. Turner, *Strumenti*, p. 120 (both sides).

Two astrolabic plates for a single latitude in Florence and Salzburg are related to work attributed to Dorn, if not actually by him.

astrolabic plate dated 1468 and serving the latitude of a locality in the Italian Alps.

i. *The «Regiomontanus-Type» Astrolabes*

Regiomontanus-type astrolabes can be classified as follows:

Groups 1-4: Astrolabes with Type A retes

Group 1: Astrolabes with a lunar volvelle on the back

a) Nuremberg – dated 1457, gothic script;[53]

b) Paris – undated, gothic script as on 1a on the plates and back, and antiqua as on 2b on the rete.[54]

Group 2: Astrolabes with an *organum Ptolemei* on the back

a) Paris – undated (with some early-sixteenth-century additions on the back of the movable plate), gothic script resembling that on 1a and 1b, some antiqua majuscules as on 1b and 2b;[55]

b) London – dated 1462, signed by Ioannes (Regiomontanus), dedicated to Cardinal Bessarion, antiqua script as on the rete of 1b.

Group 3: Standard astrolabes

a) Darmstadt – dated 1466, gothic script, some antiqua capitals;[56]

b) Munich – undated, gothic script;[57]

c) Strasbourg – dated 1481, gothic script as on 3d;[58]

---

53 IC 549: Nuremberg, Germanisches Nationalmuseum, inv. no. WI 129. See ZINNER, *Regiomontanus*, p. 222, and plate 40, figs. 70-71 (front and back); ZINNER, *Instrumente*, pp. 150, 465, and plate 50.2 (back only); PILZ, *Astronomie*, p. 54; Nuremberg 1976-1977 *Catalogue*, p. 71 (no. 15); Linz 1990 *Catalogue*, vol. 2, p. 32 (no. 29), (illustration of the back only); Nuremberg 1992-1993 *Catalogue*, item 1.75 (detailed description).

54 IC 452: Currently in a private collection; formerly in the Greppin Collection in Brussels, sold at Galerie Köller in Zurich on 17.11.1975. See Brussels 1957 *Catalogue*, p. 9 (no. A13); and GREPPIN *Catalogue*, p. 32 (no. 3122: «probablement par Georg Hartman [*sic*], vers 1460»), with an illustration of the back only on plate 13. We are grateful to the present owner for allowing us to study this instrument.

55 IC 256: Currently in a private collection, formerly in the Nachet Collection. See NACHET *Catalogue*, p. 44 (no. 126), illustration of the front on plate 14, facing p. 42; GUNTHER, *Astrolabes*, p. 434 (IC 256), and also KING, «Report», p. 7, fig. 5 (illustration of throne and inscription on mater). The inscriptions recorded by Gunther postdate the construction of the instrument by about 75 years. We are grateful to the present owner for allowing us to study this instrument.

56 IC 592: Darmstadt, Hessisches Landesmuseum, inv. no Kg. 25.15. Unpublished.

57 IC 620: Munich, Deutsches Museum, inv. no. 70678. See ZINNER, *Instrumente*, plate 41, fig. 1 (front only: «Gotisches Astrolab»).

58 IC 530: Strasbourg, Musée Historique, inv. no. IL 74. See POULLE, «Fabrication des astrolabes», p. 118 (illustrations of front and back); ZINNER, *Instrumente*, pp. 150, 421 (s.v. Arnold Kunig); Strasbourg, 1987 *Catalogue*, p. 1 (illustration of detail of plate), front cover (colour illustration of back), and back cover (colour illustration of front).

d) Cracow – dated Schmalkalden, 1482, gothic script as on 3c.[59]

Group 4: Astrolabes with unusual plates and three different styles of inscription

a) Paris – undated;[60]
b) New York – undated.[61]

Group 5: Astrolabe with Type B rete

a) Cracow – undated rete and plates, engraved in the same gothic hand as 1a, fitted to a fourteenth-century Italian mater.[62]

Table 2 - Size, number of stars on rete, and date of vernal equinox on the Regiomontanus-type astrolabes.

No. location	IC no.	size	stars	equinox	date
1a Nuremberg GNM	549	129/6	30	10 1/3	1457
1b Paris PC [Greppin]	452	137/7	30	11	–
2a Paris PC	256	168/9	27	11	–
2b London PC	640	115/4	30	10 1/2	1462
3a Darmstadt HLM	592	144/7	27	10 1/2	1466
3b Munich DM	620	132/6	27	10 1/3	–
3c Strasbourg MH	530	170/4	27	13	1481
3d Cracow MN*	4501	152/6	27	12	1482
4a Paris CNAM	516	132/5	27	10 1/4	–
4b New York CUL	545	133/7	27	10	–
5a Cracow JUM	644	**	34	**	–

* anomalous throne
** mater not original.

No: number in Groups 1-5 above
IC number: International Checklist number
Size: diameter and thickness in millimetres
Stars: number of stars on the rete
Equinox: date in March of vernal equinox as read from calendrical scale.

---

[59] IC 4501: Cracow, Muzeum Narodove, Czartoryski Collection, inv. no. XIII/753. Unpublished.

[60] IC 516: Paris, Conservatoire National des Arts et Métiers, inv. no. 3882/1. Unpublished (see the unpublished 1851 handlist available in the Museum, p. 40, no. F-b 20; brief mention).

[61] IC 545: New York, Columbia University Special Collections, inv. no. 27-256. Unpublished. See New York IM *Handlist*, p. 59 (no. 13) (brief mention).

[62] IC 644: Cracow, Jagellonian University Museum, inv. no. 4042-40/V. See ZINNER, *Regiomontanus*, figs. 74-75 (front and back); ZINNER, *Instrumente*, p. 296; BURCZYK-MARONA, «Astrolabium» (detailed description with illustrations of all parts, including all four quarters of the rete).

The inspiration for the distinctive style of the retes on these astrolabes came originally from Italy. Two Italian pieces provide the evidence: a fourteenth-century (?) astrolabe (Fig. 10),[63] and an astrolabe based on the same rete design made in Urbino and dated 1462.[64] The latter is signed «KP», initials more likely to have represented a German (or Greek) name than an Italian one. One of these pieces, or another with identical details, including the shape of the individual star-pointers, and also the rete pattern, is depicted on the intarsia in the study of Archduke Ferdinand of Urbino.[65]

The most intricate of the Type A retes is that on the 1457 piece. The Type B rete, whose sole surviving example is somewhat larger than any of the Type A ones, is also to be found later on the even larger astrolabes associated with Hans Dorn, dated 1483 and 1486.

The astrolabes of Type A provided some of the inspiration for Hartmann's *Volkastrolabien*, made in Nuremberg in the first half of the sixteenth century.[66] Some two dozen of his works survive, and with one exception, they all resemble at first sight the Regiomontanus Type A instruments. Hartmann simply used the same retes that had been popular for more than half a century. The rete form may have come to Hartmann by way of his earlier contemporary Johannes Stöffler (1452-1531), whose retes are still closer to the Italian originals mentioned above.[67] Hartmann also used rosettes on the thrones of his instruments, again imitating the works of his predecessors. In fact he even copied the stars used on the earlier instruments, sometimes taking over 20, and at other times 30 stars. These instruments are sufficiently well-documented in the modern literature that no detailed description is necessary here.[68] However, no detailed comparative study has ever been conducted.

---

[63] IC 493: Florence, Museo di Storia della Scienza, inv. no. 1107. See Florence IMSS 1954 *Catalogue*, p. 62 (no. 1107); Florence IMSS 1991 *Catalogue*, p. 8 (no. I.6).

[64] IC 4506: Moulins (Allier), Musée Départemental - stolen. See Paris BN 1973 *Catalogue*, p. 75 (no. 176), with illustrations of the front and back on p. 57.

[65] Illustrated most recently in CARDINI & FUMAGALLI, *Antiche Università*, p. 37.

[66] On Hartmann, see GUNTHER, *Astrolabes*, pp. 436-440; ZINNER, *Instrumente*, pp. 357-368; A. J. TURNER, *Instruments*, pp. 39-40, 45; KLEMM, *Hartmann*; KING, *Catalogue*, Section 7.2.

[67] See, for example, GUNTHER, *Astrolabes*, p. 428, for an illustration.

[68] For detailed descriptions of individual Hartmann astrolabes, see Rockford TM *Catalogue*, pp. 128-131 (no. 17); Washington NMAH *Catalogue*, pp. 146-150 (no. 262); Ecouen MR *Catalogue*, pp. 106-107 (no. 82); and Nuremberg 1992-1993 *Catalogue*, items 1.69 and 1.70. See also GORDON, «Metallography» and «Technology».

Some other early sixteenth-century German astrolabes continued the tradition of the Regiomontanus and Dorn astrolabes, namely:

a) four undated astrolabes by the same maker with distinctive Y-shaped frames inside the ecliptic and below (one – the only one known to Gunther – was misidentified as Italian and dated c. 1500; a later dating may be more appropriate);[69]

b) two undated astrolabes by the same maker with a vaguely rectangular frame on the rete developed from Type B.[70]

ii. *Some common features of the «Regiomontanus-type» astrolabes*

The eleven associated astrolabes display competent workmanship throughout, with various hands in evidence in the engravings. Only the distinctive gothic of 1a, 1b (plates and mater) and 5a, and the distinctive antiqua of 1b (rete) and 2b, are unmistakable. We are dealing with the productions of a series of individuals, possibly in a single workshop.

The diameters of the ten astrolabes with Type A retes vary between 115mm (2b) and 170mm (3c), with the single Type B rete (5a) fitting into an earlier mater of diameter 252mm. A diameter of 132 or 133mm is found on three instruments (3b, 4a, 4b); otherwise the diameters are all different.

The basic design of the thrones on all but one piece (3d) is identical. Each instrument has a throne consisting of three rosettes, the lower two supported by a trilobed motif at either side and the suspensory apparatus being attached to the upper one. The predominant decoration for these rosettes is a five-petal rose, which also appears occasionally on the pin holding the astrolabe together

---

[69] IC 173: Oxford, Museum of the History of Science, inv. no. LE 2065, diameter 197 mm. See GUNTHER, *Astrolabes*, pp. 327-328, and plate 73 (front and back).

IC 453: Point Lookout, N. Y., Collection of Leonard Linton, formerly Brussels, Greppin Collection (no. 3124), sold at Galerie Köller in Zurich on 17.11.1975, diameter 197 mm. See GREPPIN, *Catalogue*, p. 33 (no. 3124: «français vers 1500»), with illustrations of the front and back on p. 14; LINTON, *Catalogue*, p. 133 (no. 186), colour plate of front and back following.

IC 4505: with leather case, probably original, dated 1542: Munich, Deutsches Museum: 1991-223, 1 (astrolabe) + 2 (case). Purchased in 1991 from private owner in Jena, diameter 197 mm. Unpublished.

IC 4543: Rome, Osservatorio Astronomico, diameter not known.

[70] IC 491: Florence, Museo di Storia della Scienza, inv. no. 1110, diameter 134 mm. See Florence IMSS 1954 *Catalogue*, p. 62 (no. 1110); Florence IMMS 1991, *Catalogue*, p. 8 (no. I.7); POULLE, «Instruments astronomiques», pp. 20-21 (illustration of front).

IC 4544: Stuttgart, Württembergisches Landesmuseum (loan of the Fürstlich Fürstenbergische Sammlungen): L1985-200, diameter 187 mm. Unpublished.

(see below). A star (2b, 3a) or representations of the sun and moon (1b) sometimes replace the rosette. On three pieces rosettes are replaced by roundels with coats of arms (arms of Vienna on 1a, arms of von Wallenstein on 1b,[71] empty on 3a), and on one (2a) a magnetic compass.[72] The anomalous throne on 3d consists of a single rosette supported by a lattice with foliate motifs.

The basic design of the retes is very similar, with one larger and more developed than the rest (Type B, 5a) and one of the remainder (1a) at an intermediary stage of development. There is a roughly rectangular frame inside the southern ecliptic, formed by two arcs standing on the sides of the ecliptic and four circular arcs «parallel» to the equinoctial bar. A trilobed motif features at the middle of the upper side on the two astrolabes 1b and 2b (Bessarion's). The rete of 1a is more sophisticated than the rest of Type A, featuring elegant details such as a vertical bar shaped with a beast's head above the central disc. On the Type A retes there is a roughly Y-shaped frame below the ecliptic, formed by a series of circular frames, sometimes with a central part modified from a half-quatrefoil (1a and 2b). On several of the retes (1b, 2b, 3a, 4a, 4d) there is a small circular appendage attached to the pointer for Vega to indicate the centre of the ecliptic. On others (1a, 1b) there is a rectangular frame outside the ecliptic circle at the winter solstice. Type B has six arcs of circles along the top of the «rectangular» frame inside the ecliptic and a vertical solstitial bar spanning the ecliptic which is counter-changed three times above the central circular disc. There is a prominent trefoil design supporting the frame below the ecliptic.

The equinoctial bars on the Type A retes are all counter-changed according to the same pattern, here stylized:

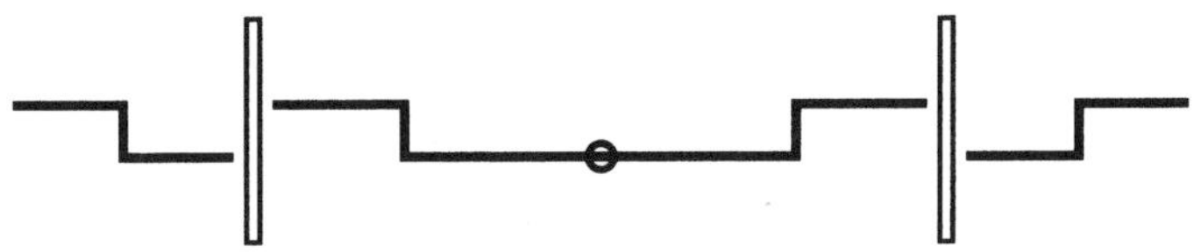

That on Type B is the same but upside down.

---

[71] The engraving reads AEW, apparently for Albrecht Václav Eusebius von Wallenstein (1583-1634), Bohemian soldier and statesman, commanding general of the armies of the Holy Roman Emperor, Ferdinand II, during the Thirty Years' War. See, for example, *Encyclopedia Britannica*, XII, pp. 470-471.

[72] See KING, «Report», p. 7, fig. 5, for an illustration.

The basic design of the star-pointers is identical. A wavy pointer is attached to a circular base, often with one or two «collars», and occasionally to a base made up of two circles. The base of the pointer for Regulus is distinguished by an asterism around the circular base on 1b, 2b, and 3a. The reason is simply that a crown is intended; the name Regulus means «little king».

The letter forms are often gothic, but in two cases, clearly involving a single engraver, antiqua is used. The gothic inscriptions are characterized by a prolonged final S on its side (Figs. 13, 17, 19, 21); and a «1» which is higher than the «0». The «2» is the typical German form resembling a «Z» (Figs. 13, 21). The elongated S was also used on the simplified Schmalkalden astrolabes and the larger instruments attributed to Hans Dorn. Some of the inscriptions are stamped rather than engraved. For example, on 4a all the lettering and numbering on three of the plates is punched. The letters are antiqua majuscules, and the numbers are all modern. The letters resemble those used by Hartmann, so it may be the case that three plates were added or substituted (or even modified) in the sixteenth century in this older astrolabe. Regiomontanus also gradually changed from the old gothic numerals to the new forms in his manuscript works.[73] His contacts with Italy also led him to abandon the gothic script for the antiqua. On the astrolabes of Hartmann, the inscriptions are stamped not engraved, and are in antiqua; however, in his manuscripts Hartmann used a gothic script.

On the Strasbourg instrument (3c) the engraving is at first sight different; for example, a different form of the «2» is used and some capital N's are reversed. The astronomical markings are achieved with less precision; in particular the calendrical scale is inaccurate, with a date for the vernal equinox that is centuries out-of-date (March 13 instead of March 10 1/2). The instruments 4a and 4b require special consideration because of anomalous plates and varied hands.

The star-names on all the instruments are not identical, but are closely related. They are of particular historical interest because they reflect an important trend of the time: the abandonment of Latinized Arabic names and the adoption of Latin names. This trend

---

[73] ZINNER, *Regiomontanus*, p. 49; G. L'E. TURNER, *The Regiomontanus Astrolabe*, p. 11. On the development of medieval and Renaissance numerals, see HILL, *Arabic Numerals*, an important work based, however, mainly on manuscripts. An investigation of the development of numeral forms as attested by instruments is a task for the future.

was not irreversible: on later instruments we witness the gradual but persistent reversion to Arabic names. It is as a result of this late trend that we still use some Arabic star-names to this day. The choice of stars and the forms of the Arabic names used for them on the Type A retes indicate the influence of the star-table of John of Gmunden (d. Vienna, 1442).[74] In particular, *alhes* for α *Crateris* derives from a corruption of Arabic *al-ka's* for «drinking glass», and *alkair* for α *Aquilae* from a Germanized rendering of *alcair*, derived from *altair* (Arabic *al-tā'ir*, «flying (eagle)») through confusion of c and t in gothic script.

Preliminary investigations of the positions of the star-pointers by Burkhard Stautz of Frankfurt indicate that the positions correspond to contemporaneous star catalogues. These investigations are still in progress. The same catalogues were used by Hartmann decades later, which explains why the star positions on Hartmann's astrolabes correspond to a date several decades before his time.

We have already commented on the lunar volvelles which feature on the backs of two pieces (1a and 1b), and the *organa Ptolemei* on two others (2a, 2b). On the former pair the shadow scale is found inside the calendrical scale. On the latter pair it is ingeniously divided into four rectilinear parts in the annulus between the scale for the lunar mansions and the outer scales. With the exception of 1a, 1b, 2a, and 2b, all instruments have a double universal horary quadrant and a double shadow square on the back.

Two pieces (3b, 4a) have a set of horizons for each 10° engraved on the mater. The latitudes used for the plates are not identical, but there are favoured patterns. The increments of 3° between 30° and 54°, which are a common feature, are intended to make the astrolabes «universal». The specific latitudes then correspond roughly to the boundaries and middles of the climates of Antiquity (with extensions to serve Northern Europe); the continued explicit and implicit use of the climates throughout the Middle Ages and the early Renaissance is only now being recognized.[75] The latitudes served by the plates on the eleven astrolabes are:

---

[74] See KUNITZSCH, *Arabs and Stars*, chapter XXII, and also KUNITZSCH, *Sternverzeichnisse*, pp. 111-116 («Typ XVII»). On John of Gmunden see, for example, ZINNER, *Instrumente*, pp. 322-323; *Dict. Sci. Biog.* VII, pp. 117-122.

[75] On climates, see HARLEY & WOODWARD, *History of Cartography*, I, especially pp. 182-183; Nuremberg, 1992-1993 *Catalogue*, I, p. 107; KING «Astronomical Secrets».

Table 3 - The latitudes engraved on the plates

1a:	32°	37°	48°	51°	(one plate missing)			
1b:	32°	35°	37°	42°	45°	48°	51°	
2a:	30°	36°	39°	42°	45°	54°	(one plate missing)	
2b:	39°	42°(R)	45°	48°	51°	54°		
3a:	36°	39°	42°	45°	48°	51°	54°	57°
3b:	39°	42°	45°	48°	51°	54°		
3c:	42°	45°	48°	51°				
3d:	42°	45°	48°	51°				
4a:	0°	3°	6°	9°	12°	51°	54°	90°
4b:	39°	42°	51°	54°	*			
5a:	48°	49°	50°	51°	52°	67°		

R = ROMAE; * two sets of markings not yet explained.

Only the latitudes used on the Bessarion astrolabe (2b) and the undated Munich piece (3b) correspond precisely to those on Regiomontanus' astrolabes which Hartmann had seen (see above). The latitudes used by Regiomontanus for Vienna in his written works were 47° 46′ (from John of Gmunden) and later 48° 0′, 48° 15′, 48° 22′, and 48° 27′; for Nuremberg he used 49° 0';[76] it was typical of medieval and Renaissance instrument-making to use integral degrees. On several instruments (including 1a, 1b, 3a, 3b, 4a) there are markings on the lugs of the plates, either a reminder of the latitude or marks to distinguish one plate from another. The latitudes used on 4a are unique in surviving European instruments; they were clearly chosen for pedagogic purposes.

Those original alidades that survive are elegantly fashioned. Some (2a, 3b, 3c, 3d) have folding sights. One (1a) has ends decorated with beasts' heads (also found on the retes of 1a and 5a). None bears any astronomical markings. Several pieces have a radial rule for use on the front; two (1a, 2b) have a diametral rule. (The radial rule without markings on 3c and the diametral rule without markings on 2a are replacements). The northern and southern declinations are indicated by Septentrionalis (3a), Latitv Septen (3b), Septemtrionalis (1a, 1b), or Septemtrio (2b), and Meridionalis (1a), Lati Meri (3b), or Meridies (2b) respectively.

Two instruments aside from the Bessarion astrolabe bear marks of ownership. On 2a are the date 1535 and two names (Fig. 19)

[76] Zinner, *Regiomontanus*, pp. 39, 60, and plate 30, fig. 55.

engraved on the blank back of the mater over which the *organum Ptolemei* rotates; on the back of the latter are inscribed the date 1537 and initials. On the second (3d), the initials SWIII (= SWM?, or the third instrument produced by SW?) occur inside the shadow square.

One instrument aside from the Bessarion astrolabe bears mention of a location. On Bessarion's astrolabe (2b) Rome is mentioned, most probably as the place where the astrolabe was presented. The 1482 piece with the anomalous throne (3d) bears the name Schmalkalden (near Suhl in Thuringia) by the date. We can be fairly certain that of the eleven astrolabes only this one was made there. One could speculate that the maker was visiting Hans Bamman, several of whose astrolabes, all contemporaneous with this piece, are marked with the name of this town.

iii. *Concluding remarks*

We are dealing here with a newly-identified corpus of historical material. Only recently (since 1989) have most of the items been identified in the course of the Frankfurt cataloguing project. Still more recently (since 1991) have they been inspected by us consistently, paying particular attention to the engraving and construction marks. But we have not been able to compare any two of these instruments with each other except by using photographs. We lack reliable information on the workshop(s) where they were made. It is possible that the location of the workshop was in Vienna. We know that Peuerbach, Regiomontanus, and Dorn devised or made astrolabes. In the earliest dated piece (1a from 1457) Peuerbach may have played a role. One could conjecture, and this is pure speculation, that Peuerbach started the piece in two different hands (1b) and Regiomontanus, still in Vienna and still favouring the old Arabic star-names, finished it. Dorn is the only one of the three who was mainly in Vienna throughout the 1450s and 1460s, and he may have had a role in the later productions. But the instruments attributed to him (of which he signed only one) are quite different in character from the ones we have described here.

Inevitably the corpus raises as many questions as its discovery solves, but these questions are now on a different plane. Two of these astrolabes, the 1457 piece and the Bessarion astrolabe, were exhibited together at the Germanisches Nationalmuseum in

Nuremberg in 1992-1993,[77] and the Bessarion astrolabe in the Bibliothek Otto Schäfer in Schweinfurt in 1993-1994.[78] If sufficient interest in this remarkable corpus of instruments can be aroused, it would be worthwhile to exhibit all of them together. Ideally, the entire corpus of German astrolabes up to about 1550, a significant date when new trends are detectable, should be studied in detail, along with the available texts.

Acknowledgements

The authors are grateful to the directors and staff of all those museums that have allowed instruments to be examined and photographed. They are also indebted to a number of private collectors for similar facilities. They wish especially to thank Dr. W. F. Ryan and the late Jon Darius for their comments on the text.

Bibliography

Ameisenowa, *Bylika Globe*
Zofia Ameisenowa, *The Globe of Martin Bylika of Olkusz and Celestial Maps in the East and the West*, translated by Andrzej Potocki, Wrocław, Cracow, Warsaw: Polska Akademia Nauk, 1959.

Anonymous (London)
«Under the protection of the divine Bessarion called Cardinal I arise in Rome... 1462», *Antiques Trade Gazette*, 28 October, 1989, p. 27.

Brownsword, «Metalwork»
R. Brownsword, «X-ray Fluorescence Analysis on Non-ferrous Archaeological Metalwork: Part 2. Archaeological Applications», *Analytical Proceedings*, 22 (1985), pp. 73-75.

Brussels 1957 *Catalogue*
[H. Michel, *et al.*], *Exposition: Art, Histoire et Sciences. Catalogue*, Brussels: Musées Royaux d'Art et d'Histoire, 1957.

Burczyk-Marona, «Astrolabium»
D. Burczyk-Marona, «Zegar słoneczny i astrolabium profesora Jana

[77] See Nuremberg 1992-1993 *Catalogue*, items 1.75 (the Nuremberg GNM astrolabe of 1457) and 1.76 (the Bessarion astrolabe).

[78] Schweinfurt, 1993-1994, *Catalogue*, item 179.

Brozka w zbiorach Muzeum Uniwersytetu Jagiellońskiego», *Opuscula Musealia* (*Zeszytu Naukowe Uniwersytetu Jagiellońskiego*), [Warsaw and Cracow], 3 (1988).

CARDINI & FUMAGALLI, *Antiche Università*
Franco Cardini and M. T. Fumagalli Beonio-Brocchieri, *Antiche Università d'Europa: Storia e personaggi degli Atenei nel Medio Evo*, Milano: Editoriale Mondadori, 1991.

Dict. Sci. Biog.
Dictionary of Scientific Biography, 16 vols, New York: Charles Scribner's Sons, 1970-1980.

DIECKHOFF, «Arsenius-Astrolabium»
R. DIECKHOFF, «Cosmographia planisphaeria: Ein 'Arsenius'-Astrolabium des späten 16. Jahrhunderts in Kölnischen Stadtmuseum», *Kölner Museums-Bulletin: Berichte und Forschungen aus den Museen der Stadt Köln*, 2 (1990), pp. 23-44.

EADE, *Astrology*
J. C. EADE, *The Forgotten Sky: A Guide to Astrology in English Literature*, Oxford: Clarendon Press, 1984.

Ecouen MR *Catalogue*
ADOLPHE CHAPIRO, CHANTAL MESLIN-PERRIER, and ANTHONY TURNER, *Musée national de la Renaissance, Château d'Ecouen: Catalogue de l'horlogerie et des instruments de précision du début du XVIe au milieu du XVIIe siècle*, Paris: Editions de la Réunion des musées nationaux, 1989.

Encyclopaedia Britannica
The New Encyclopaedia Britannica, 15th edn, Chicago: University of Chicago Press, 1986.

FLEGG, *Numbers*
GRAHAM FLEGG, *Numbers through the Ages*, Basingstoke: Macmillan, 1989.

Florence IMSS 1954 *Catalogue*
M. L. BONELLI, ed., *Catalogo degli strumenti del Museo di Storia della Scienza*, Firenze: Olscki, 1954.

Florence MSS 1991 *Catalogue*
M. MINIATI, ed., *Museo di Storia della Scienza: Catalogo*, Firenze: Istituto e Museo di Storia della Scienza, 1991.

GOLDSTEIN, «Transversal Scale»
BERNARD R. GOLDSTEIN, «Levi ben Gerson: On Instrumental Errors and the Transversal Scale», *Journal of the History of Astronomy*, 8 (1977), pp. 102-112.

GORDON, «Technology»
R. B. GORDON, «Sixteenth-Century Metalworking Technology used in

the Manufacture of two German Astrolabes», *Annals of Science*, 44 (1987), pp. 71-84.

GORDON, «Metallography»
R. B. GORDON, «Metallography of Brass in a 16th-century Astrolabe», *Journal of the Historical Metallurgy Society*, 20 (1986), pp. 93-96.

GREPPIN, *Catalogue*
La Collection Greppin, Instruments scientifiques, Zurich: Galerie Koller, 1975. [Catalogue of auction, 17 Nov. 1975].

GUNTHER, *Astrolabes*
R. T. GUNTHER, *Astrolabes of the World based upon the Series of Instruments in the Lewis Evans Collection... with Notes on Astrolabes in... other Public and Private Collections*, 2 vols [continuously paginated], Oxford: privately printed by the University Press, 1932; reprinted bound as one, London: The Holland Press, 1976.

HARLEY & WOODWARD, *History of Cartography*
J. B. HARLEY & DAVID WOODWARD, *The History of Cartography Volume I: Cartography in Prehistoric, Ancient, and Medieval Europe and the Mediterranean*, Chicago & London: Chicago University Press, 1987.

HARTNER, «Astrolabe»
WILLY HARTNER, «The Principle and Use of the Astrolabe», chapter in Arthur U. Pope, *A Survey of Persian Art*, III, London: Oxford University Press, 1939; reprinted with original plates in *Astrolabica 1*, Paris: Société Internationale de l'Astrolabe, 1978, and the article «Asturlāb» in *Encyclopedia of Islam*, I, pp. 722-728. Both reprinted in W. HARTNER, *Oriens-Occidens*, Hildesheim: Georg Olms, 1968, pp. 287-318.

HILL, *Arabic Numerals*
G. F. HILL, *The Development of Arabic Numerals in Europe Exhibited in Sixty-Four Tables*, Oxford: Clarendon Press, 1915.

KING, «Report»
DAVID A. KING, «Medieval Astronomical Instruments: A Catalogue in Progress», *Bulletin of the Scientific Instrument Society*, no. 31 (1991), pp. 3-7.

KING, *Catalogue*
DAVID A. KING, *A Catalogue of Medieval Astronomical Instruments: Astrolabes, Quadrants, and Sundials*, parts 1-5, Eastern (Islamic and Indian), parts 6-10 European: in preparation.

KING, «Medieval Picard Astrolabe»
DAVID A. KING, «Rewriting History through Instruments: The Secrets of a Medieval Astrolabe from Picardy», in R. G. W. ANDERSON, J. A. BENNET & W. F. RYAN, eds, *Making Instruments Count: Studies in Honour of Gerard Turner*, Aldershot: Variorum, 1993.

KING, «Secrets»
DAVID A. KING, «Some Medieval Astronomical Instruments and their Secrets», in Renato G. Mazzolini, ed., *Non-verbal Communication in Science prior to 1900*, (Biblioteca di «Nuncius», Studi e testi), Firenze: Leo S. Olschki, 1993.

KING, «Astronomical Secrets»
in HARRY KÜHNEL, ed., *Kommunikation zwischen Orient und Okzident, Sitzungsberichte der Österreichischen Akademie der Wissenschaften, Phil.-Hist. Klasse, Veröffentlichungen des Instituts für Realienkunde des Mittelalters und der frühen Neuzeit*, Bd. 17, Vienna: Verlag der Österreichischen Akademie der Wissenschaften, 1994.

KLEMM, *Hartmann*
HANS GUNTHER KLEMM, *Georg Hartmann aus Eggolsheim (1489-1564): Leben und Werk eines fränkischen Mathematikers und Ingenieurs* (Wissenschaftliche und künstlerische Beiträge, Ehrenbürg-Gymnasium Forchheim, Heft 8), Forchheim: Gürtler Druck, 1990. [Revision of catalogue of exhibition at Eggolsheim, 1989, for 500th anniversary of Hartmann's birth.].

KUNITZSCH, *Sternverzeichnisse*
PAUL KUNITZSCH, *Typen von Sternverzeichnissen in astronomischen Handschriften des zehnten bis viersehnten Jahrhunderts*, Wiesbaden: Otto Harrassowitz, 1966.

KUNITZSCH, *Arabs and Stars*
PAUL KUNITZSCH, *The Arabs and the Stars: Texts and Traditions on the Fixed Stars and their Influence in Medieval Europe*, Variorum Collected Studies Series CS307, Northampton: Variorum, 1989. [Includes «The Star Catalogue commonly appended to the Alphonsine Tables», *Journal for the History of Astronomy*, 17 (1986), pp. 89-98].

LINTON *Catalogue*
Scientific Instruments, Rare Books: 1 - Leonard Linton Collection, Paris: Alain Brieux, 1980. [Catalogue of the Linton Collection offered for sale at Nouveau Drouot, Paris, 9-10 October 1980].

Linz 1990 *Catalogue*
WILFRIED SEIPEL, ed., *Mensch und Kosmos: Die Heraufkunst des modernen naturwissenschaftlichen Weltbildes (Katalog: Oberösterreichische Landesausstellung 1990)*, Linz: Landesmuseum, 1990. [Exhibition in Schloßmuseum, 7 May-4 Nov. 1990].

London BM *Catalogue*
F. A. B. WARD, *A Catalogue of European Scientific Instruments in the Department of Medieval and Later Antiquities of the British Museum*, London: British Museum Publications, 1981. [See Price, «Review»].

London NMM *Astrolabe*
The Planispheric Astrolabe, London: National Maritime Museum, 1976.

MADDISON, «Rojas Projection»
FRANCIS R. MADDISON, «Hugo Helt and the Rojas Astrolabe Projection», Separata da *Rivista da Faculdade de Ciências* [Coimbra], 39 (1966), 61 pp. and one foldout [pp. 195-251 of the orignal].

MENNINGER, *Zahlwort und Ziffer*
KARL MENNINGER, *Zahlwort und Ziffer: Eine Kulturgeschichte der Zahl, Band II: Zahlschrift und Rechnen*, 2nd edn, Göttingen: Vandenhoeck & Ruprecht, 1959; translated as *Number Words and Number Symbols*, Cambridge, Mass.: MIT Press, 1970.

MICHEL, *Astrolabe*
HENRI MICHEL, *Traité de l'Astrolabe*, Paris: Cauthiers-Villars, 1947; reprinted Paris: Librairie Alain Brieux, 1976.

Milan MAA *Catalogue*
TULLIO TOMBA and GIUSEPPE BRUSA, *Museo d'Arti Applicati: Strumenti scientifici-orologi*, Milano: Electa Editrice, 1983.

Milan PA *Catalogue*
TULLIO TOMBA, «Gli astrolabi della Collezione Settala nella Pinacoteca Ambrosiana», *Atti della Fondazione Giorgio Ronchi*, 33, part 2 (1978), pp. 297-314, continued in 33, part 4 (1978), pp. 3-14.

MORISON, *Printing*
STANLEY MORISON, *Four Centuries of Fine Printing*, 4th edn, London: Ernest Benn, 1960.

MORTIMER, «X-ray Fluorescence Analysis»
C. MORTIMER, «X-ray Fluorescence Analysis of Early Scientific Instruments», *Proceedings of 1987 Archaeometry Symposium*, pp. 311-317.

NACHET, *Catalogue*
A. NACHET, *Collections Nachet: Instruments scientifiques et livres anciens*, Paris, 1929; reprinted Paris: Librarie Alain Brieux, 1976.

New York IM *Handlist*
Anonymous, *The Industrial Museum of New York, Exhibit of Early Astronomical and Mathematical Instruments (February 1930)*, New York: Museums of the Peaceful Arts [1930].

NORTH, «Astrolabe»
J. D. NORTH, «The Astrolabe», *Scientific American*, 230, no. 1 (January, 1974), pp. 96-106; reprinted with modifications in J. D. NORTH, *Stars, Minds, and Fate: Essays in Ancient and Medieval Cosmology*, London and Ronceverte, West Virginia: The Hambledon Press, 1989, pp. 211-220.

Nuremberg 1976-1977 Catalogue
L. Hennig, ed., *500 Jahre Regiomontan – 500 Jahre Astronomie*, Nuremberg: Germanisches Nationalmuseum [1976].

Nuremberg 1983 *Catalogue*
J. Willers and K. Holzamer, *Treasures of Astronomy: Arabic and German Instruments of the German National Museum*, Nuremberg: Germanisches Nationalmuseum, 1983 [In English, German, Arabic].

Nuremberg 1992-1993 *Catalogue*
Gerhard Bott and Johannes Willers, eds, *Focus Behaim-Globus*, Nuremberg: Germanisches Nationalmuseum, 1992.

Osborne, *Decorative Arts*
Harold Osborne, ed., *The Oxford Companion to the Decorative Arts*, Oxford: Clarendon Press, 1975.

Panicali, *Orologi*
Roberto Panicali, *Orologi e orologiai del Rinascimento Italiano: La scuola Urbinate*, Urbino: Quattro Venti, 1988. [In Italian, French, English].

Paris BN *Catalogue*
Bibliothèque Nationale: Nicolas Copernic ou La révolution astronomique, Paris: Bibliothèque Nationale, 1973.

Patterson, *Life*
Elizabeth C. Patterson, *Mary Somerville 1780-1872*, Oxford, 1979. [Booklet of 44 pp.].

Pilz, *Astronomie*
Kurt Pilz, *600 Jahre Astronomie in Nürnberg*, Nuremberg: Verlag Hans Carl, 1977.

Pollard, «Authenticity of Brass Objects»
A. M. Pollard, «Authenticity of Brass Objects by Major Element Analysis», unpublished proceedings of the 1984 Archaeometry Symposium.

Poulle, «Fabrication des Astrolabes»
Emmanuel Poulle, «La Fabrication des astrolabes au Moyen Age», *Techniques et Civilisations*, no. 22 (vol. 4, part. 4, 1955), pp. 117-124.

Poulle, «Instruments astronomiques»
Emmanuel Poulle, «Les Instruments astronomiques du Moyen Age», Museum of the History of Science, Oxford, Selected Off-prints no. 7 (1969); reprinted from *Le Ruban rouge*, 32 (March 1967).

Price, *Equatorie*
Derek J. Price, *The Equatorie of the Planetis: Edited from Peterhouse MS 75.1*, Cambridge: Cambridge University Press, 1955.

PRICE, «First Renaissance Instrument»
DEREK J. DE S. PRICE, «The First Scientific Instrument of the Renaissance», *Physis: Rivista di Storia della Scienza*, 1, pt. 1 (1959), pp. 26-30.

PRICE, *Checklist*
SHARON L. GIBBS, JANICE A. HENDERSON, and DEREK de SOLLA PRICE, *A Computerized Checklist of Astrolabes*, [New Haven, Conn.: Yale University], November 1973.

PRICE, «Review»
DEREK J. DE SOLLA PRICE, «Instruments in the British Museum» [review of London BM *Catalogue*], *Journal for the History of Astronomy*, 13 (1982), pp. 129-132.

ROCKFORD TM *Catalogue*
BRUCE CHANDLER, ed., *The Time Museum Catalogue of the Collection Vol. 1: Time Measuring Instruments, Part 1: Astrolabes, Astrolabe Related Instruments*, by A. J. Turner, Rockford, Illinois: The Time Museum, 1985.

ROSE, *Italian Mathematics*
PAUL LAWRENCE ROSE, *The Italian Renaissance of Mathematics: Studies of Humanists and Mathematicians from Petrarch to Galileo* (Travaux d'Humanisme et Renaissance CXLV), Geneva: Libraire Droz, 1975.

ROY (Paris)
JOSEPH ROY (pseudonym) «Sous le signe des étoiles», *L'Express*, no. 1997, 13-19 October, 1989.

Schweinfurt 1993-1994 *Catalogue*
UWE MÜLLER, ed., *450 Jahre Copernicus* De revolutionibus: *Astronomische und mathematische Bücher aus Schweinfurter Bibliotheken*, Schweinfurt: Stadtarchiv (Veröffentlichung Nr. 9), 1993.

STEINBERG, *Printing*
S. H. STEINBERG, *Five Hundred Years of Printing*, 3rd edn, Harmondsworth: Penguin Books, 1974.

Strasbourg 1987 *Catalogue*
FRANCIS DEBAUVAIS, *Catalogue des livres et objects scientifiques exposés à la Bibliothèque Nationale et Universitaire de Strasbourg, Section des Sciences et Techniques, à l'occasion du Congrès de la Société Française de Physique, du 6 au 10 juillet 1987*, Strasbourg, 1987.

Trento, *Astrolabio*
PAOLO TRENTO, *L'astrolabio, storia, funzioni, costruzione*, Rome: Biblioteca del Vascello, 1989.

A. J. TURNER, *Instruments*
ANTHONY J. TURNER, *Early Scientific Instruments: Europe 1400-1800*, London: Sotheby's Publications, 1987.

A. J. TURNER, see also Rockford TM *Catalogue*.

G. L'E. TURNER, *The Regiomontanus Astrolabe*
GERARD L'E. TURNER, *The Regiomontanus Astrolabe dated 1462*, Supplement to *Fine Scientific and Philosophical Instruments*, 28 Setptember, 1989, London: Christie's South Kensington, 1989.

G. L'E. TURNER, *Strumenti*
GERARD L'E. TURNER, *Gli Strumenti*, vol. 1 of *Storia della Scienza*, general editor Paolo Galluzzi, Torino: Giulio Einaudi editore, 1991.

WARD, «Sundials»
F. A. B. WARD, «The Earliest String-Gnomon Sundials», *Antiquarian Horology*, June, 1976, pp. 800-801.

Washington NMAH *Catalogue*
SHARON GIBBS and GEORGE SALIBA, *Planispheric Astrolabes from the National Museum of American History*, Washington, D.C.: Smithsonian Institution Press, 1984.

Washington NGA *Catalogue*
JAY A. LEVENSON, ed., *Circa 1492: Art in the Age of Exploration*, Washington, D.C.: National Gallery of Art, and New Haven: Yale University Press, 1991. [Exhibition at National Gallery of Art, 1992.]

ZINNER, *Instrumente*
ERNST ZINNER, *Deutsche und niederländische astronomische Instrumente des 11.-18. Jahrhunderts*, Munich: Verlag C.H. Beck, 2nd edn 1967; reprinted 1979.

ZINNER, *Regiomontanus*
ERNST ZINNER, *Leben und Wirken des Joh. Müller von Königsberg genannt Regiomontanus*, 2nd edn, Osnabrück: Otto Zeller, 1968; translated by Ezra Brown as *Regiomontanus: His Life and Work*, Amsterdam: North-Holland, 1990. [NB. References in the text are to the German edition.]

Fig. 1 - IC 640; 2b Bessarion's astrolabe, front.

Fig. 2 - IC 640; 2b Bessarion's astrolabe, back.

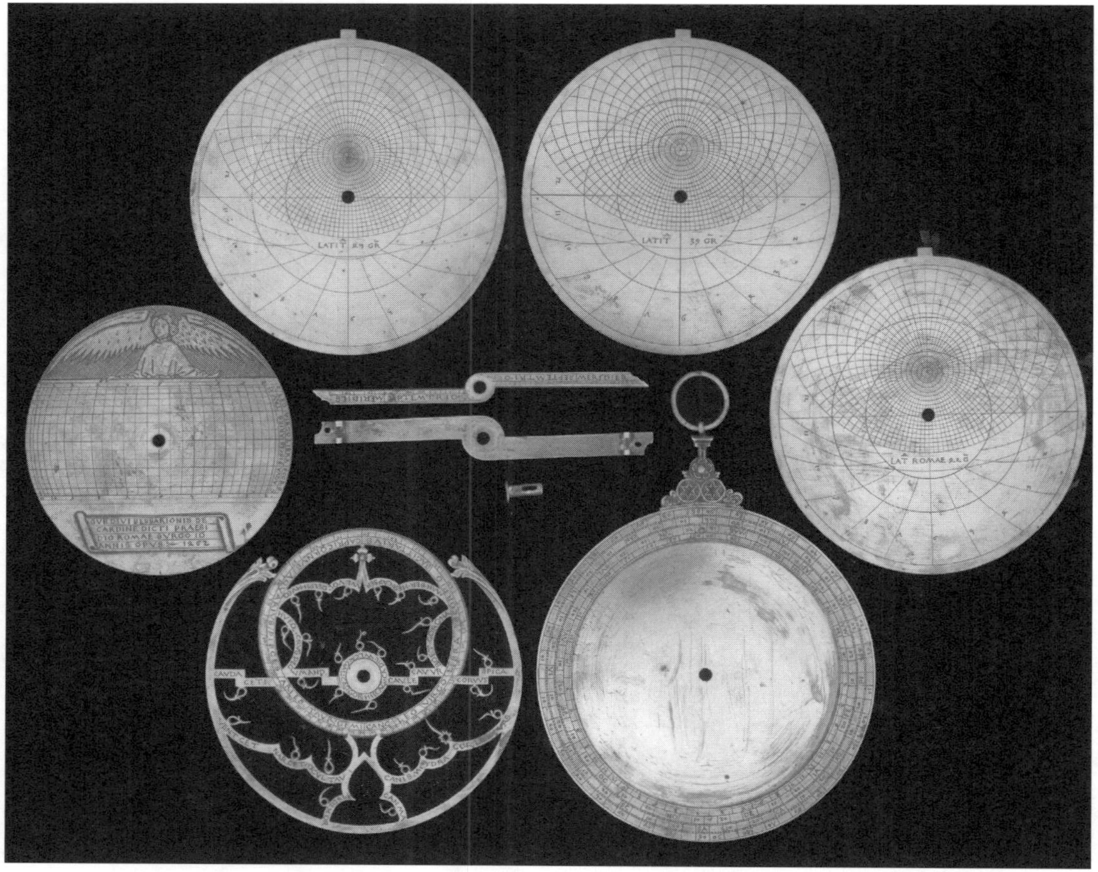

Fig. 3 - IC 640; 2b Bessarion's astrolabe, constituent parts: mater, plates, rete, disc with dedication, rule, alidade, pin.

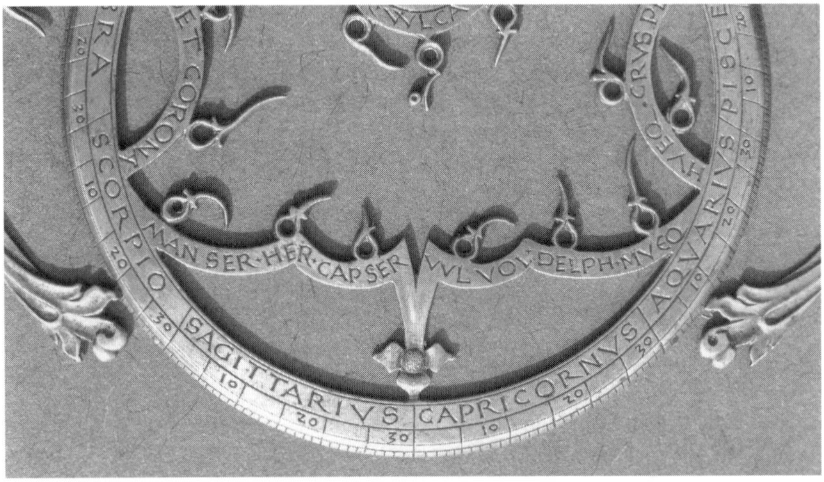

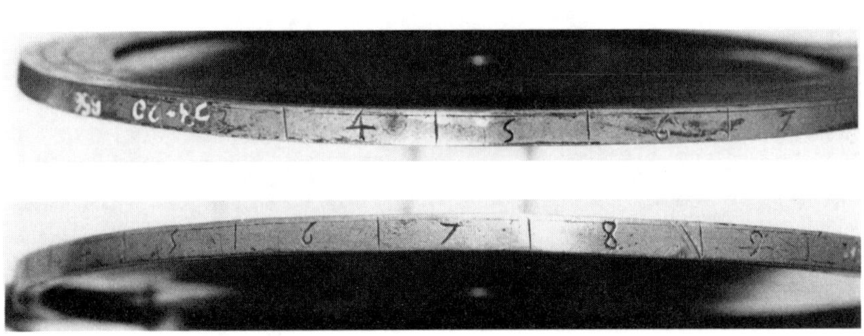

Fig. 4 - IC 640; 2b Bessarion's astrolabe, letter forms on the rete.

Fig. 5 - IC 640; 2b Bessarion's astrolabe, letter and number forms on disc.

Fig. 6 - IC 640; 2b Bessarion's astrolabe, number forms around outer edge of rim.

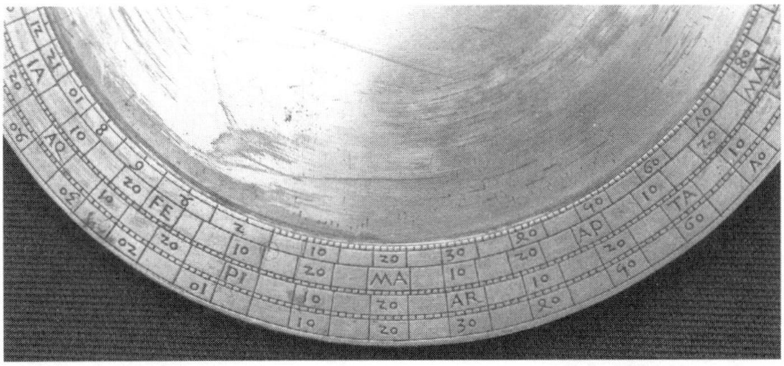

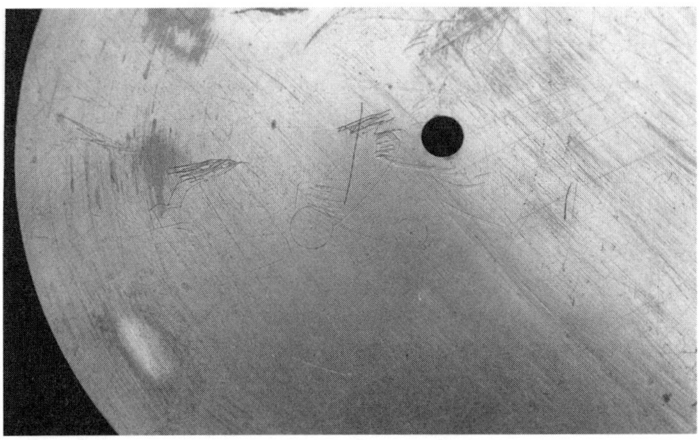

Fig. 7 - IC 640; 2b Bessarion's astrolabe, back of the astrolabe showing the burin jabs produced after scribing degree lines.

Fig. 8 - IC 640; 2b Bessarion's astrolabe, back of rete, showing 5-degree layout lines on the ecliptic circle.

Fig. 9 - IC 640; 2b Bessarion's astrolabe, back of disc, showing a scribed arc with degree dots, scratches, and doodles.

Fig. 10 - IC 493; Florence IMSS inv. no. 1107; fourteenth-century (?) astrolabe rete.

Fig. 11 - IC 549; 1a Rete; note beasts' heads and appendage to side of star-pointer for Wega, marking centre of ecliptic.

Fig. 12 - IC 549; 1a Front of throne, with roundels containing the arms of Vienna.

Fig. 13 - IC 549; 1a Back of astrolabe, showing the throne with two roundels bearing blank shields, and date 1457. Note calendar scale with elongated S.

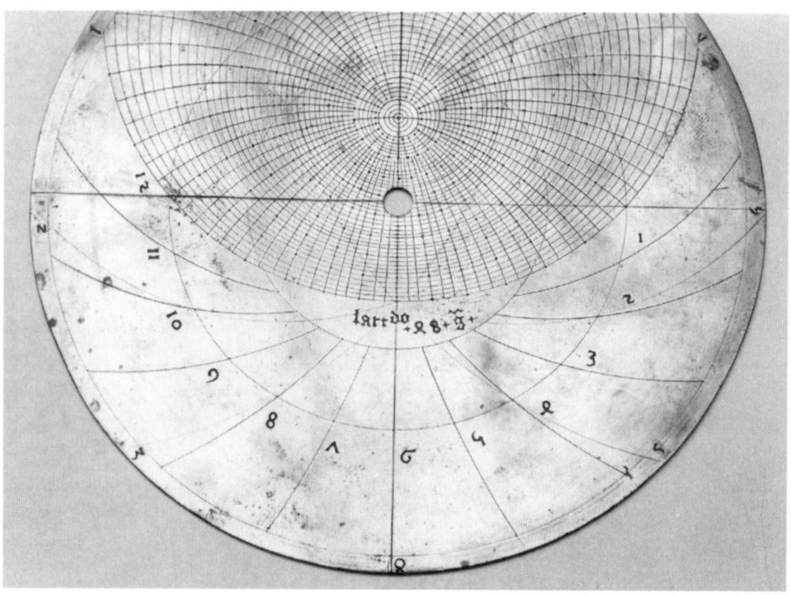

Fig. 14 - IC 452; 1b Rete, showing identical features and antiqua lettering by the same hand as on 2b, Bessarion's astrolabe.

Fig. 15 - IC 452; 1b Part of plate for latitude 48°, showing gothic script.

Fig. 16 - IC 452; 1b Front of throne, with roundels containing the Sun and the Moon.

Fig. 17 - IC 452; 1b Back of throne, showing roundels with shields.

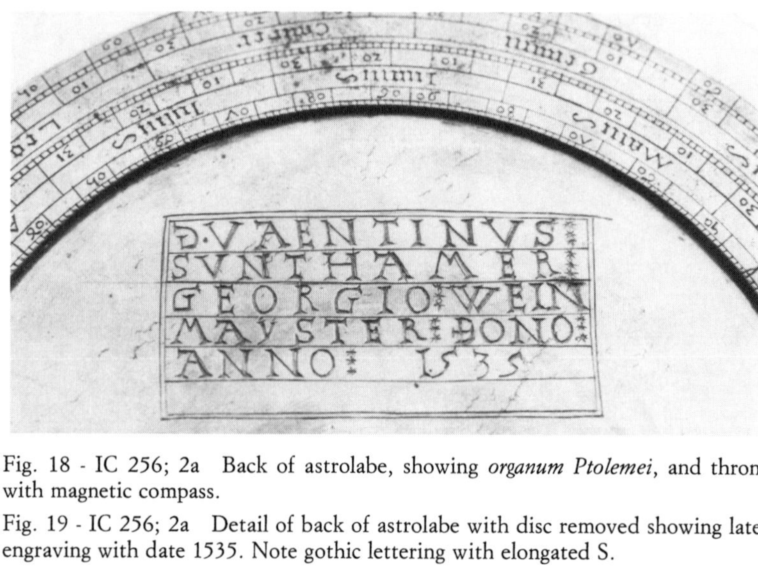

Fig. 18 - IC 256; 2a Back of astrolabe, showing *organum Ptolemei*, and throne with magnetic compass.

Fig. 19 - IC 256; 2a Detail of back of astrolabe with disc removed showing later engraving with date 1535. Note gothic lettering with elongated S.

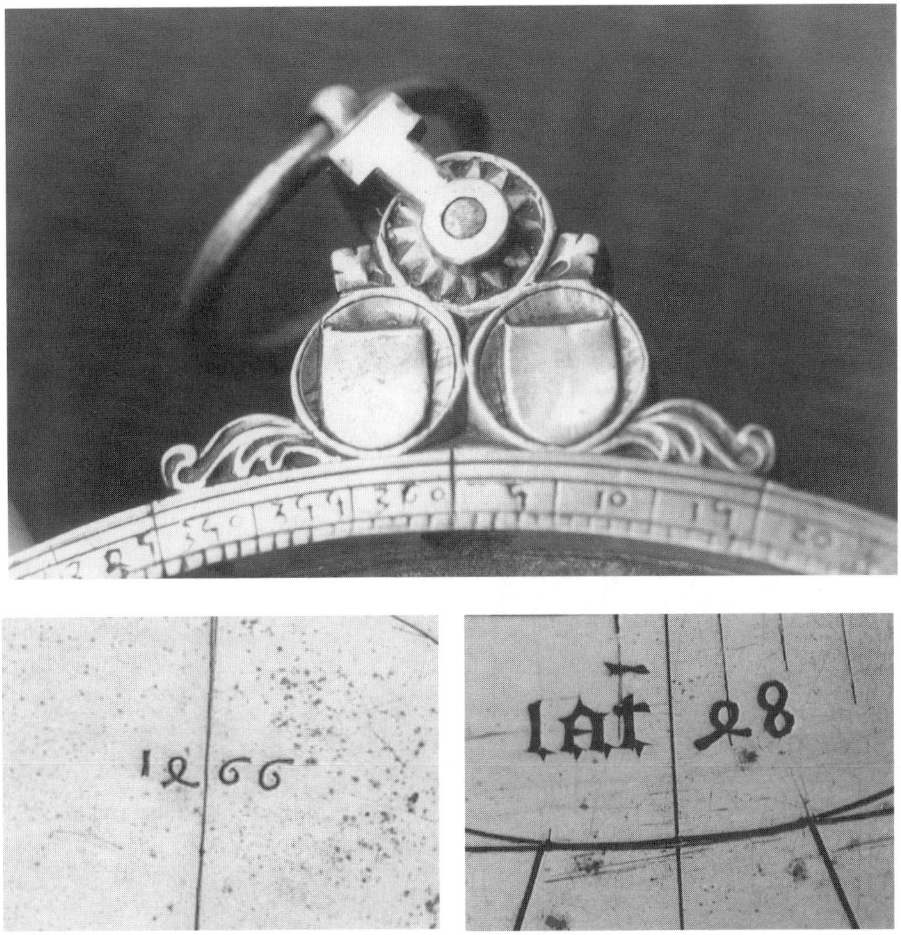

Fig. 20 - IC 592; 3a Details: front of throne; the date, 1466; plate for latitude 48°.

Fig. 21 - IC 620; 3b Back of astrolabe.

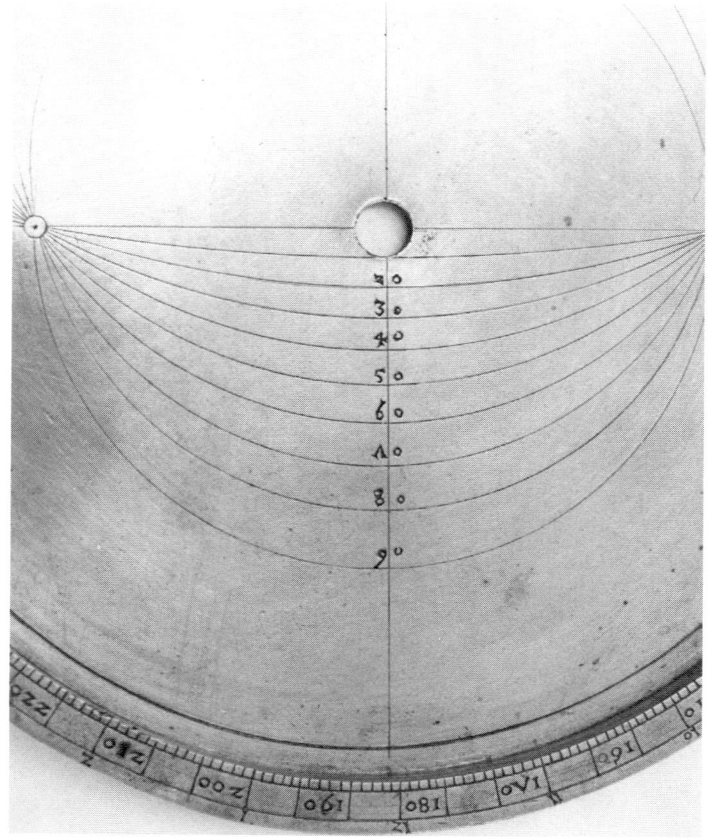

Fig. 22 - IC 620; 3b Front of astrolabe, showing part of mater inscribed with tablet of horizons.

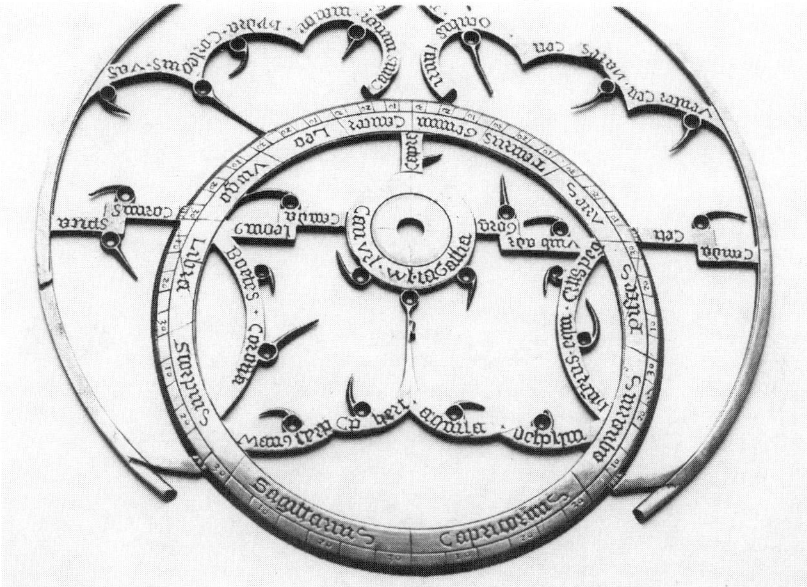

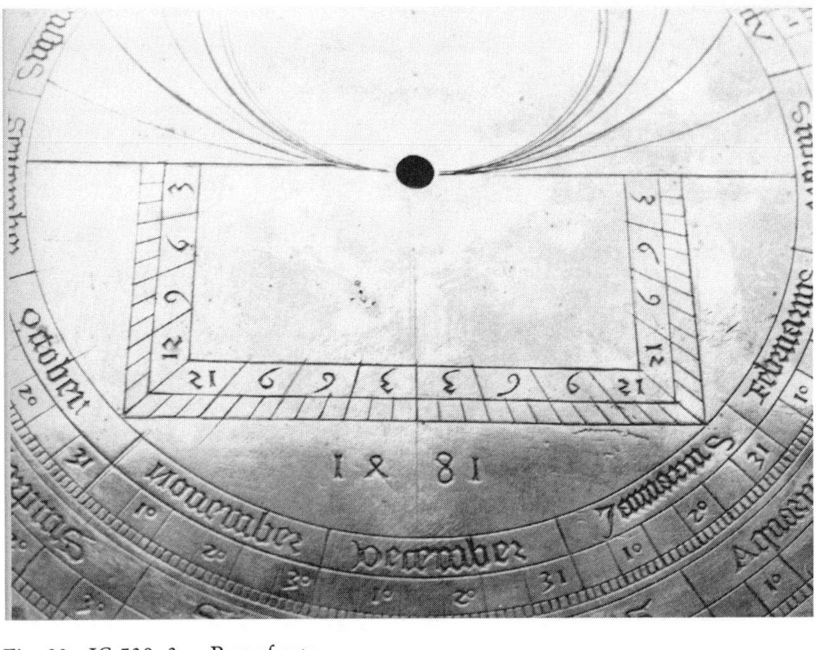

Fig. 23 - IC 530; 3c Part of rete.

Fig. 24 - IC 530; 3c Back of astrolabe, showing date 1481.

Fig. 25 - IC 4501; 3d Front of astrolabe, showing throne with variant decoration including one rosette.

Fig. 26 - IC 4501; 3d Back of astrolabe, showing date 1482.

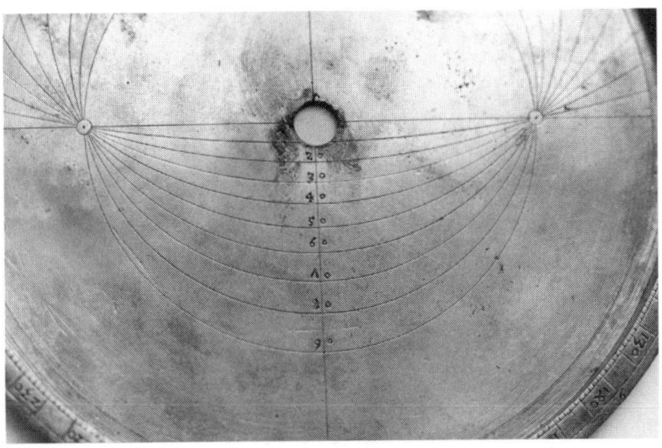

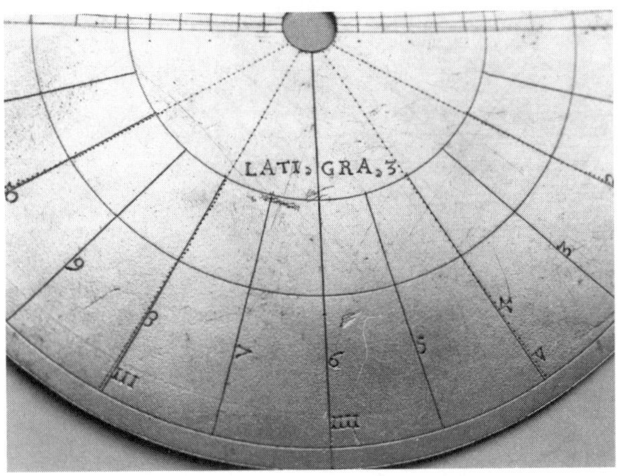

Fig. 27 - IC 516; 4a Detail of back of astrolabe with throne.

Fig. 28 - IC 516; 4a Detail of mater, showing tablet of horizons.

Fig. 29 - IC 516; 4a Detail of lettering on plate for 3°, lettters and numbers punched.

Fig. 30 - IC 644; 5a Astrolabe with rete of Type B.

V

AN ASTROLABE BELONGING TO GALILEO?

1. The Facts

The National Museum of American History has an astrolabe of modest size (diameter 160 mm) that is signed and dated: GEORGIVS HARTMAN NORENBERGE FACIEBAT ANNO M D XXXVII (Fig. 1).[1] It is similar in most respects to the many Hartman astrolabes scattered over many collections.[2] There are the usual three plates engraved for six latitudes: 39°/42°, 45°/48°, 51°/54°; the letter and numeral punches are characteristic of Hartman's workshop. A fourth plate is somewhat thinner, is quite blank on one side, while the other is cut with the equator and tropics, a set of almucantars at 2° intervals (no azimuths) and a set of 24 curved lines for Italian hours which lie between the tropics (Fig. 2). The brass is not very good, and shows several cracks; probably it is thin. As near as may be judged, the almucantars are drawn for the latitude of 43°. There is no number to show the latitude, and the point of intersection of the

---

1 Sharon Gibbs with George Saliba, *Planispheric Astrolabes from the National Museum of American History* (Washington, DC: Smithsonian Institution Press, 1984), pp. 146-150. IC 262, NMAH no. 336117. Previously in the Samuel Verplanck Hoffman collection, no. 33.

2 This astrolabe, among others by Hartman, is described and illustrated in R. T. Gunther, *The Astrolabes of the World* (Oxford, 1932), No. 262, pp. 437-438, plate CVIII.

latitude line with the North-South line is in the middle of the central hole in the plate. The almucantars are labelled every 10° by number punches, which are not the same as those used on the rest of the instrument. This extra plate was not made by Hartman.

A latitude of 43° is found on smaller astrolabes made for Florence, which othewise is usually given on instruments as 43° 40′.[3] An examination of the non-Hartman punches used for the numerals shows that they come from one of the sizes employed by Giovan Battista Giusti, who was responsible for a large skilled workshop at Florence from about 1556 to 1580:[4] the end date is not firm. Giusti had three sets of numbers, and the smallest set (N3) has been used on the plate; they average 1.5 mm in height (Fig. 3). Among the 26 instruments described, of which there are eight different types and eight astrolabes, there is an extra plate made by Giusti to fit on a large astrolabe manufactured in the Duisburg workshop of Gerard Mercator *c.* 1570.[5] This plate is of the same thickness as its companions, and it carries on both sides the projection for the latitude of 43°. A horary quadrant by Volpaia has a previously blank back engraved by Giusti with a shadow square.

As could be expected, an instrument maker may add to or repair an instrument brought in to him. Somebody possessed in Florence in the 1570s, or thereabouts, a Hartman astrolabe of 1537. Not being satisfied with the old-fashioned lines of unequal (planetary) hours, he caused the Giusti workshop to produce a plate special to his requirements.

2. An Inscription

The mater of the Hartman astrolabe is marked with the tropics, equator, and the N-S and E-W lines, the rest being blank. Around the middle is some handwriting, which has been described as: «the following inscribed phrase is barely visible between the equator and the Tropic of Cancer: «Appartenne a Galileo Galilei».[6] This phrase is difficult to see

---

[3] For example, astrolabes marked 43°: IC 179, IC 186, IC 490, IC 4006; marked 43 2/3°: IC 181, IC 458, IC 4007; marked 43° 40′: IC 180, IC 484.

[4] Gerard L'E. Turner, «The Florentine Workshop of Giovan Battista Giusti, 1556-*c.* 1575», *Nuncius: Annali di Storia della Scienza*, x (1995), fasc. 1, pp. 131-172.

[5] Gerard L'E. Turner and Elly Dekker, «An Astrolabe attributed to Gerard Mercator *c.* 1570», *Annals of Science*, 50 (1993), 403-443; Gerard L'E. Turner, «The Three Astrolabes of Gerard Mercator», *Annals of Science*, 51 (1994), 329-353.

[6] Gibbs, note 1, p. 146. Gunther, note 2, p. 437, refers to this «graffito» with the reading: «*Appartene [sic] a Galileo Galilei*».

and very hard to photograph. It is actually lightly *scratched* on the metal in a prominent place, but quite large (height of G = 10 mm). With the throne of the astrolabe at the top, the name is disposed equally between the upper quadrants, while the verb is in the lower left quadrant. The letters forming the verb are not quite as large as those forming the name, so from their size and from the location of the words, they could have been added at a later time.

Silvio Bedini has kindly examined the inscription, and he writes that it is scratched «so lightly that only bits and pieces are visible to the naked eye, or even when examined with a relatively high powered glass».[7] From a photograph taken with the best lighting for the purpose, I have made a tracing (Fig. 4). It is this inscription that connects the Hartman astrolabe, with its Giusti addition, to Galileo himself. The prominence and placing of the name is consistent with Galileo having marked his possession himself, irrespective of the preliminary words.

3. Speculation

The handwriting does appear close to the script and signatures of Galileo (Fig. 5). Specimens can be seen in the Collected Works, and in the monograph published by the Burndy Library, where a complete letter written by Galileo is reproduced in facsimile.[8] Although this letter is dated 1635, the signature has not changed from 1613. Galileo's hand appears to have been schooled through the models of chancery cursive presented by G. F. Cresci, *Il perfetto cancellaresco corsivo* (Rome, 1579).[9] The upper case G is just as Galileo wrote it, and the lower case letters are the same, allowing for the small personal variations that are to be expected. When discussing the inscription on the astrolabe allowance must be made for the different effect produced by laborious scratching and by fluid writing with a pen. On the astrolabe the two names are positioned between two circles 15 mm apart.

---

[7] Silvio A. Bedini, private communication, 3 May 1996.

[8] Antonio Favaro, ed., *Le opere di Galileo Galilei*, 20 vols (Florence, 1890-1909; reprinted with additions, Florence, 1929-1939, 1965). Bern Dibner and Stillman Drake, *A Letter from Galileo* (Norwalk, Conn.: Burndy Library, 1967). Many examples of Galileo's hand, but not his signature, are in Stillman Drake, *Galileo's Notes on Motion*, Monografia N. 3, supplemento agli *Annali dell'Istituto e Museo di Storia della Scienza* (1979).

[9] A. S. Osley, *Luminario: An Introduction to the Italian Writing-books of the Sixteenth and Seventeenth Centuries* (Nieuwkoop: Miland Publishers, 1972), pp. 82-83.

It is certain that the astrolabe was in Florence in around 1580. If the writing in the mater is Galileo's holograph, then when and why did he so mark the instrument?

A possible reason is that Galileo ordered the new plate from Giusti and marked the astrolabe for identification while it was in the workshop. But this is not convincing, for Giusti is unlikely to have had another with which to confuse it. It is more likely that, when one of a group of students, or teaching a class of his own, Galileo marked his astrolabe as his personal property. The astrolabe was a most versatile instrument, much used in medicine and its then sister study, astrology. There are many examples of astrolabes with astrological features, and others of the sixteenth century are wholly astrological.[10] So any student of medicine, astrology, astronomy, or mathematics would have much use for an astrolabe, and could probably have afforded to possess one. Hartman had a workshop using specialist craftsmen working on a batch production line of small brass astrolabes. Much less expensive card astrolabes, with the surfaces of printed paper, were also on the market, including those produced by Hartman.[11]

Galileo was born at Pisa in 1564, and his father had settled the family in Florence by 1575. Galileo was schooled as a novice in 1578, but in 1581 he became a medical student at Pisa university. He could have had need of an astrolabe at this time, and this old German one may have been relatively inexpensive. Two years later he was studying mathematics, and in 1585 he was back in Florence teaching mathematics privately. So 1580 to 1585 seems the most likely period for Galileo to have owned this astrolabe.

Could there, in a later age, have been an attempt to enhance the value of a rough German astrolabe by imitating Galileo's handwriting? Well yes, but it would have been pointless, since the value of such an astrolabe at the end of last century would have been only £5,[12] and to give an instrument the dignity of having belonged to the eminent Galileo would have demanded a piece of much higher quality. After all, the great astrolabe at Florence is commonly called «Galileo's astrolabe».[13] It was made *c.*

---

[10] For example, GUNTHER, note 2, No. 257, pp. 434-435, plate CVI; GERHARD BOTT, editor, *Focus Behaim Globus*, Teil 2: Katalog (Germanisches Nationalmuseum: Nuremberg, 1993), p. 592, and another at p. 611.

[11] GUNTHER, note 2, No. 263, p. 438, plate CIX. Lewis Evans collection, LE 2024, Museum of the History of Science, Oxford.

[12] Lewis Evans, who was the founding benefactor of the Museum of the History of Science, Oxford, in around 1900 paid between £3 and £30 for his German astrolabes.

[13] In the Uffizi inventory of 1654 it is recorded as: «un astrolabio di ottone, come fosse un

1570 by Giusti, and is Europe's largest astrolabe with a diameter of 831 mm.[14]

4. Conclusion

The additional plate, made by Giusti for this instrument, establishes the period when the astrolabe was in Florence, that is, around 1580. Galileo was in Florence at that time as a young man, aged between 17 and 21, studying mathematics, and teaching. The inscription is in a hand of that period, and it could be that of Galileo himself. There is, of course, the possibility that the scratched writing was added by him or by another person as late as the seventeenth century, but the contemporary hypothesis seems more likely. The symmetrical disposition and prominence of the name suggest deliberate identification of a possession. The words *Appartenne a* (had belonged to) in the inscription signify that the instrument had been in Galileo's possession. The astrolabe may have been given by Galileo to one of his students, who wrote *Appartenne a* on it to mark the gift.

Acknowledgements

I am particularlyu grateful to Silvio A. Bedini for obtaining photographs and confirming the reading of the inscription in the mater of the Hartman astrolabe, and to Dr Peggy Kidwell, Curator of Mathematical Instruments at the National Museum of American History for facilitating the study and photography of the instrument. Eric Long, Smithsonian Photographic Laboratories, was able to make visible the light scratches of the inscription. I thank Dr Stefano U. Baldassari, Yale University, for confirming the interpretation of the inscription, and Dr Mara Miniati for sending facsimile signatures of Galileo, and A. V. Simcock for comments on a draft.

---

tavolo da gioco, del Galileo»; see Guglielmo Righini, «Il grande Astrolabio del Museo di Storia della Scienza di Firenze», *Annali dell'Istituto e Museo di Storia della Scienza di Firenze*, II (1977), fasc. 2, p. 46. The instrument is called Galileo's throughout M. Felli, *L'Astrolabio di Galileo* (Florence, 1983).

[14] IC 484. See Turner, note 4, pp. 157-160.

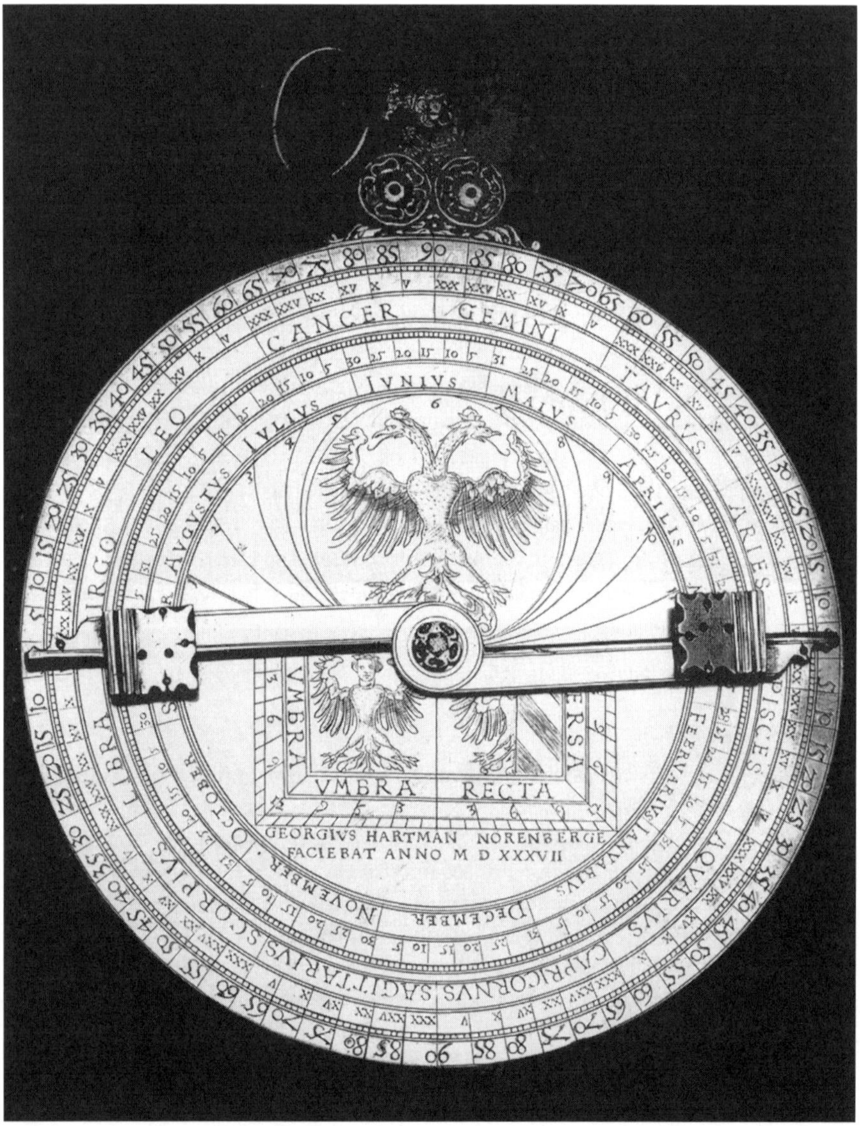

Fig. 1. - The back of the Hartman, 1537 astrolabe (IC 262) in the National Museum of American History, Washington, D.C. (note 1). Copyright Smithsonian Institution.

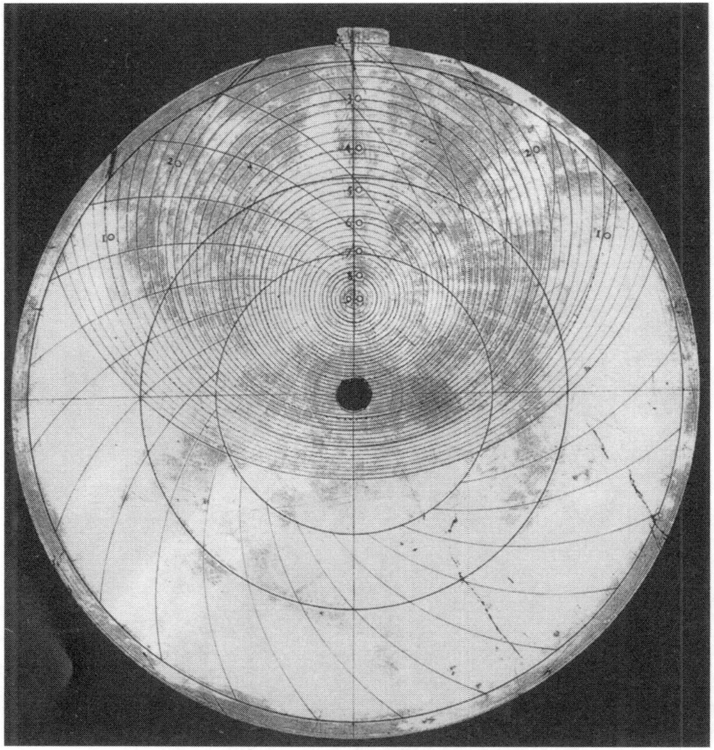

Fig. 2. - The extra plate with the Hartman, 1537 astrolabe (IC 262), made by Giovan Battista Giusti of Florence. Copyright Smithsoniam Institution.

Fig. 3. - The punched numbers on the rule and ecliptic circle are Giusti's size N3. All the other numbers are Giusti's size N2. Astrolabe (IC 458) at the Musée de la Vie Wallonne, Liège, inv. no. 36 (Turner, note 4, pp. 167-168). Photograph by the author.

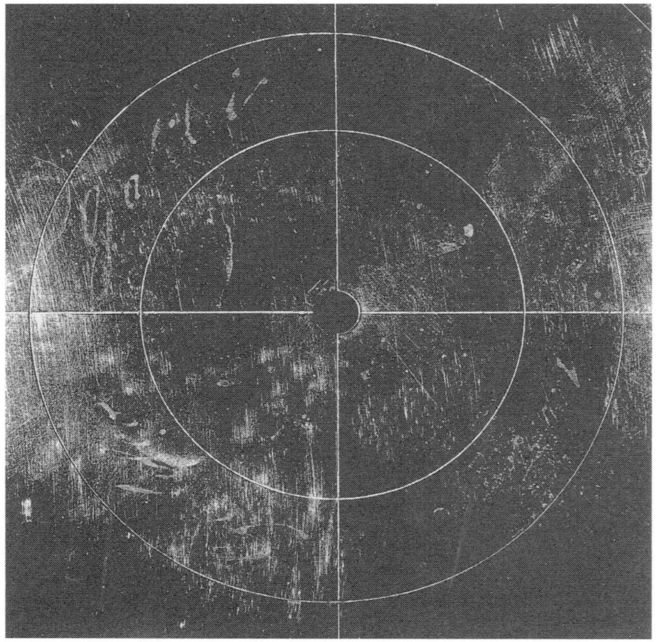

Fig. 4. - *a*) The scratched inscription on the mater of the Hartman astrolabe. The radial distance between the two circles is 15 mm. Courtesy of the Smithsonian Institution Photographic Laboratories.

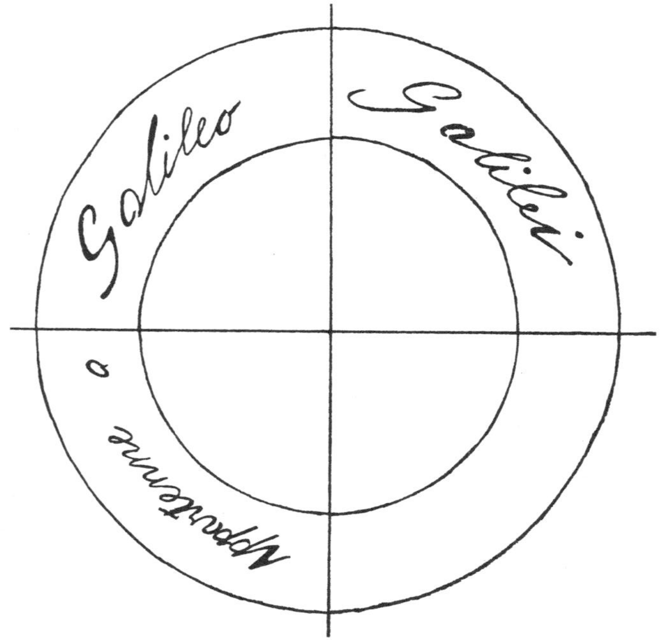

Fig. 4. - *b*) Tracing of the writing on the mater taken from the photograph. The scratches are hard to see even with a lens, so this tracing cannot be an exact replica of the original scratches.

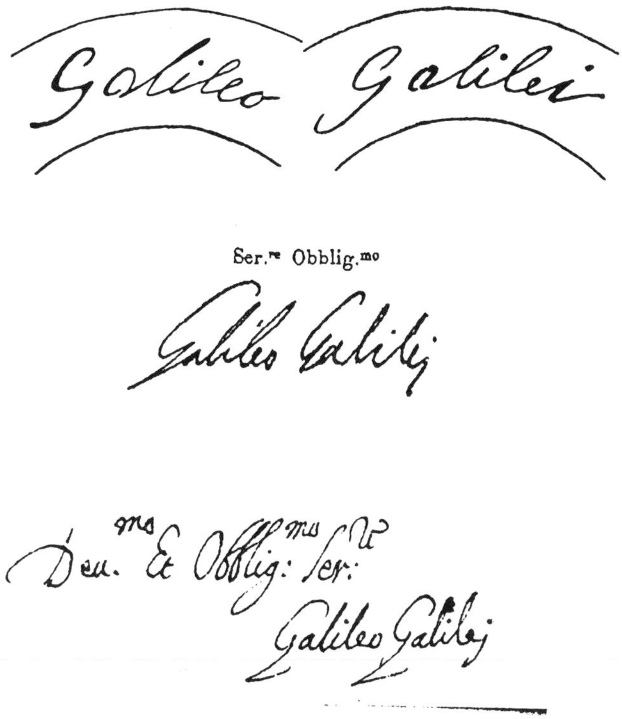

Fig. 5. - The inscription, and Galileo's signatures from documents dated 1613 (Favaro), and 1635 (Dibner and Drake). See note 8 for locations.

VI

THE FLORENTINE WORKSHOP OF GIOVAN BATTISTA GIUSTI, 1556-*c*. 1575

1. Introduction

This paper originated in February 1983 while I was Maria Luisa Righini Bonelli Visiting Professor at the Istituto e Museo di Storia della Scienza at Florence. I had examined all the microscopes, and had turned my attention to the English Elizabethan instruments taken to Florence in 1606 by Sir Robert Dudley. Some of these were signed, but most were not. By a careful examination of the engraved letter and number forms it was possible to identify 20 instruments as English and belonging to the Elizabethan time. It was necessary to scan all the cases in the museum that contained likely instruments. While on this search I looked at other instruments for comparison, and one was the large astrolabe (XV)[1] commonly said to be Galileo's, and said to be made by Egnazio Danti.[2] It was not believable that Danti would be a craftsman; it was more likely that he commissioned it. The astrolabe has punched numbers and letters on the rete only, the rest being engraved. The punches meant that there was a chance of finding them on other instruments in the museum. This proved to be true, and soon three horary quadrants (II-IV) were found, two dated (1565 and 1568), showing the use of the same punches, and signed by Giovan Battista Giusti. On the *primum mobile* (XIV, 1568) the larger letters and numbers appear to good effect, also on the medium-sized astrolabe in the museum (XVII, 1568).

Here was a good beginning, but more was needed. The Oxford and Dresden (now lost) astrolabes illustrated by R. T. Gunther in his *The Astrolabes of the World* (1932), were easy to add, but there were still only seven instruments. Gradually more instruments were identified, both in Florence and in other parts of the world. A signed nocturnal was sold at Sotheby's in 1990, but its present whereabouts is not known. With twenty instruments or parts of them it was time to make an assessment and to write a paper. Then suddenly there was a flush. In May 1994 I saw a device of eight connected bars labelled as a *grafometro* of the seventeenth century by Carlo Doni (XIII). It turned out to be by Giusti, Doni being the owner, and the bars could be arranged to form the *radio latino* described by Egnazio Danti in his book of 1583. In London in June, I was shown a device that I could see had the characteristic punched letters of Giusti, and which turned out to be a declinatory for constructing vertical declining sundials (XII). Then in July, while searching the literature, I recognized a quadrant from a small picture in an exhibition catalogue (Va). When tracing its location I was shown a photograph from a 1986 Parisian sale catalogue of yet another identical quadrant (Vb). The total is twenty-six at the time of writing, and may be higher by the time this paper is published (see note 18).

Of these twenty-four, only three quadrants and the nocturnal (II-IV, VII) are signed by Giusti. These are fairly straightforward to

---

[1] The roman numerals give the number of the description in Section 3 of this paper.

[2] M. L. Righini Bonelli, *Il Museo di Storia della Scienza a Firenze* (Milan, 1968), p. 163, no. 103. M. Miniati, editor, *Museo di Storia della Scienza: Catalogo* (Florence, 1991), p. 44, no. 27, where the latitude of the single plate is said to be for 45° 24′ the latitude of Padua, the city where Galileo might have used it, and where the construction is attributed to Egnazio Danti. The instrument is called Galileo's throughout M. Felli, *L'Astrolabio di Galileo* (Florence, 1983), but the latitude is correctly given, 43° 40′.

make, and were in common use, so may be classed as «shop goods». The remainder are finely-crafted, special pieces, several obviously commissioned. There was no need to sign these. Indeed, there are several examples of unsigned instruments by fine makers during the sixteenth, seventeenth and eighteenth centuries.

A key piece of evidence lies with the Gerard Mercator astrolabe at Florence, which I identified as by him in February 1992.[3] It has seven plates, one of which is a polar projection of the globe, a feature found on several Giusti astrolabes, although infinitely more crudely drawn. The original, Mercator, latitude plates did not include one suitable for use in Florence, so one was provided locally with a projection for 43°. This plate was engraved to imitate those of Mercator, but fortunately one row of numbers was punched, and pointed immediately to Giusti (XX). This meant that the astrolabe was in Florence during the Giusti period, and a variety of considerations strongly suggested that the astrolabe - one of the finest ever made - was acquired for the crowning of Cosimo de' Medici as the Grand Duke of Tuscany in August 1569. In this year Danti published his treatise on the astrolabe. An astrolabe with a geographical plate is a flat, portable symbol of the heavens and the Earth; Danti, cosmographer to Cosimo, an astronomer, designer of instruments, and publisher of them, promoted the astrolabe and will have advised Giusti. Perhaps we may regard Danti as the patron of Giusti. He may have had a say in ordering the extra plate for the Mercator astrolabe, and will have commissioned the large astrolabe and *primum mobile* that bear his name or initials. The former may well have been a gift from Cosimo to his favoured servant. This does not really matter, because what is needed is to date the high peak of astrolabic activity. This does seem to be 1568 to 1572 (the astrolabes with dates), with the high point at the time of the crowning of Cosimo.

If Giusti were sufficient of a craftsman to make the large and very large astrolabes, then why could he not have made a version of the Mercator astrolabe? Simply, there was no other workshop in the world with the skill and reputation of Gerard Mercator at Duisburg. His astrolabe incorporates the Copernican revisions, and is accurate. Giusti is old-fashioned and none too accurate in positioning his stars.

---

[3] G. L'E. TURNER and E. DEKKER, «An Astrolabe attributed to Gerard Mercator *c.* 1570», *Annals of Science*, 50 (1993), pp. 403-443; G. L'E. TURNER, «The Three Astrolabes of Gerard Mercator», *Annals of Science*, 51 (1994), pp. 329-353.

In fact, there seems little rhyme or reason in Giusti's choice of stars, and variations abound. Perhaps the purchaser made his own selection – personalized the rete. No attempt has been made in this paper to consider the stars. They are listed because it is important to gather together information to do with craft practice. This is why even the breaks between words are noted, and where there are solid or open dots.

The «Galileo» astrolabe (XV) is attributed to Egnazio Danti, and it is clear that it was under his patronage that Giusti produced several of the more important instruments in the group being considered. Danti (1536-1586) was born in Perugia, of an educated family, both his father and his grandfather being interested in astronomy.[4] Egnazio entered the Dominican order, and soon acquired a reputation for learning that led to his being ordered by Duke Cosimo II, in 1562 when only 26 years of age, to prepare maps, and a terrestrial globe. Designated *Cosmografo*, his work on decorating with maps the walls of a room in the Palazzo Vecchio described as the *Guardaroba*, or sometimes the *Sala delle Carte geografiche*, won such approval that in 1571 the Grand Duke requested the general of the Dominican order to allow Danti to reside at his court in Florence. Cosimo had in mind the reform of the calendar, and for this Danti, in 1572, installed a large quadrant on the facade of Santa Maria Novella, and inside the church a meridian.[5]

Even before his move to court, Danti had published, in 1569, his *Trattato dell'uso et della fabbrica dell'astrolabio*, and the study that this book contains found its physical embodiment in the instruments manufactured by Giusti. The *Trattato* was reprinted in 1578, with the addition of sections on various other instruments. Danti also translated Proclus into Italian, *La sfera di Proclo* (1573), and published an edition of his grandfather's translation of Sacrobosco's *De Sphaera* (1579), which he annotated, thus producing the first two astronomical works in his native language.

It is interesting to compare Danti's *Trattato* with Geoffrey Chaucer's *Treatise on the Astrolabe* (1391), which was the first technical study of an instrument to be written in English. Chaucer's

---

4 *Dizionario Biografico degli Italiani* (Rome, 1960-), Danti, Egnazio, XXXII (1986), pp. 659-662.

5 M. L. RIGHINI BONELLI and T. B. SETTLE, «Egnatio Danti's Great Astronomical Quadrant», *Annali dell'Istituto e Museo di Storia della Scienza di Firenze*, 4, fasc. 2 (1979), pp. 3-13.

text was not the work of a mathematician, and was probably inspired by an existing astrolabe. Danti, on the other hand, was clearly inspired by the work of Rojas, whose projection had been published less than twenty years earlier in 1550 (see Section 3 below).

Danti fell from favour with the Medici court on the death of his patron in 1574. He was immediately despatched to Bologna, to become professor of mathematics in 1575. In 1577, he returned for a few years to Perugia as a cartographer, producing a topographical map of the country around that city. But his map-making skills attracted the notice of Pope Gregory XIII (Ugo Buoncompagno, Pope 1572-1585) who ordered Danti to map all the papal states. The smallest quadrant (II) is said to have belonged to Gregory, but this may be just a guess by the compiler of the inventory. In 1580, Danti was called to Rome to reform the calendar, and the Gregorian calendar was introduced from 1582. Danti ended his life as Bishop of Alatri.

The extant instruments attributable to Giovan Battista Giusti fall naturally into two groups, designated A and B. The first group (A) comprises horary quadrants, a nocturnal, declinatory, *radio latino*, and surveyor's simple theodolites. Four are signed by Giusti, of which three are dated. Then there is a quadrant made of *lignum vitae*, dated 1556 (I). It is similar to the other quadrants, and is inscribed: «Cosimo M. R. P. Flo Duc II». It is difficult to match a wooden instrument into a group of metal objects, because, although punches were used, punching in wood gives a less defined edge, and the impress was filled with a white substance to provide contrast. This has now fallen out or become blurred.

The second group (B) comprises astrolabes, of which there are seven complete, plus two separate plates, and a *primum mobile* instrument (XIV). This and one astrolabe (XVII) are dated 1568, and one other astrolabe is dated 1572 (XXIII). The *primum mobile* and one large astrolabe (XVI) bear the initials or full name and arms of Egnazio Danti. It is safe to assume that the astrolabes were made by Giusti as a result of Danti's interest in this instrument. The reason for the choice of Giusti as craftsman we shall never know for certain, but on the evidence of the extant instruments, he was highly skilled, certainly more so than his contemporaries. When Danti wrote the last section of his *Trattato*, the craftsman he most likely had in mind was Giusti, because he demanded a diligent artificer. He also strongly recommended making the instrument as large as possible, since the errors in reading the scales would be less. The passage reads:[6]

Non manchero gia di ricordare due cose a quelli, che desiderono fabbricare gl'Astrolabii: L'una che cerchino di hauere gli Artefici diligenti, perche tutte le volte, che no si usera gran diligenza, gl'instrumenti saranno poco giusti, tanto e facile a chi non e piu che accurato in tal fabbrica pigliare errore. L'altra che si ingegnino di far tali strumenti di quella maggior grandezza, che possibile, perche quanto saranno maggiori, tanto piu minutamente saranno diuisi, & tanto piu ageuolmente, & bene si operera con essi.

We find here a reason for the very large astrolabe at Florence (XV) and the large one at Oxford (XVI), arguably the finest of all Italian astrolabes. It is these large instruments that show the exact work, whereas on the small astrolabes, of which three are known to exist, the workmanship is less good, and sometimes sloppy. It may be that these were made to a price, requiring no engraving apart from the mathematical lines; with the use of punches the labelling could be left to a less skilled workman.

While considering sizes, it should be pointed out that there is a definite ratio in the diameters of the four models of astrolabe, and the diameters can be sized in terms of the length standard of the period. The very large, «Galileo», astrolabe is described in detail by Guglielmo Righini (1977), and he quotes the Uffizi inventory, *Giornaletto della Galleria*, no. 62, for the period 1642-1688, under the date 1654:[7]

Un tavolino di noce ottagono, sopra il quale vi è una faccia d'astrolabio di piastra d'ottone di diametro B^a 1 e 3/4 con sua rete, diottra e traguardi, ferma col suo piè di noce mastiettato per alzarlo per taglio il detto tavolino.

Righini took the Florentine *braccio* to measure 583.6 mm, with the result that 1 3/4 *braccia* measure 1021 mm, as against the actual diameter of the astrolabe, which is 831 mm. Righini continues to rationalise the measure, the inventory, and the astrolabe, but he chose the *wrong braccio*. There were two different measures, *terra* and *pan-*

---

[6] E. DANTI, *Trattato dell'Vso et della Fabbrica dell'Astrolabio* (Florence, 1569), p. 38.

[7] G. RIGHINI, «Il Grande Astrolabio del Museo di Storia della Scienza di Firenze», *Annali dell'Istituto e Museo di Storia della Scienza di Firenze*, 2, fasc. 2 (1977), 45-66. The case for Galileo and the use at Padua is dismissed, pp. 63-65. It is, however, dated as 1563 by a falacious method (see XV below), and: «The fact that Danti was precipitously expelled from Florence by Francesco I, could account for its being unfinished and unsigned». This paper queries both these points.

no. Angelo Martini says the law of 1782 abolished the *braccio a terra* (551 mm) and standardized the other (583.6 mm).[8] By taking the *braccio a terra*, one finds that the very large astrolabe measures just one-and-a-half *braccia*. It is true that this still leaves a discrepancy with the one and three-quarters of the inventory, but the inventory may have been read incorrectly, or perhaps the writing was unclear.

To establish that this is not a trivial point, a table has been made of the diameters of the astrolabes, the figures converted to *braccia* and to *soldi* (20 *soldi* = 1 *braccio*). There seems to be a perfectly reasonable progression. What is surprising is the consistency in the diameters of the astrolabes (see descriptions in Section 3), and the inference that a standard measure was decided upon, and adhered to, is plain.

TABLE 1. - The sizes of Giusti's astrolabes based on the Florentine *braccio a terra*.

Size	*Diameter mm*	*braccia*	*soldi*
Very large	831	1.5	30
Large	410	0.74	15
Medium	248	0.45	9
Small	177	0.31	6

The Florentine *braccio a terra* is 551 mm. It is sub-divided into 20 *soldi* of 27.6 mm. In this table the important figures are the *soldi*. The middle column could give the impression that the diameters are meant to be one half or one third, but just miss out by ten percent. The truth is the diameters progress in round numbers of *soldi* (6/20 = 0.3; 9/20 = 0.45), with the familiar steps of multiples of three.

The instruments in Groups A and B have letters and numbers punched, but with some exceptions. The wooden quadrant (I) has some markings in ink, while others are engraved in parts. One of the quadrants (III, 1565) has punches and punched letters touched up by engraving. The small, gilded quadrant (II) has punched letters and engraved numbers. The surveying instruments are punched, and the instruments in Group B are nearly all punched, with the exception of the base plate of the very large astrolabe (XV), which is all engraved because of the size of lettering and numbering needed. The rete, on the other hand, is all punched. The astrolabes are also linked by the use of distinctive, parallel shading.

[8] A. MARTINI, *Manuale di Metrologia* (Turin, 1883), pp. 206-207.

The use of punches as a supplement to engraving is found on Volpaia instruments as well as those by Giusti. They obviously speeded up production, and made it possible to employ less skilled people. From studies of these and other instruments, it is clear that the laying out of mathematical lines, and the lettering, can be done by different hands, and to do so was economically sensible.

i. *The claims for Egnazio Danti as a maker*

The *primum mobile* (XIV) and one astrolabe (XVI) have been inscribed with the initials or name of Danti: F.E.D.P.F., and: F. Egnatio Dantes, respectively. Of the latter, R. T. Gunther wrote: «made by Egnatio Danti about 1580».[9] He also assumed that the Dresden astrolabe (XXII) was by Danti. The initials were construed by Boffito as: Frater Egnatius Dantes Perusinus faciebat,[10] but for Bonelli it is: Frater Egnatius Danti Praedicatorum Fecit.[11] In his *Trattato* (1569), we find Danti referred to as: «F. Egnatio Danti dell'Or. di S. Domenico», and as «F. Egnatio Danti Perugino». In Danti's *La Prospettiva di Euclide* (1573) are these titles: «R[everendo] P[adre] M[aestro] Egnatio Danti Cosmografo del Seren[issimo] Gran Duca di Toscana», and «M. Egnatio Danti Perugino Frate di San Domenico». The initials are inevitably open to various interpretations. However, a more likely construction is: Frate Egnatio Danti Perugino a Firenze.

Danti as the maker of instruments is now embedded in the literature, in spite of what is known of his career. That Danti, an aristocrat and scholar, could be said to have «made» an instrument is improbable, and in his book he says very clearly that one has to search for a diligent craftsman.[12] Now these two instruments are grouped with twenty-four others, and it has been shown that all originated in the same workshop. It is no longer reasonable to study an instrument in isolation; it is the workshops that are important.

---

[9] R. T. Gunther, *The Astrolabes of the World* (Oxford, 1932), p. 332; for the Dresden astrolabe, see p. 333.

[10] G. Boffito, *Gli Strumenti della Scienza degli Strumenti* (Rome, 1929), p. 220.

[11] M. L. Righini Bonelli, *Il Museo di Storia della Scienza a Firenze* (Milan, 1968), p. 12.

[12] See passage quoted and note 4 above.

ii. *Who was Giusti?*

Nothing is known about Giusti the mathematical instrument maker, although one would not be surprised if his background were not rather similar to that of Egnazio Danti. I am grateful to Dr Silvio A. Bedini for informing me that he had made a concerted effort to find out about Giusti because of the quadrant (II) he made for the Pope. Although he found nothing about the craftsman, he did find a man of the same name, and I quote accordingly: [13]

Giovan Battista Giusti was the son of Alberto and Ginevra (Viviani) Giusti born January 25, 1553. The family originated in Colle Valdelsa but moved to Florence where eventually they obtained Florentine citizenship. In 1571 G. B. Giusti became a Knight of the Order of San Stefano and in 1579 he married Poliziana Acciaioli, and they became the parents of seven children. He served as a member of the Council of Two Hundred and many other public offices both in Florence and in Colle. His last known public office was that of superintendent of the jails in Florence in 1629. The date of his death is not known.

2. Some features of the Astrolabes

i. *The Rojas astrolabe projection*

All the Giusti astrolabes, except the very large one (XV), which has a blank back, are engraved with a Rojas projection on the back. This was made much of by Danti in his *Trattato*. This is a universal projection that has the advantage of not requiring a plate engraved for each latitude. It is an orthographic projection of the celestial sphere onto the plane of the solstices. Thus the lines of latitude project as straight lines, and the lines of longitude project as semi-ellipses. Although not his invention, this projection has acquired its name from the two editions of Juan de Rojas, *Commentariorum in Astrolabium quod Planisphœrium vocant* (Paris, 1550; second edition Paris, 1551).[14] Danti illustrates the projection on an astrolabe copied from the book of Rojas, and he also takes over what I have called the multiple feature plate, which Rojas had on

[13] Private communication, 28 September 1990.

[14] For a full discussion on the projection, illustrations, and a checklist, see F. R. Maddison, *Hugo Helt and the Rojas Astrolabe Projection* (Agrupamento de Estudos de Cartografia antiga XII) (Coimbra, 1966).

the front of his astrolabe. For Danti-Giusti this has to be a plate, because the rete is kept, perhaps out of a sense of tradition, even though a universal projection is on the back. Amongst the descriptions below are comments on the inaccurate placing of some stars.

ii. *The map plates*

In the sixteenth century, geographical map plates are a not uncommon addition to astrolabes. An early account of such a circular world map in polar stereographic projection was published by Peter Apian (1495-1552) in his *Cosmographia* of 1524.[15] Through the efforts of Gemma Frisius (1508-1555), the well-known mathematician and physician, the book was reprinted and translated into many languages. The introduction of geographical plates reflects the growing interest in geography that was first stimulated by the many printed editions of Ptolemy's *Geography* from 1472 onwards, and was further encouraged by the voyages of exploration. Until the sixteenth century, the astrolabe had been a purely astronomical instrument, serving various purposes related to astronomical phenomena. The addition of a geographical map plate converted the astrolabe into an image of the world as a whole, in a way comparable to a pair of globes. It is, therefore, not surprising to find a geographical plate (five are known) among the plates with some of Giusti's astrolabes.

iii. *Sizes and form of the punched letters and numbers*

Although some of the instruments described here have been engraved, and the hand is correctly described as calligraphic, most of the lettering and numbering has been performed by punches. Indeed, it is through the characteristics of the punches used that the identification to Giusti's workshop has been made.

The punches used by instrument makers are the same as those used by letter-cutters for printing. For the printer the punches are used to make the matrix, from which the lead type is cast. The act of punching the matrix is exactly the same as punching letters to label an instrument.

The English globe and instrument maker, Joseph Moxon

---

[15] P. APIANUS, *Cosmographicus Liber* (Landshut, 1524). There were 29 editions. For an illustration, see R. W. SHIRLEY, *The Mapping of the World: Early Printed World Maps 1472-1700* (London, 1984; reprinted with additions, 1993), p. 57, plate 50.

(1627-1700), describes the making of punches in his *Mechanick Exercises on the Whole Art of Printing* (1683). He says: «The making of *Steel Punches* is a Branch of the *Smith's* Trade. ...The *Black-Smith's* Trade comprehends all Trades that use either Forge or File, from the *Anchor-Smith*, to the *Watch-maker*».[16] Moxon continues with a full description of the making of punches.

Some of the earlier instruments by Giusti show evidence of an imperfect set of punches, because the A may be an inverted V with an engraved bar. Later he has a complete set. This may have been because in his early days the expense of a full set was great, and he was not yet in a good way of business. Well-made steel punches would be costly, and Giusti eventually had sets for three sizes of letters and three sets of numbers. Punching into brass would cause more stress on the punch that when used on copper, so the leading edge could begin to break up. Moxon is aware of this even with letter punches used on copper matrices. If a punch is struck lightly by the hammer then the defects in the leading edge can be seen, but if the hammer blow is hard, the punch will sink further into the metal, and the shoulder will leave a more continuous outline.

For the examination of the letter forms on instruments it is essential to use a microscope with a magnification of about x 20. Better still is a binocular microscope, which enables one to see into the cut, when the form of the leading edge is revealed. When examining the instruments at Florence, as well as some of the others, I used an Olympus SZ stereo zoom microscope with a camera attachment. By this means it was possible to link the various instruments to Giusti's workshop. It also removed from consideration the punches used by Volpaia and other Florentine makers, whose letter forms, although superficially similar, are not the same.

It is tempting to look for wear, and for the eventual replacement of one or more punches, with the aim of placing the instruments in date order of their manufacture. However, it might be that both old and new punches were in use for a period at the same time; it would depend on how discriminating Giusti was and how well he controlled his workmen.

The letters and numbers are each in three sizes, and these are designated, Letters: L1, L2, L3; Numbers: N1, N2, N3. The respective

---

[16] J. Moxon, *Mechanick Exercises on the Whole Art of Printing*, edited by Herbert Davis and Harry Carter (London: Oxford University Press, 1958), p. 88.

average heights are, Letters: 30, 20, 11 tenths of a millimetre; Numbers: 31, 20, 15 tenths of a millimetre. The height as measured on the instruments will vary according to how hard the punch has been struck, and therefore the depth the shoulder has penetrated. The measurements do help to ensure that one is comparing like with like. All letters are majuscules and the height of each is nearly the same. The numbers have risers and descenders, so the average is not the height of a given number except on occasion (e.g., 8). The letter forms are roman in the Italian old style, resembling Centaur or Bembo.

iv. *Changes in the use of N1 punches*

The same set of large number punches was used in the making of instruments bearing the name of Giusti, and of instruments (all the astrolabes) that do not bear his name. Moreover, the same set of punches appear on instruments of which the earliest recorded date is 1556 (I), and the latest 1572 (XXIII). Table 2 marks changes in the

TABLE 2. - Changes in the use of N1 Punches: Old and New.

No.	*Name*	*0*	*1*	*2*	*3*	*4*	*5*	*6*	*7*	*8*
I	Q: 1556	o	o	o	o	o	o	o	o	o
IV	Q: 1568	o	o	o	o	n	o	o	n	o
V	Q	–	n	n	n	–	–	n	–	–
VI	Q*	–	n	n	n	–	–	o	–	–
IXa	ST	–	n	n	n	–	–	n	–	–
X	ST	–	o	o	o	–	–	o	–	–
XI	ST	o	n	n	n	n	n	n	n	o
XIV	PM: 1568	o	o	o	o	o	o	o	o	o
XV	A:*vl*	o	o	o	o	n	o	o	n	o
XVI	A:*l*	o	o	o	o	n	o	o	n	o
XVII	A:*m* 1568	o	o	o	o	o	o	o	o	o
XVIII	A:*m*	o	o	o	o	o	o	o	o	o
XIX	APL	o	o	o	o	o	o	o	o	o
XX	APL:*mer*	n	n	n	n	n	n	n	n	n
XXI	A:*s*: back	o	o	o	o	n	o	o	n	o
	plate	n	n	n	n	n	n	n	n	n
XXIII	A:*s*: 1572	–	o	–	o	n	–	–	–	–

No. roman number refer to the descriptions in Section 3
Q quadrant ST simple theodolite PM *primum mobile*
A astrolabe; *vl* very large; *l* large; *m* medium; *s* small
APL astrolabe plate; *mer* plate with Mercator astrolabe
* back with Giusti punched letters and numbers front by Camillo della Volpaia.

number punches that produced N1 numerals on sixteen instruments. (Not all instruments have N1 numerals.)

In Table 3, the N1 numbers from 0 to 8 (9≡6) are arranged in rows for six pieces, linking signed and unsigned instruments. Two instruments dated 1568, *primum mobile* (XIV) and the astrolabe (XVII), show numbers from the same, old, set of punches. The quadrant (IV) signed: **IOANNES BAPTISTA IVSTI FLO FAC A ·D· MCCCCCLXVIII** shows the same punches *except* for 4 and 7. The old 4 and 7 were certainly breaking up; if the dates are meaningful, then the quadrant could have been made after the others, and up to March, the beginning of the new year 1569 in the Tuscan calendar. The very large astrolabe (XV), sometimes referred to as «Galileo's», shows the same punches as the quadrant, and perhaps it was made in 1569. The same may be said for the large astrolabe (XVI) at Oxford that bears the name: F. EGNATIUS DANTES. The simple theodolites (IXa, XI) show punches of the same style and size, but all are new with the probable exception of 8. Other simple theodolites (IXb, X), however, have the older punches for 1, 2, 3, 6, the only ones needed for a shadow square.

Commentary

1. All the instruments considered are linked by characteristic punched letters and numerals, as well as by other features.
2. The punches link signed instruments with unsigned.
3. It is proposed that Giusti is the owner/custodian of these punches, and that this implies the identity of a workshop. As no personal information on Giusti has so far come to light, and as virtually nothing is yet known about the structure of the instrument trade in Florence in the sixteenth century, it is impossible to say whether he was himself a craftsman, an entrepreneur, or a retailer.
4. It is tempting to see a date sequence marked by the introduction of new punches, but this would require the detailed examination of punches in all six sequences on all the instruments, and has so far been undertaken only in sufficient detail to establish the connections shown in this paper.
5. Among the unsigned instruments in Group B, the attribution to Egnazio Danti of two instruments bearing his name has to be reconsidered. It is more likely that he was the patron who commissioned them, or was the recipient of a gift.
6. It has already been proposed that the Mercator astrolabe in

TABLE 3. - The N1 punches on six instruments.

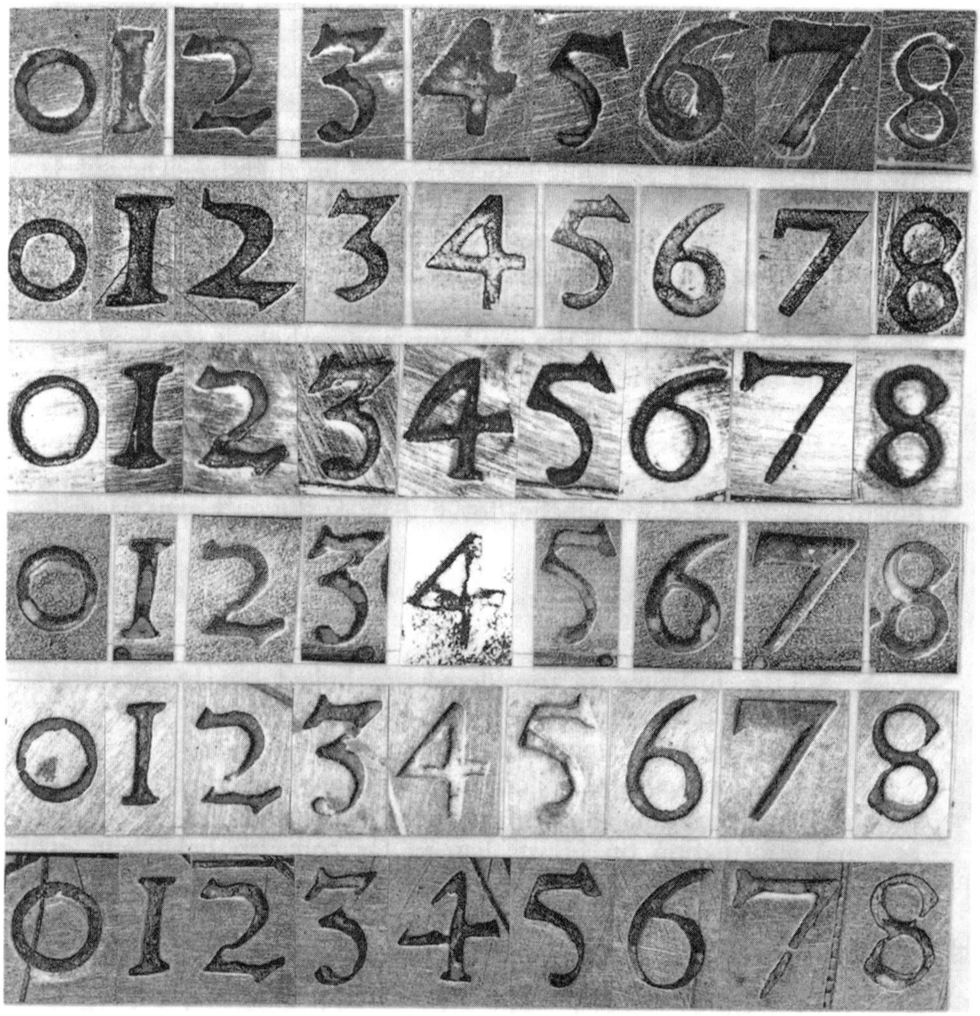

Top to bottom:
IV: Quadrant, signed by Giusti, 1568.
XI: Simple theodolite.
XIV: *Primum mobile*, Egnazio Danti's initials, 1568.
XV: Very large astrolabe (IC 484).
XVI: Large astrolabe (IC 180), Egnazio Danti's name.
XVII: Medium astrolabe (IC 4006), 1568.

Florence, of which one plate (XX) was made in Florence and belongs to the corpus of instruments under consideration here, can be dated to *c.* 1570 (but probably 1569) from the political situation of the Medici, and the status and publications of Egnazio Danti. What may be described as the astrolabe vogue at the Medici court around 1569 is supported by the appearance of two new punches (4, 7) on most of the astrolabes. This provides a more secure dating for the «Galileo» astrolabe than 1563, as put forward by Righini.[17]

3. The descriptions of the instruments

There follow descriptions of twenty-six instruments or parts of separate instruments that are signed (four) or attributed to the workshop of Giusti, so revealing for the first time the range and quality of his production.[18] There are eight distinct types of instrument, including the uncommon *primum mobile*, declinatory, and *radio latino*. These descriptions are not intended to say all that could be said about the instruments, because the purpose of the paper is to establish a workshop. Much more can be said about the metal work and constructional details, about errors, erasures, over-punching, and over-engraving.

The star names have been transcribed as nearly to the original style as is possible. They show how variable is the choice of star and the way it is named. The conventional word to confirm a spelling – *sic* – is not used because it would appear so many times. The reader may be reassured that the copy has been frequently checked against the original, so the transcriptions are as accurate as humanly possible.

Because some instruments or parts of instruments were engraved while others were punched, the convention is used of emboldening the typography (**PONENTE**) for punches, as against ordinary roman capitals for engraving (PONENTE).

---

[17] Righini, *op. cit.*, note 7.

[18] Two further instruments were found in early 1995 when this paper was in proof. Textural additions have been made under IXb, theodolite, and XXIV, clinometer.

TABLE 4. - Instruments attributed to Giovan Battista Giusti of Florence, 1556-*c.*1575.

No.	*Name*	*Collection*	*Inventory no.*	*Inscription*
I	Quadrant	Florence	2521	COSIMO M.R.P.FLO DUC II (1556)
II	Quadrant	Florence	2525	IOANNES BATIS I
III	Quadrant	Florence	2524	IOANNES BAPTISTA IVSTI FLOE FACIEBAT (1565)
IV	Quadrant	Florence	2520	IOANNES BAPTISTA IVSTI FLO FAC (1568)
Va	Quadrant	PC Belgium		
Vb	Quadrant	Nouveau Drouot		
VI	Quadrant	Florence	3628	C.V.F.F. [Camillo Volpaia]. Back by Giusti
VII	Nocturnal	Sotheby's		IOANNES BAPTISTA IVSTI FE (1565)
VIII	Theodolite	Florence	2506	MATTEO E GIOVANBATISTA BOTTI [owners]
IXa	Theodolite	Florence	2457*r*	
IXb	Theodolite	Florence	1295	
X	Theodolite	Florence	2507	
XI	Theodolite	Florence	2508	
XII	Declinatory	Florence	3822	
XIII	*Radio latino*	Florence	619	CARLO DONI [owner]
XIV	*Primum mobile*	Florence	2643	F.E.D.P.F. 1568
XV	Astrolabe	Florence	3361 (IC 484)	
XVI	Astrolabe	Oxford	2072 (IC 180)	F.EGNATIVS DANTES
XVII	Astrolabe	Florence	1285 (IC 4006)	1568
XVIII	Astrolabe	Smithsonian	316756 (IC 2005)	
XIX	Plate, map	Liverpool	1982.545.45 (IC 4005)	
XX	Plate, 43°	Florence	1098 (IC 490)	
XXI	Astrolabe	Liège	36 (IC 458)	
XXII	Astrolabe	Dresden	(IC 181)	[destroyed]
XXIII	Astrolabe	London SM	A629481 (IC 4007)	DON COLV̄BINVS DE ALFIANO (1572)
XXIV	Clinometer	Florence	1311	

i. *Horary quadrants*

I Dated 1556

Florence Istituto e Museo di Storia della Scienza, inv. no. 2521.

Horary quadrant backed by a perpetual calendar. Made in lignum vitae. Radius 168 mm; sides 176, 187 mm; thickness 11 mm.

Inscribed in large incised and painted letters QVADRANS HORARVM AD LAT· GRA· XLIII· MIN·XLV·; on the back dated: MDLVI. Along one edge is the inscription: COSIMO M.R.P.FLO DUC II.

Lignum vitae is a very dense wood, with heart wood that is nearly black, and sap wood that is a light tan colour. Here the board has sap wood on two sides. The sights are part of the block, and are decorated with a small scroll motif that is to be found on the thrones of the astrolabes described in this paper. The plumb-line is missing. Apart from the titling, the numbers and letters are punched, so indicating the Giusti workshop.

The front has a degree scale divided in ones and marked in fives to 90°. The hour lines are flanked by calendar and zodiac scales. Time is read in Italian hours (counting from sunset) and the lines are distinguished by colours according to the half of the day: LINEAE ARGENTEÆ ANTE MERID AVR POST. A small shadow square is in the angle. On the back are nine concentric circles forming a perpetual calendar, with the following elements: solar cycle, dominical letter, the same for leap years with the year (1556) commencing the sequence, golden number, epact, zodiac symbols, months, indiction. Two corners have foliate decoration.

Sources: Identification may be made in an inventory: «Inventario del 1574 delle robe gia del Gran Duca Cosimo... et hoggi del Ser. Gran Duca Francesco», folio 9*r* reads: «Quadrante uno d'ebano in sua busta di cuoio nero e dentro scritto per lettere 1556». G. Boffito, *Gli Strumenti della Scienza e la Scienza degli Strumenti* (Florence, 1929), p. 199; here Boffito confuses Cosimo Duke II (from 1569 Grand Duke I) with Cosimo Grand Duke II, so misdating.

II No date

Florence Istituto e Museo di Storia della Scienza, inv. no. 2525.

Horary quadrant backed by a magnetic compass and sundial. Gilded brass, with shaped and tooled red leather case, lining claret velvet; radius 80mm; sides 85, 95mm; thickness 4.4 mm.

Signed: IOANNES BATIS I.

Inscribed: **QVADRANS HORARIV' AD LATITVDINEM GRAD XLII** (the latitude of Rome).

Degree scale and Italian hour diagram as before (I). On this brass instrument, the lines after noon are distinguished by dotted lines: **LINEA PVNTORVM POS MERIDIES**. In the corner is a magic square of the third order. It has been shown that this can refer to the Moon. The sights are supported by scrolls; the plumb-line is missing.

The back has a magnetic compass, which is complete, and a pin-gnomon dial for Italian hours, with cover, hinged in the beard of a grotesque face. The eight winds are marked by the first two letters of their names.

Sources: Identified in the inventory of Granduca Ferdinando I: «Quadrante d'ottone dorato ordinario con oriuolo italiano che ere di Pape Gregorio [XIII] in custodia di corame». Also identifiable in the Uffizi inventory of 1654. Vladimír Karpenko, «Between Magic and Science: Numerical Magic Squares», *Ambix*, 40 (1993), pp. 121-128.

III Dated 1565

Florence Istituto e Museo di Storia della Scienza, inv. no. 2524.

Horary quadrant backed by a wind rose. Bronze; radius 178 mm; sides 190, 204 mm; thickness 5.5 mm.

Signed: IOANNES BAPTISTA IVSTI FLOR FACIEBAT MDLXV.

Inscribed: QVADRANS HORARIVM AD LATITVDINEM GRADVVM XLIII XL MINVTORV[3] / LINEA PVMTOR POS MERIDI.

Degree scale with half degree dots; tropics and equator positioned with their degrees. Italian hour diagram and a shadow square too small to be more that a rough guide. Within this square is a TABVLA SOLIS. The circles give the initial letter of the sign of the zodiac, the initial of the month, and the Italian hour of noon for that month. There has been difficulty in inscribing this face, probably because bronze is hard. Many of the punched letters have been improved with a burin, sometimes the complete letter is engraved. The regular punch for 7 has not been used (perhaps it was too frail; see other instruments), and a V is used on its side.

The back has a windrose, and there is evidence for a lost magnetic compass that will have been held by solder. The winds are: OSTRO, LIBECC, PONE, MAESTR, TRAMON, CRECHO, LEVANT, SYROCC. The spelling and the contractions are not typical of the practice found on the other instruments of Giusti.

Sources: G. Boffito, *Gli Strumenti della Scienza e la Scienza degli Strumenti* (Florence, 1929), p. 200; identified in the 1776 inventory.

IV Dated 1568

Florence Istituto e Museo di Storia della Scienza, inv. no. 2520.

Horary quadrant backed by a shadow square. Hard brass; radius 227 mm; sides 231, 250 mm; thickness 4.2-5 mm.

Signed: **IOANNES BAPTISTA IVSTI FLO FAC A·D· MCCCCCLXVIII.**

Inscribed: QVADRANS HORARIVM AD LATITVDINEM GRAD: XXXXIII: M: XL MINVTORV[3].

Degree scale, and Italian hour diagram as before, with explanation for dotted and continuous lines. In the angle is a table for: ROTA MEDII MOTVS SOLIS QVANTITAS DIEI. This comprises the first two letters of the zodiac, first three of the months, hours and fractions of daylight, the initials of eight winds. Also on this face is a third-order magic square, with the same arrangement as a previous quadrant (II).

The back has a large shadow square, marked in half divisions to 12 at the apex. Named and numbered in punched letters and numerals. The plumb-line is located in a notch.

Va No date

Belgium Private collection.

Horary quadrant backed by shadow square. Brass; radius 84 mm; sides 90 mm; thickness 1.5 mm.

This is a smaller version of quadrants III and IV above, with a strap-work cartouche giving the latitude of Florence, ·ADLATITVDINEM·POLI·A·GRADI·XXXXIII·M·XXXX·. The quadrant for Italian hours has the explanation for the dotted lines (afternoon), and across the middle the line is labelled: EQVINOCTIALE. In the apex is a circular table for the entry of the Sun in the signs of the zodiac. The back has a shadow square, which is marked out and labelled as on the back of VI, with letters and numbers of the same size. There remains the short swivel arm to which the plumb-line is attached, and when used at the back the line is held in a notch at the apex.

The front is illustrated with a brief description in *La Mesure du Temps dans les Collections Belges*, where it is attributed to the workshop of Volpaia. The three initials, S.P.F., at the bottom of the cartouche are interpreted as: Senatus Populusque Firenze.

Sources: La Mesure du Temps dans les Collections Belges (Exhibition held at Brussels, 26 Jan.-1 July 1984) (Brussels: Société Générale de Banque, 1984), p. 57, No. 32.

Vb No date

Paris Nouveau Drouot, rue Rossini, Paris.

Horary quadrant backed by shadow square. Said to be of brass, with radius of 84 mm.

This instrument has not been examined, and the only photograph is of the front in the sale catalogue. This shows that it appears to be identical to Va, with the same initials: S.P.F.

Sources: Nouveau Drouot, 12, rue Rossini, Paris, catalogue for sale 18-19 December 1986, lot 233, plate XXI.

ii. *Additional horary quadrant*

VI — No date

Florence Istituto e Museo di Storia della Scienza, inv. no. 3628.

Horary quadrant by Volpaia backed by shadow square using Giusti punches. Brass; radius 170 mm; sides 182, 185 mm.

Signed: C·V·F·F·; which signifies: Camillo della Volpaia Florentino Fecit.

The front bears a horary quadrant for Italian hours at the indicated latitude of 43° 30′, a table for the Sun's motion, and a cartouche with the six balls of the Medici surmounted by a crown.

The back may have originally been bare. Engraved on it, however, is a large shadow square which is labelled and numbered with punches as used by Giusti, and not like those used on the front. The sides and diagonal are labelled thus: ∘VM∘RECTA∘; ∘VM∘VERSA∘; ∘VM∘MEDIA∘. The notch at the apex is to locate the plumb-line for use at the back.

Camillo della Volpaia died in April 1560, and the firm continued under his sons, Lorenzo (died 1566) and Girolamo (1530-1611). The Giusti-type punches (L1, N1) suggest a date of about 1570 for the shadow square; see Table 2 above.

Sources: Carlo Maccagni, «Notizie sugli Artigiani della Famiglia della Volpaia», *Rassegna periodica di informazioni del Comune di Pisa*, 3, no. 8 (August 1967), 13 pp.

iii. *Nocturnal*

VII — Dated 1565

London Sotheby's New Bond Street, London.

Nocturnal backed by horary quadrant. Brass; diameter 110mm; radius of quadrant 82 mm.

Signed: **IOANNES BAPTISTA IVSTI FA MCCCCLXV.**

Not examined; description from catalogue and photographs. A typical nocturnal, labelled by punching with some engraving. The quadrant on the back is similar to the small quadrant (II) described above. The device is inscribed: **QVADRANS HORARIVM ADLATIT CRADW XLIIII M XL**. As no detailed examination has been possible, conclusions regarding the letter forms can only be tentative. The set of letters clearly lacks G, since one finds these spellings: VIRCO, CEMINI, CRAD. A and T are partly broken, and 7 appears to be a V on its side. The quadrant III, also 1565, has these particularities, and so does IV (1568), but to a much lesser extent. The latter is large and requires larger numbers, which are the N1 type, and a new set of letter punches are employed.

Sources: Sotheby's, New Bond Street, London, *Catalogue* for sale of Clocks, Watches,... and Scientific Instruments, 4-5 October 1990, lot 515; colour picture p. 173. Photographs were provided by courtesy of Sotheby's.

iv. *Simple theodolites*

VIII No date

Florence Istituto e Museo di Storia della Scienza, inv. no. 2506.

Simple theodolite with plane table rule. Brass; diameter 199 mm; thickness 1.65-1.95 mm; length of rule 195 mm.

Inscribed with the names of the owners: ∘M[RE]∘MATTEO∘E∘GIO-VANBATTISTA∘BOTTI∘.

The outer edge of the circular plate is divided in single degrees and numbered in eight sections of 45°. The names of eight winds are marked: **TRAMONTANO, GRECO, LEVANTE, SCIROCCO, OSTRO, LIBECCIO, PONENTE, MAESTRO**. The southern part of the plate is engraved with a shadow square, each quarter divided to 12, each unit roughly sub-divided in fifths.

Across the northern half of the plate is a horizontal sundial, marked out for Italian hours. To use the sundial, the alidade is turned to lie along the East-West line when the pin-gnomon, which is attached to the alidade, is on the North-South, or meridian, line marked on the plate. By means of the magnetic compass, the instrument is aligned on the meridian, when the shadow cast by the pin-gnomon will point to the inscribed hour line appropriate to the moment of use; the hours are marked **10** to **23**. An alidade, bearing a pair of sights (edge and pinhole), rotates round the compass box. One arm carries a pair of brackets into which slips a tangential bar.

The magnetic compass at the centre has a needle with arrow point and feathered tail. Magnetic north is 5° East. The glass cover is full of bubbles, so probably original. On the under side is a screw-in fixing bracket and the punched names of the owners.

At the «Libeccio» position there is an obvious repair; a piece of brass soldered into place. The decorated supports on the alidade have a shape resembling the thrones on the three small astrolabes described below (XXI-XXIII). There are horizontal lines for shading, typical of Giusti's style.

Apart from the lines, there is no engraving; letter and number punches do not «show through» to the under side, but those giving the owner's name do show through to the top surface, and were therefore punched in after the instrument was finished (otherwise raised parts would have been filed and polished). The name inscription uses the same punches as the rest.

IXa — No date

Florence Istituto e Museo di Storia della Scienza, on loan, reg. no. 2457.

Simple theodolite with plane table rule. Brass; diameter 184 mm; thickness 1.86-2.05 mm.

The description of this instrument follows closely that of the previous one. The tangential rule is missing, but the bracket to which it was secured is present. The magnetic compass has lost its glass and needle; magnetic north is offset some 5° East.

The outer edge of the circular plate is marked in single degrees 0-90°-0, 0-90°-0. The names of eight winds are marked as before, except for the spelling of: **TRAMONTANA**. The shadow square has each unit roughly sub-divided in thirds. The alidade rotates round the compass box, and is held in place by a circular ring which fits on with a triple, slotted catch.

The horizontal sundial is marked out for Italian hours; the hours are marked **X** to **XXIII**. Also marked is: **◦ALT◦POL◦G◦43 2/3**. This is the latitude of the city of Florence. The equinoctial line is marked by the symbols for Aries and Libra.

There is no fixing underneath the plate for mounting. There is a hole and ring at the top for vertical use with the shadow square.

IXb No date
Florence Istituto e Museo di Storia della Scienza, inv. no. 1295.

Simple theodolite with alidade (modified and incomplete). Brass; diameter 168 mm; thickness 1.67-1.87 mm.

The outer edge is divided in single degrees and numbered in quarters to 90°. The names of the winds are marked as on IXa.

The description follows that of IXa, with the same scale of degrees, shadow square, winds, and horizontal dial. Points of difference are OMBRA; no latitude given; Italian hours are marked in arabic numerals; the pin-gnomon (lost) is not carried on a bracket attached to the alidade, but slots into a small hole. Since the present alidade is unlikely to be original, this hole, which has rough edges, is probably a later modification. The compass box is empty. There is no fixing underneath the plate.

X No date
Florence Istituto e Museo di Storia della Scienza, inv. no. 2507.

Simple theodolite with plane table rule. Brass; diameter 168 mm; thickness 1.6-1.8 mm; length of rule 170 mm.

Similar, but much simpler than the previous three. Divided as above (IXa, IXb), and with the same spelling for eight winds. Shadow square division one unit. Like one other (IXb), the scale is labelled: OMBRA RETTA and OMBRA VERSA rather than VMBRA RECTA and VMBRA VERSA, an interesting and uncommon use of Italian. The compass needle is lost, but the glass is probably original. Tangential rule is in good condition. The alidade is without decoration.

The underside is completely flat with no attachments.

XI No date
Florence Istituto e Museo di Storia della Scienza, inv. no. 2508.

Simple theodolite, incorporating a rule. Brass; square plate, 209x209 mm; thickness 1.9-2.1 mm.

This functions in the same way as the previous four instruments, but it is square, and is without a sundial. The compass is complete, and has an arrow needle, and East variation. The plate is marked with a circle (diameter 180 mm) divided into four 90° sections. Eight winds are indicated, but not all spelt in the same way as on the other simple theodolites: **TRAMONTANA, GRECO, LEVANTE, SCIROCHO, OSTRO, PONENTE, MAESTRO**. The alidade is decorated in a manner similar to two other simple theodolites.

On each edge is a straight measure, running to 100 with half units crudely added. These 100 units measure 180 mm. Shadow square with Latin wording; divided to a single unit. A hole at the North point is for a ring (as with the others) for suspension when using the shadow square, for which purpose the compass and alidade have to be removed. There are pinhole and edge sights.

Underneath is an attachment for the compass box with a nut.

These five simple theodolites resemble the *theodelitus* of Leonard Digges, shown in his *A Geometrical Practise Named Pantometria* (London, 1571).

v. *Dialling Instrument*

XII — No date

Florence Istituto e Museo di Storia della Scienza, inv. no. 3822.

Declinatory. Brass; square plate 170x170 mm, thickness 1.9-2.3 mm; radial arm 148 mm, thickness 1.8 mm.

The plate has pivoted from one corner a radial arm with a point that transverses a quadrant divided in degrees, and numbered each way from 0° to 90°. Incorporated at the lower end of the radial arm is a magnetic compass, which is complete (diameter 22.6 mm). The sides of the plate are inscribed with the names of the compass directions. To the left of the pivot point for the arm: **·LEVANTE ·A·MERIDIANO·**; to the right: **·PONENTE·A·MERIDIANO·** The other sides are inscribed: **·SEPT·A·LEVANTE·**; and: **·SEPT·A·PONENTE·**. The underside of the plate is completely without markings.

The declinatory is a device used by makers of sundials when designing a vertical declining sundial. For any wall that does not face due South, or due East or West, the exact angle that the wall makes with the meridian has to be measured so that the hour lines on the dial may be calculated and drawn out to fit the situation. In the case of the present instrument, the edge of the plate is held horizontal and placed against the wall. The arm is then moved until the compass needle shows that the arm is aligned with the meridian, when the pointer gives the angle measured from the scale. The correct edge of the plate is chosen for a wall declining East or West.

The instrument may also be used for inclining dials, that is for dials which are not in the plane of the horizon. Then the plate is held vertical, and the pivoted arc, weighted by the compass box, hangs in the vertical plane and allows the inclination to be read from the scale. The present example is a very early model. It can be shown that the labelling uses punches L1 and N2 characteristic of the Giusti workshop of Florence, *c.* 1570.

Sources: The declinatory is described in N. Bion (Paris, 1723), and in Edmund Stone's translation of Bion (*The Construction and Principal uses of Mathematical Instruments*, 2nd edn. (London, 1758), Book VIII, chap. 2). For other examples, see D. Syndram, *Wissenschaftliche Instrumente und Sonnenuhren* (Munich: Callwey, 1989), p. 145, No. 57, plate p. 142 (instrument signed: *N. Bion A Paris*); Elly Dekker, «No Additions Please, We are Serious!: A Sundial Delineator by Anthony Sneewins», *Bulletin of the Scientific Instrument Society*, No. 34 (1992), pp. 3-5.

vi. Radio latino, *or gunnery instrument*

XIII — No date

Florence Istituto e Museo di Storia della Scienza, inv. no. 619.

Radio latino. Brass; eight bars pivoted together, opened in line length 1,455 mm.

Inscribed: **CARLO DONI**.

The instrument comprises four bars with linear measure, followed by one with degrees and a shadow-square scale, then by one with degrees, and finally one with linear measure. The eighth bar is an alidade, which can be attached at an appropriate point, and there are four sights and two crescent-headed screws, whose shape allows sightings to be made over the top. Also provided are two steel pins and a steel inkpen.

Bars 1, 2, 3 are brought together in line, and the two angle-divided bars to a right angle, which leaves two linear bars to form the hypotenuse of the right-angled triangle. The four sights can go in the three corners of the triangle and the mid-point of the hypotenuse. At this mid-point the alidade can be attached, in which case the instrument measures the inclination. Sighting from hypotenuse to apex gives a level of 0°. Other configurations can be employed, e.g., place the junction of the arms with degree scale so that it can move along bars 1-3 set in line; thus there is formed a quadrilateral of variable angle with sights at the corners. Now the instrument functions in the manner of a forestaff. A magnetic compass (complete) is on a mounting that can slide over the «handle» formed from the first bars.

When turned over, the other scales are displayed. With hypotenuse at the top, on left: **DIVISIO CIRCULI**; on right: **DIVISIO LINEE**; these give the measure for inscribed polyhedra (3-15), and division of a line (1-10). The personal name, Carlo Doni, is then just left of centre. Central is the first bar displaying powder and shot density: **POLVERE**; **FERRO**. The scale for lead and stone shot is on the neighbouring edges of the bar: **PIOMBO**; **PIETRA**.

Bars 1, 2, 3, in line, give two linear measures. One is to 19 (464 mm), subdivided in tenths of 2.4 mm. This equals 1 *denaro* (2.43 mm), 1/240 of a Florentine *braccio a panno* (583 mm). The other is to 9 (465 mm) divided in tenths of 5.17 mm, doubtless a town measure.

This type of instrument was developed by Egnazio Danti, and published by him at Rome in 1583 and 1586. As he said on the title page, it was derived from the regula of Ptolemy and the ancient *radio astronomico*. He said the current instrument was: «inventato dall'Ill.mo et eccell.mo signor Latino Orsini». Danti annotates the text thoroughly, and illustrates an instrument just like that described above. He shows it in many of its configurations for fortification, surveying, and gunnery. Such all-in-one instruments are complicated to use, which can explain why so few are known.

Sources: [Egnazio Danti], *Trattato del Radio Latino: instrumento giustissimo & facile...* (Rome, 1583). G. Boffito, *Gli Strumenti della Scienza e la Scienza degli Strumenti* (Florence, 1929), p. 92, plate 40. A similar model is in the Conservatoire national des Arts et Métiers, Paris, illustrated in Maurice Daumas, *Les instruments scientifiques aux XVIIe et XVIIIe Siècles* (Paris, 1953), plate 4, fig. 11. J. J. Roche, «The Radius Astronomicus in England», *Annals of Science*, 38 (1981), pp. 1-32.

vii. Primum mobile

XIV Dated 1568
Florence Istituto e Museo di Storia della Scienza, inv. no. 2643.

Model after Peter Apian. Brass; height 279 mm, width 348 mm, thickness 0.9-1.05 mm.

Named as: **·INSTRVMENTV·PRIMI·MOBILIS·** Below a cartouche with the Medician balls surmounted by a crown, is an oval showing arms: in chief three olive leaves on a stalk, party two bends sinister azure (see also XVI). Above is the date **1568**, and below are the initials **∘F∘E∘D∘P∘F∘**. The interpretation of this as an attribution to Egnazio Danti is discussed in the Introduction above. The back is blank.

The letters and numbers are typical of those found on most of Giusti's products. The close parallel line shading is also typical, and the foliate decoration is also used by him.

This instrument is regarded as a mathematical quadrant, used for the calculation of sines and cosines. The arc is divided in single degrees, and the semicircles represent Sinus versus and Sinus rectus. These scales are marked in units of 1000. The conception is that of Peter Apian (Petrus Apianus, 1495-1552), the German mathematician, astronomer, and inventor of instruments. The device is illustrated in Apian's book (1534), where are printed for the first time sines for every minute using a decimal radius.

Sources: Peter Apian, *Instrumentum Primi Mobilis, a Petro Apiano nunc primum et inventum et in lucem editum* (Nuremberg, 1534). G. Boffito, *Gli Strumenti della Scienza e la Scienza degli Strumenti* (Florence, 1929), pp. 77-78, 220, plates 24-26.

viii. *The very large astrolabe*

XV No date
Florence IC 484; Istituto e Museo di Storia della Scienza, inv. no. 3361.

Brass; diameter 831 mm.

Such a very large astrolabe, inscribed on one side only of a brass disc (the underside is only rough finished), was intended as an accurate analogue computer. The stereographic projection is for the latitude of Florence, being marked: .G.**43** M.**40**. The projection includes the astrological houses, unequal hours, the twilight line, the tropics, and the equator. Around the rim, starting from the outside, are eight winds (OSTRO, LIBECCIO, PONENTE, MAESTRO, TRAMONTANA, GRECO, LEVANTE, SCIROCCO), the 24 hours, degrees of the zodiac subdivided by transversals to 3 minutes of arc, the symbols for the signs of the zodiac, a shadow scale (here curved, not square), and the calendar (eccentric). The alidade is counterchanged, and is devoid of markings. The sights are pinholes. On the whole of the plate the letters and numbers are engraved, except for 43° 40′ indicating the latitude. On the rete, however, all letters and numbers are punched.

The engraved letters on this astrolabe and that at Oxford (XVI) are clearly by the same hand, and the punches used on both are from the same set. It is, of course, the punches in particular that link all the instruments described in this paper.

The rete is to the pattern customary for Giusti, made up of circles: equator, ecliptic, and two tropics (for comparison, see the Oxford astrolabe that follows). Because of the large diameter, there is a pattern of strapwork linking the circles, forming twists and loops. Some of the flamboyant pointers are plain, while others have Giusti's characteristic lined shading. This astrolabe has been fully described by Guglielmo Righini (1977), but his list of 41 stars is not entirely accurate and is not complete. Giusti made his rete for 50 stars, but some pointers have been replaced, others are missing entirely. Professor Righini listed 41 stars that were «reperibili», and these numbers are included in the list below. Many of the star pointers are attached to the strapwork by a dovetail joint, and a few have fallen out. Other pointers have broken off, rough treatment being inevitable for a large device that was table-mounted and displayed in the *Sala delle carte geografiche* of the Palazzo Vecchio.

Being so large and prominent, this instrument has received much attention, but mostly casual. It is now customary to refer to it as «Galileo's Astrolabe», and to claim that its *maker* was Egnazio Danti, the *Cosmographer* to Cosimo de' Medici. It is said that tradition has it that Galileo used it when he was at Padua. The myth keeps recurring, and a recent manifestation is in d'Hollander, *L'Astrolabe* (1993): «avec tympan unique pour la latitude de *Padoue*, qui aurait servi à *Galilée* pour ses calculs»; «E. Danti a fabriqué notamment un très grand astrolabe qui aurait servi à Galilée».

Guglielmo Righini has «dated» the astrolabe to 1563 from the first point of Aries: this treatment is invalid. Comparing this instrument with the others now described here, several of which are dated, gives a more secure date near to 1570. For other considerations, see the Introduction above.

Sources: Ettore Allegri & Alessandro Cecchi, *Palazzo Vecchio e i Medici* (Florence, 1980), pp. 303-304; Maria Luisa Righini Bonelli, *Il Museo di Storia della Scienza a Firenze* (Milan, 1968), pp. 13, 163, plate 4; Guglielmo Righini, «Il grande Astrolabio del Museo di Storia della Scienza di Firenze», *Annali dell'Istituto e Museo di Storia della Scienza di Firenze*, 2, fasc. 2 (1977), pp. 45-66. For a general description and a card model, see Marcello Felli, *L'astrolabio di Galileo* (Florence, 1983), 32 pp. and R. d'Hollander, *L'Astrolabe* (Toulouse: Musée Paul Dupuy, 1993), p. 135, C2; p. 137, Ee9.

Star Names on the very large Astrolabe IC 484

Righini numbers	*Engraved name*	
1	PET·CA·	
2	·ANDROMEDA·	
	·CAVDA·CETI·	[tip broken]
3	·ARIETIS·COMN·PRECE·E·P·8·OR	[dovetailed]
4	·INASPICE·TRIANGVLI·	[dovetailed]
5	·CETI·NARES·	
6	·CAPVT·MEDVSE·	
7	·PERSEI·DES·LA·	
	·P·S·PLEADVM·	
8	·OCVLVS·TAVRI·	
9	·HIRCVS·	
10	·PES·ORIONIS·	[dovetailed]
11	·D·HV·ORIONIS·	
12	·CANIS·MAI·	[dovetailed]
13	CASTOR·	
14	·CANIS·MINOR·	[dovetailed]
15	POLLVX	
16	·ASELLVS·AVS·	
17	·VR·MA·DOR·	
18	·LVCIDA·HIDRE·	[broken]
19	·COR·LEONIS·	
20	·CERVIX·LEONIS·	
21	·CAVDE·PRINC·	
22	·MED·CAVD·VR·MA·	
23	·DORSVM·LEONIS·	
24	[·VR·MA.] EX·CAVDE·	[shared with 22]
25	·CAVDA·LEONIS·	
26	·SPICA·VIRG·	
27	·CORVI·ALA·DEXTRA·	
28	·BOOTIS·SINISTER·HVM·	
	·ARTVRVS·IN·CRV·	[pointer on bar]

Righini numbers	*Engraved name*	
29	·EXTREMA·CHELE·BOREA·	
30	·LVCIDA·CORONE·SEPTEN·	[dovetailed]
	·EXT·CHELE·AVSTRINA·	[no pointer]
34	·COR·SCORPII·	[broken]
31	·CAPVT·HERCVLIS·	
32	C·SERPENTARII	
33	·CA·DRACONIS·	[dovetailed]
	·CA·O	[on strap]
35	·LIRA·	
36	·AQVILA·VEI·VVLTVR·VOLANS·	
37	·SAGITTA·	[dovetailed]
38	·CAVDA·DELFINI·	
39	·RICTVS·PEGASEI·	
	·CAVDE·CAPRIC·PRECEDENS·	
40	·TIBIA·AQVARII·	
41	HVMERVS·PEGASEI·	[tip missing]
	·CRVX·PEGASEI·	[tip missing]
	·GALLINA·	[broken]
	·CEPHRI·D·HV·	[dovetailed pointer missing]

Note: CANIS MINOR: the word MINOR has been written over a mistake.

ix. *The large astrolabe*

XVI No date

Oxford IC 180; Museum of the History of Science, Oxford, inv. no. 2072.

Brass; diameter 410 mm.

Inscribed: ·F·EGNATIVS DANTES, under a coat of arms: Per fesse Azure three olive leaves on a stalk in pale Gold, and party Silver and Silver two bends sinister Azure, on a barred helm the Crest: an olive stock couped and sprouting. Mr J. A. Goodall writes: «This is a distinctive coat which could well be Italian in origin, the barred helm indicating that the bearer was of a noble, or perhaps a patrician family aggregated to the nobility». Simplified arms, with Danti's initials, are on the *Primum mobile* (XIV), which is dated 1568. After considering the wear in Giusti's N1 punches as shown in Tables 2 and 3 for the three instruments dated 1568, it is reasonable to date this astrolabe to late 1568 or early 1569.

Limb. The limb is riveted to the back plate. The outer circle is divided in 10° intervals, numbered in each quarter, with four cardinal winds engraved in Italian. There follows an hour scale, divided to 4 minutes of time, and numbered in engraved roman numerals I-XII twice. Just inside the limb is a rotating annulus that is divided to 4 minutes of time and numbered in engraved arabic numerals 1-24, for registering Italian hours.

Mater. The stereographic projection is engraved on the mater, and no plates are allowed for. The projection is for Florence, the marked latitude being 43° 40′. A projection of this quality could not have been engraved inside a mater with an integral limb. In the present case a brass plate received the engraving and was then riveted to the limb.

Back. Rojas projection with 17 stars. The regulus and cursor are present, the connecting piece being decorated with a grotesque face. The stars marked and labelled are:

LIRA	**AQVILA**	**CIGNVS**	**ARCTVRVS**
EXT·CAVDA·VRS·MAI	**CAVDA LEONIS**	**COR·LEO**	**OCVLVS TAVRI**
CANIS·MINOR	**HIRCVS**	**CAP·GEMI·ANTEC**	**DEX·HV ORIONIS**
COR SCORPII	**CAVDA CAPRICORNI**	**SPICA VIRGINIS**	**LVCIDA HIDRA**
CANIS·MAIOR			

Plate. There is no separate plate; the projection is on the mater. It is very carefully drawn, with almucantars every 2°, and azimuths every 5°. Also engraved are the astrological houses, numbered I-XII, the planetary hours punched **1-12**, and the twilight line.

Rete. The design is typical for Giusti, a simple frame made up of circles, the equator, ecliptic, and tropic of cancer, held within the tropic of capricorn. The zodiac names and the star names are all punched. The zodiac signs are also present, showing the typical Giusti shapes. There are 31 flamboyant pointers, shaded with distinctive Giusti parallel lines. One end of the outer circle, at 7° Aquarius, is broken off.

CAVDA∘CETI	**∘COR∘LEONIS∘**
ANDROMEDÆ∘SC	**CORVI∘ROSTRVM**
ARIE∘CORN∘PREC	**VENDEMIATOR**
VENTER∘CETI	**SPICA∘VIRGINIS**
CAPUT∘MEDUSA	**VRS∘MA∘CAV∘VLTIMA**
PELE∘BOR	**ARCTVRVS**
HIAD∘IN∘∘NARIBS∘TA∘	**CHELE∘BOR**
∘OCVULVS∘TAVRIS	**COR∘SCORPII**
ORIONIS∘SINIS∘PES	**CAP∘ERCULIS**

HIRCVS	**LIRA**
HVM DEX∘PERSEII	**AQVILA**
∘C∘MA∘SIRIOS	**CVSPIS∘SAGGITE**
CAP [Gem] PRECED	**CAVDA∘DELFINI**
∘CI∘MI∘PROCION∘	**CAVDA∘CIGNI**
ASELLVS∘BOREVS	**CRVS∘PEGASEI**
∘HYDRIE∘LVCIDA∘	

Alidade. The alidade, with slit and pinhole sights, is fitted on the front. It has some floral decoration, but no scales.

Throne. The small throne is continuous with the back plate, which is a reason for its being small. It has a pair of S-shaped supports for the shackle and ring.

Sources: R. T. Gunther, *The Astrolabes of the World* (Oxford, 1932), pp. 332-333, plate LXXIV. It was purchased by Lewis Evans on 13 March, 1911, at the Roussel sale in Paris, Hôtel Drouot sale catalogue lot 70. For advice on the coat of arms I am indebited to Mr John A. Goodall, private communication 18 August 1994, through Dr W. F. Ryan.

x. *The medium astrolabes*

XVII Dated 1568

Florence IC 4006; Istituto e Museo di Storia della Scienza, inv. no. 1285.

Brass; diameter 248 mm.

Limb. Divided in single degrees, numbered tens to 90° in each quarter; 24 hours marked in roman numerals, **I-XII** twice.

Mater. The base is blank, and the surface is very rough through lathe judder and beating. At the bottom is a peg to fix the plates in position.

Back. Rojas projection, with regula, but cursor missing. There are 17 stars named:

LIRA	**AQVILA**	**CIGNUS**	**ARCTVRVS**
VRSA MAI	**CAVDA** [Leo]	**COR** [Leo]	**OCVLVS** [Tau]
CANIS MIN	**HIRCUS**	**CAPVT** [Gem]**ANTEC**	
COR [Sco]	**CAVDA** [Cap]	**SPICA VIR**	**HIDRA**
CAN MAIOR	**CANOPVS**		

4 more stars are not named.

Plates. Three plates: 41°/43°; 45°/48°; North polar projection of the Earth/multiple features. These features are: 24 hours, marked **1-12** in arabic numerals, twice; degree scale divided to 2° and marked to 90° in quarters; the zodiac, the signs indicated by symbols not words; the calendar, months indicated by the first three letters; upper left and lower right quadrants, conversion between equal and unequal hours based on times of sunrise and sunset; lower left quadrant, shadow square; upper right blank. At the bottom of the plate is a small hole to fix the position in the mater. The latitude plates have almucantars at 3°, and azimuths at 10°.

The map plate (diameter 215 mm) has a rough outline of the continents, and they, with the sea and tropics, are labelled by engraved letters. CANAR, EVROPE, SPA, AFFRICA, ARABIA, ASIA, TRAPOBANA, MOLVCCHE, INDIA, NVOVA SPAGNA, AMERICA, OCEANO, TROPICO DEL CANCRO, ÆQVINOCTIALE, TROP DEL CAPR. An attempt to punch the last has been made adjacent to 100°, but abandoned after **TRO**. The second position, at 180°, has started with smaller letters that have been over engraved with larger ones.

Rete. The pattern is a simple one, two overlapping circles, those of the equator and ecliptic, held within a partial outer circle, the tropic of capricorn. The thin flamboyant pointers mark 27 stars, but there are only 26 engraved names because CAUDA CAPRICORNI has pointers for both γ Cap and δ Cap. On this rete all the names are engraved, while all the numbers are punched. The pointer for Libra septentriones is broken.

CAVDA CETI
PEC CASSIOP
PERSES^{S}
PLEIADES
OCVLVS TAURI
HIRCVS
PES.ORIONIS
CANIS MAIOR
CAST
CAN^{S} MINOR
IDRA
COR LEONIS
LIBRA SEPTEN^{S}
SPICA VIRGINIS
ARTVRVS
VRSA MAIOR
COR SCORPII
CAPVT HERCVLIS
VVLTVR CADENS
LIRA
AQVILA
GALL^{A}
CAVDA CAPRICORNI
TIBIA AQVAR
CRVS AQVARII
PEGASEVS

Alidade. The alidade, with pinhole sights, fits on the front. One arm has a declination scale, the other an hour scale. All numbers are punched.

Throne. The throne is formed from open scrolls, chased on the front, flat on the back. It is set into the limb by a dovetail. There is a simple shackle holding the ring. Between the scrolls is a magnetic compass (complete), with a cover on the front. At the back, the compass box has an engraved cartouche, and at the top of the scrolls is the date: **1568**.

Sources: This astrolabe was acquired by the Museo through the Viviani bequest (1895), see Mara Miniati, *Museo di Storia della Scienza: Catalogo* (Florence, 1991), p. 44, no. 29.

XVIII — No date

Washington IC 2005; National Museum of American History, Washington, DC, inv. no. 316756

Brass; diameter 248 mm.

The description of this astrolabe is virtually the same as that for the Florence instrument (XVII). On the limb, the hours are marked by arabic numerals, not roman. The mater is blank, and shows lathe marks. A pin at the bottom of the mater locks the plates. The back bears a Rojas projection, with regulus but no cursor. There are 6 stars all named: **AQVILA, CIGVS, COR LEONIS, OCVLVS TAVRI, CADE CAPRICOR, CANIS MINOR**. The names show carelessness in punching, and the star positions do not match well the corresponding ones on the Florence astrolabe. For instance, Aquila differs in position by 10°.

There is one plate, 45°/multiple features. The latter corresponds to the similar Florence plate, but March, instead of being to the left, is at the bottom, by the locking hole. The months are indicated by their initial letter except for **MARZO**. The throne is the same shape as Florence, but not chased. The compass box has a swing lid, but no glass or needle.

The rete is to the same pattern as XVII, with only very minor variations. Remarkably, this astrolabe has the same placing and number of star pointers (27), and the same names, engraved and identically lettered. From evidence given in Table 2 above, this astrolabe could have been made at the same time as XVII, which is dated 1568.

Sources: Sharon Gibbs with George Saliba, *Planispheric Astrolabes from the National Museum of American History* (Washington, D.C.: Smithsonian Institution Press, 1984), pp. 151-153. The instrument was no. 40 in the Samuel Verplanck Hoffman (1866-1942) collection (pp. v, 61).

XIX No date

Liverpool IC 4005; National Museums & Galleries on Merseyside, Liverpool, inv. no. 1982.545.45.

Plate only. Brass; diameter 215 mm.

This plate is of exactly the same diameter as the map plate on the Florence astrolabe, which is 215 mm. The North polar projection is backed by a stereographic projection for 48°, which is the same as the other medium-sized plates. The map is the same, but the wording is punched, and there are some errors. The visible words are: **CANAR, SPA[GNA], EVROPA, AFFRICA, ARABIA, ASIA, TRAPOBANA, MOLVCHE, INDIA, AMERICA, OCEANO, TCO DI CANCRO, EQATOR, TROPICO DEL CAPR.** Although **OCEANO** is visible only at 100°, it may be made out faintly at 10°, where it had been engraved. NOVA SPAGNA is also barely visible, and outlines can be seen elsewhere: AEQ at **EQATOR**, AME at **AMERICA**, and CANAR and SPA at the respective locations. There is a mess at Tropico di Cancro. The outline of TROPICO can be seen behind the stamped letters **TCO DI**. It is probable that **CANCRO** was punched first, and then came the realization that there was no space for the word **DI**, hence the forced contraction for Tropico. The outlines were probably made by the master for the journeyman to follow with punches, but he failed to over-punch two words, and correctly to spell others.

The hole to lock the plate on the pin in the mater is at 220° on the map, and at the South point on the 48° plate.

Sources: Formerly in the Wellcome Historical Medical Museum, inv. no. R 16014. It is said that it was acquired from Ruegg, a dealer at Lausanne, before 23 April 1925 when it was catalogued by the Wellcome Museum. *Bulletin of the Scientific Instrument Society*, No. 33 (1992), p. 17, Fig. 3.

xi. *Additional astrolabe plate*

XX Date *c.* 1570

Florence IC 490; Istituto e Museo di Storia della Scienza, inv. no. 1098.

Gilded brass; diameter 293 mm.

This plate is inscribed on both sides with a stereographic projection for 43°. On one side the azimuths are at intervals of 5°, and on the other $2^1/_2$°; the almucantars on both sides are at 2°. The plate was constructed by Giusti to fit the astrolabe made by Gerard Mercator that was newly acquired for the Grand Duke Cosimo I. There were six original plates, which included projections for 42° and 48°, but not one suited to Florence (modern figure 43° 47′). It is curious that the projection is in a whole number of degrees when the three known small astrolabes have plates for $43^2/_3$°, and the large and very large astrolabes both have projections for 43° 40′. However, one medium sized astrolabe (XVII, dated 1568) has a plate for 43° exactly.

Giusti attempted to make his engraved letters and numbers match those of Mercator as closely as he could, which is why this plate is of little use for intercomparison purposes. However, there is one sequence of numbers (N1), those labelling the unequal hours, that are punched with Giusti's readily-identifiable numbers. They are similar to numbers found on the Giusti astrolabe dated 1568. It was these numbers that proved that this plate was made in Florence to match the others with the astrolabe, and therefore at about the time it was acquired. It has been shown by Turner and Dekker that the Mercator astrolabe was most likely acquired by, or was a gift to, Cosimo on the occasion of his crowning in August 1569 as the Grand Duke of Tuscany.

There is one other point of difference. Although Giusti tried to imitate Mercator, he did not copy Mercator's engraved symbol for Capricorn, but used the form he was used to, which is on a punch. Examples of this may also be seen on some other the products of Giusti's workshop described in this paper.

Sources: Gerard L'E. Turner and Elly Dekker, «An Astrolabe attributed to Gerard Mercator, *c.* 1570», *Annals of Science*, 50 (1993), pp. 403-443; Gerard L'E. Turner, «The Three Astrolabes of Gerard Mercator», *Annals of Science*, 51 (1994), pp. 329-353.

xii. *The small astrolabes*

XXI No date

Liège IC 458; Musée de la Vie Wallonne, Liège, inv. no. 36.

Brass; diameter 177 mm.

Limb. Divided in single degrees, numbered tens to 90° in each quarter; 24 hours marked in roman numerals, I-XII twice. Eight compass points punched by the outer edge, beginning from the top under the throne: **OSTRO, LIBECCIO PONENTE, MAESTRO, TRAMONTANA, GRECO, LEVANTE, SCIROCCO.**

Mater. The base is blank, and the surface is very rough through lathe judder.

Back. Rojas projection, with regula and cursor.

Plates. Two plates; 41°/43⅔°; North polar projection of the Earth/multiple features. These features are: around the edge the zodiac, each sign divided in single degrees; the calendar, divided in days; upper left quadrant in conjunction with the arc in lower right quadrant, conversion between equal and unequal hours based on times of sunrise and sunset; lower left quadrant, shadow square. The upper right quadrant is blank apart from three star names, more of which are distributed over the other quadrants, making 13 stars in all. They are arranged in order of Right Ascension, and the disc forms a ready reckoner when rotated within the hour scale on the limb. It is also called a mobile zodiac. The latitude plate has almucantars at 3°, and azimuths at 10°.

The simple, North polar projection shows distorted continents and seas They are named: **CANAR, SPA, EVROPA, AFFRICA, ARABIA, ASIA, TRAPOBANA, MOLVCCHE, INDIA, AMERICA, OCEANO.** Other names are: **TROPICO DEL CANCRO, AQVINOCTIALE, TRO DEL CAPR.**

Rete. The flamboyant star pointers are hatched with parallel lines, a characteristic feature of the Giusti workshop. There are 23 stars:

·CAVDA·CETI·	**·CORVI·ROSTRVM**
·ANDROM·SC	**VRS·MA·CA·VLT**
·VENTER·CETI·	**·SPICA·VIRGINIS·**
·CAP·MEDVS	**·ARTVRVS·**
·OCVLVS·TAVRI·	**·COR·SCORPII**
·ORION·SIN·PES [over an error]	**CAP·ERCVLIS**
HIRCUS	**·CHELE·BOR·**
CAN·MAIOR	**·LIRA·**

CAN·MIN	**AQVILA**
·CAP·GEM·PREC	**·CAVD·CIGNI**
·COR·LEONIS·	**CRVS·PEGAS**
·HIDRA·	

Alidade. The alidade, with pinhole sights, fits on the front because a universal projection is on the back. One arm has a declination scale, the other an hour scale.

Throne. The throne is attached to the instrument by setting it into the limb and back plate by a dove-tail. The scroll support provides a ball swivel to the ring, and holds a magnetic compass, in original condition.

Sources: Illustrated in H. Michel, *Traité de l'Astrolabe* (Paris, 1947), p. 46, Fig. 25, map plate, and plate VIII for the front and back (said to be by Egnazio Danti, *c.* 1550). Henri Michel, *Les Cadrans solaires de Max Elskamp* (Liège: Editions du Musée Wallon, 1966), plate XXIX (colour): «Astrolabe géographique (no. 36) Louvain (?), v. 1570». *La Mesure du Temps dans les Collections Belges* (Exhibition held at Brussels, 26 Jan.-1 July 1984), pp. 41-42, No. 12, five photographs. The captions are unreliable. For the ready reckoner, or mobile zodiac, see Michel, *Traité*, p. 102, and for details of its use and a clear diagram, see H. N. Saunders, *All the Astrolabes* (Oxford: Senecio Publishing Co., 1984), pp. 70-71.

XXII — No date

Dresden IC 181; Staatlicher Mathematisch-Physikalischer Salon, Dresden.

Presumably brass; diameter 177mm. This astrolabe is now lost.

This astrolabe seems to be adequately described by R. T. Gunther in *The Astrolabes of the World*. It is the same size as that at Liège, and judging from the illustration and description given by Gunther, the description is the same. There are three plates: 35°/38°; 41°/43⅔°; North polar projection/multiple features. The regulus lacks the cursor, and the compass the glass. It was acquired by the Dresden museum in November 1911 at the Frederick Muller auction rooms, Amsterdam, lot 377. The catalogue has a long description and a good photograph of the front and of the map plate.

The map plate has the names as given above for Liège, but there is one addition, **NVOVA SPAGNA**. The rete differs only slightly in shape, and there are similar flamboyant pointers. Gunther says there are «about 26 stars», and since he does not list them, an attempt has been made to identify the stars from the illustration. It appear that there are 27 star pointers, but not all the names can be read because the alidade is above the rete. The conjectures are indicated by brackets [ ]. As only a few of the dots separating the words can be discerned, none have been recorded below. There are dots, which confirms the workshop practice.

PEGASI OP
ANDROMEDE
VENTER CETI
CAP MEDVS
OCVLVS TAVRI
S PES ORI
HIRCUS
ORIO DEX HUM
CANIS MAIOR
[CAP GEM PREC]
CANIS MI
DOR VRS MAI
LVCIDA HIDR[A]
COR LEO
CORVI ALA DESTRA
SPICA VIRG
ARCTVRVS
CHELE BOREA
CORONA SEP
COR SCORPII
CAP ERCVLIS
OS DRAGON[IS]
LIRA
AQVILA
[CAVD CIGNI]
RICTVS PEGASI
TIBIA AQVARI

Sources: R. T. Gunther, *The Astrolabes of the World* (1932), p. 333, plate LXXV. Frederick Muller & Cie, *Catalogue des Objects d'Art et de haute Curiosité... 28-30 novembre 1911* (Amsterdam, 1911), p. 26, lot 377, plate.

XXIII Dated 1572

London IC 4007; The Science Museum, London; formerly in the Wellcome Historical Medical Museum, inv. no. A.629481.

Brass; diameter 178 mm.

Inscribed at the top of the otherwise blank mater:

DON COLVBINVS DE ALFIANO MONACVS VALLIS VMBROSE VTEBATVE MDLXXII. This may or may not be the year of construction. Vallombrosa is in the Apennines, about 20 kilometres from Florence. There was a monastery founded there in 1038 by St John Gualbert, the son of a Florentine nobleman. With his disciples an order was founded under the Benedictine rule, emphasising the contemplative life. Eventually there were about sixty houses of Vallombrosians.

The astrolabe is the same size as that at Liège, and the

description is the same, except that no winds are named on the front. There are two plates: 41°/43⅔°; North polar projection/multiple features. The throne is the same, but lacks the suspension and ring; the magnetic compass is intact. The scrolls of the throne bear the words: **HOC FAC ET VIVES**. The regulus lacks the cursor.

The map plate has the same names, with the addition of **NVOVA SPAGNA**. However, **TROPICO DEL CAPRIO** is mixed with the longitude numbers, so that caprio appears as **CAPR210**.

The rete is quite distinct from those on the Liège and Dresden instruments. The frame is to a different pattern of circles, the degrees in the zodiac circle are marked in roman numerals, the star pointers are not hatched, there are 23 stars, but the choice of stars is different, and there are errors. For example, Cauda ceti (RA 13°) is engraved next to a pointer with a Right Ascension of 336°; the same position on the Dresden astrolabe is inscribed **TIBIA AQUARII**. Finally, the letters are engraved and not punched. The rest of the instrument has letters and number punches of the typical Giusti form. The star names are:

·ANDRO·VMB	·ARTVRVS·
·PECT·CASIO	·BOOTIS·S·HV·
·VENTER·CETI·	COR· [Sco]
·OCVLVS·[Tau]	·OS·DRA·
·PES·ORIO·S·	·SERP·PAL·S·
CANIS·MAI	·LIRA·
·CANIS·MINOR·	·AQVILA·
·LVCIDA·HI· [over an error]	·SAGITTA
·COR· [LEO]	·RICT·PEGAS·
M·CAVD·V·M·	·CAVDA·CETI· [Crus Aquary intended]
·CORVI·ALA·DES·	·CRVS·PE·
·SPICA·VIR·	

xiii. *Clinometer*

XXIV — No date

Florence Istituto e Museo di Storia della Scienza, inv. no. 1311.

Clinometer. Brass; diameter 210 mm; thickness 3.8-40 mm.

The outer edge is divided in single degrees and numbered

0°-90°-0°. The straight edge is 9mm in width, and is probably intended to be set to a surface whose angle can be measured by the heavy ostensor. At the back are four concentric circles divided by dots in 24 parts.

The numerals are Giusti's N 2 size.

Acknowledgements

I thank in particular Professor Paolo Galluzzi and the Comitato Scientifico dell'Istituto e Museo di Storia della Scienza, Florence, for awarding me as M. L. Righini Bonelli Visiting Professorship in 1983. At Florence, two of the most helpful people to find in any museum are Dr Mara Miniati, Curator, and Mrs. Franca Principe, who photographed superbly fifteen of the instruments. I am indebted to the Renaissance Trust for assistance in many ways, including provision of equipment. Access to instruments has been provided by Dr D. A. Robinson, Keeper of Physical Sciences, and Mr K. Johnson at the Science Museum, London; Mr M. K. Suggett, Keeper of Physical Sciences, Liverpool Museum, Liverpool; the Director, Musée de la Vie Wallone, Liège; Mr F. R. Maddison, Curator, and Mr A. V. Simcock, Librarian, Museum of the History of Science, Oxford. I thank Mr J. Baddeley, Sotheby's, New Bond Street, London, who kindly provided a photograph of the nocturnal, and Dr Peggy A. Kidwell, Smithsonian Institution, Washington D.C., for facilitating photographs of an astrolabe. Mr B. G. Archer, Oxford, Mrs J. A. Bigos, Oxford, Dr Elly Dekker, Utrecht, and Dr W. F. Ryan, Warburg Institute, London, gave their customary support.

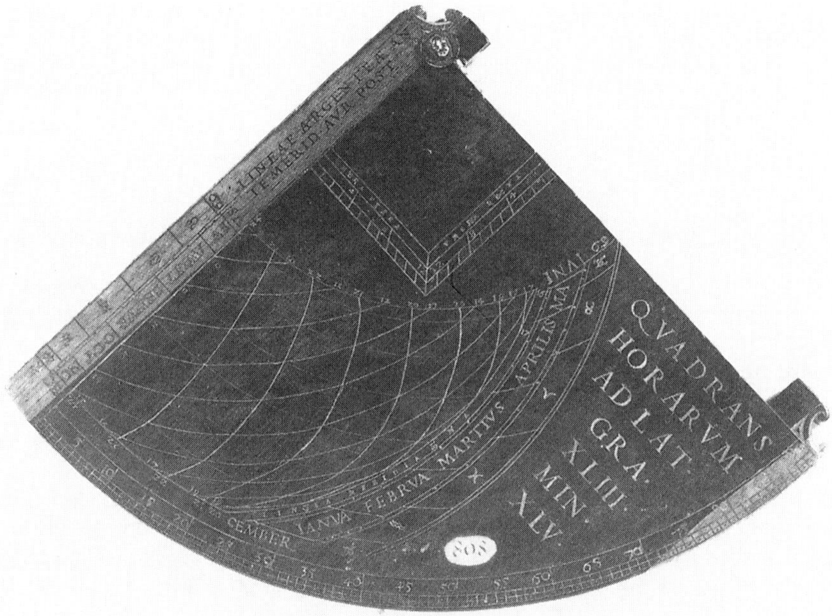

1a

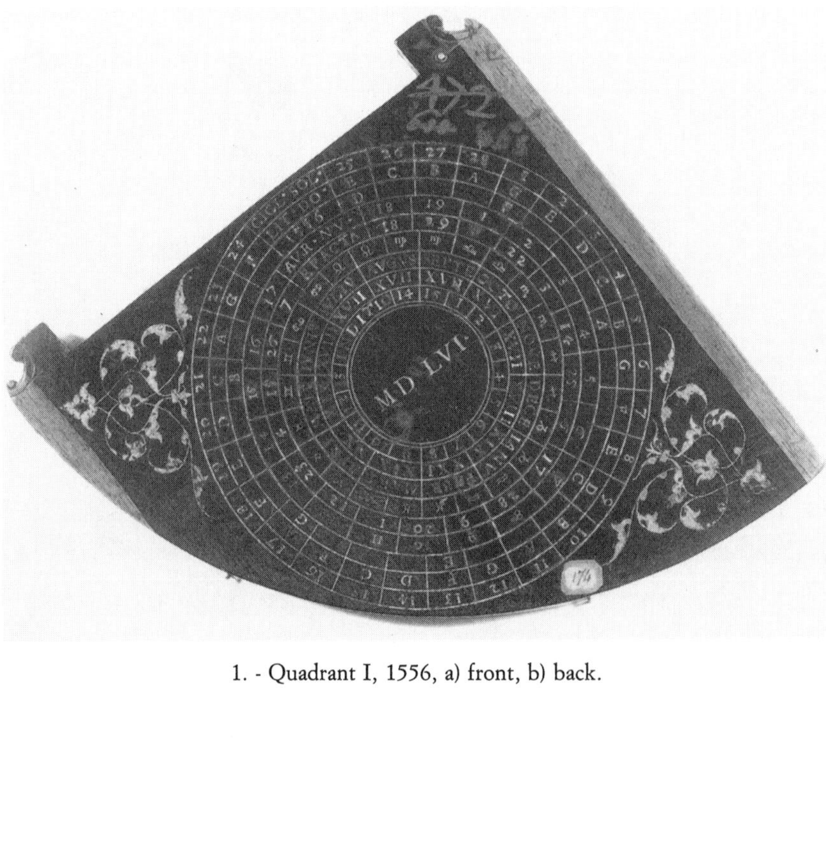

1b

1. - Quadrant I, 1556, a) front, b) back.

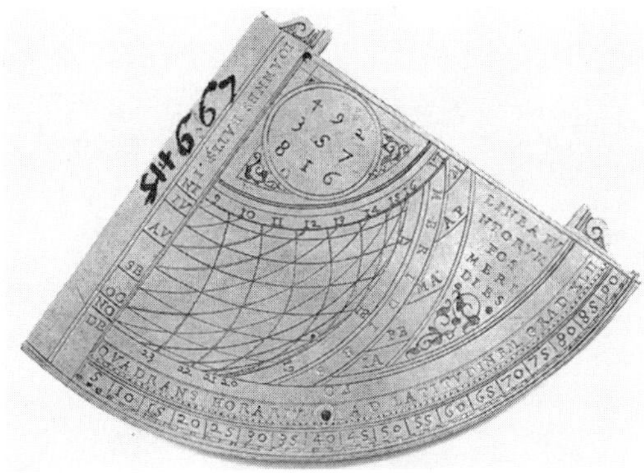

2a

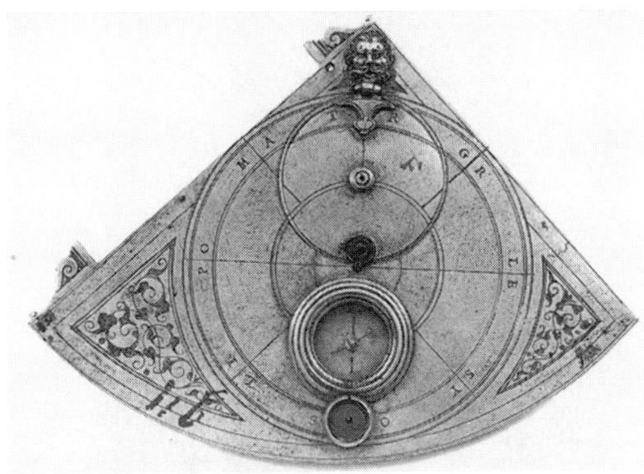

2b

2c

2. - Quadrant II, a) front, b) back, c) back showing sundial.

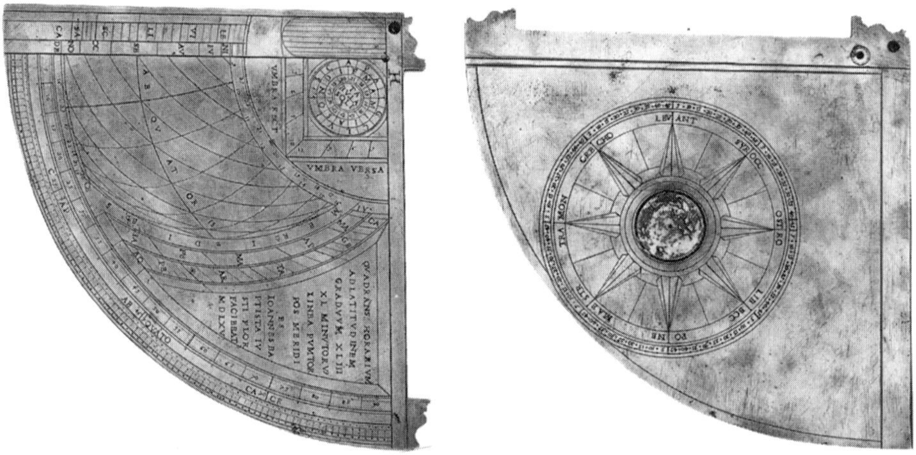

3. - Quadrant III, 1565, a) front, b) back.

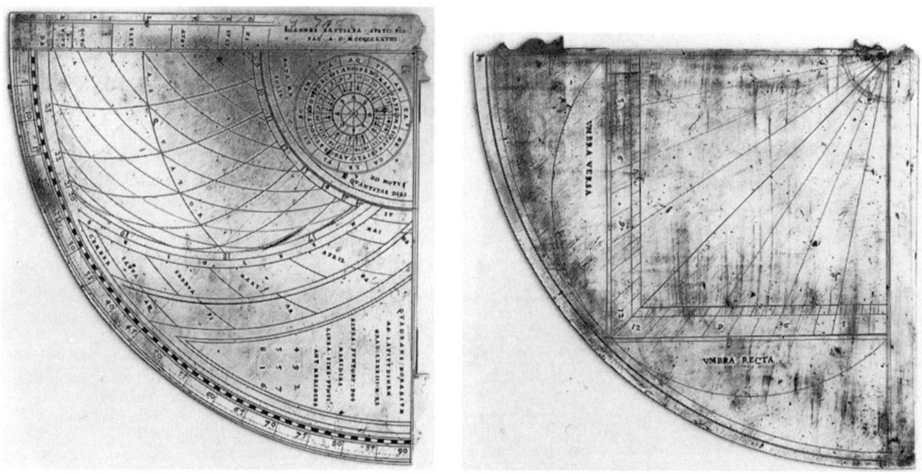

4. - Quadrant IV, 1568, a) front, b) back.

5a

5b

5. - Quadrant Va, a) front, b) back.

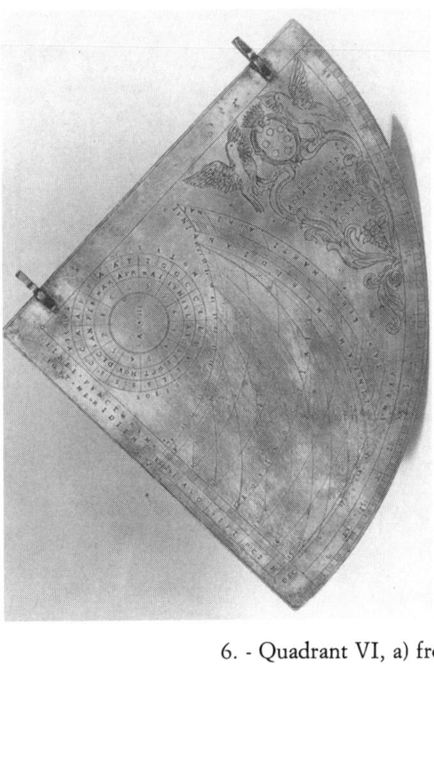

6a

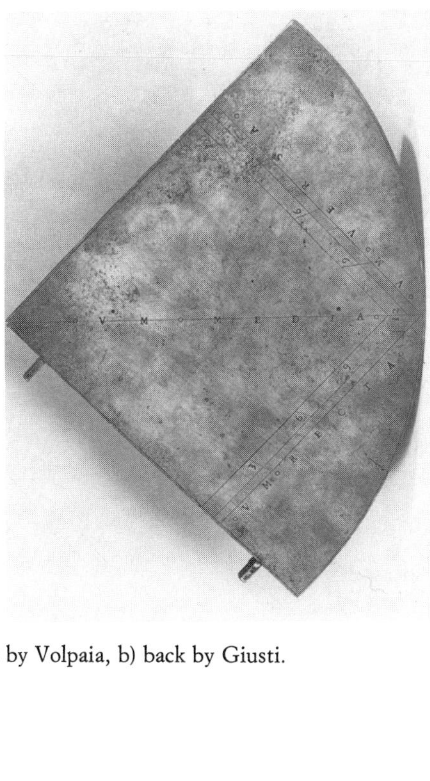

6b

6. - Quadrant VI, a) front by Volpaia, b) back by Giusti.

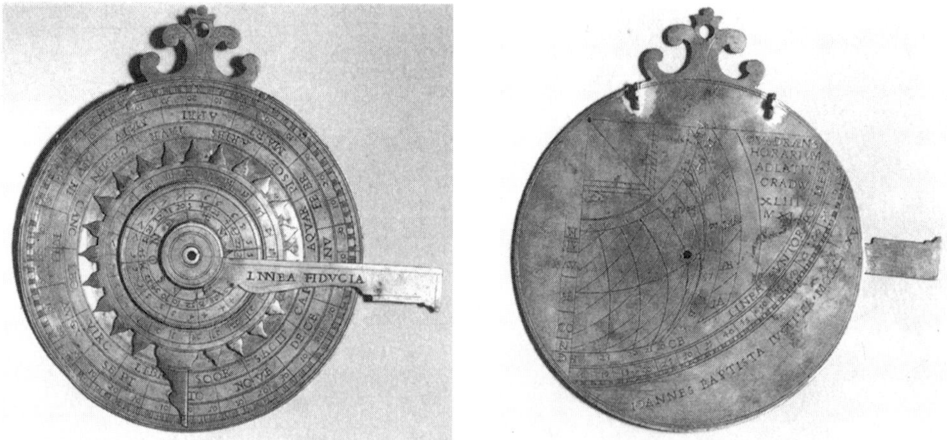

7a

7b

7. - Nocturnal VII, a) front, b) back.

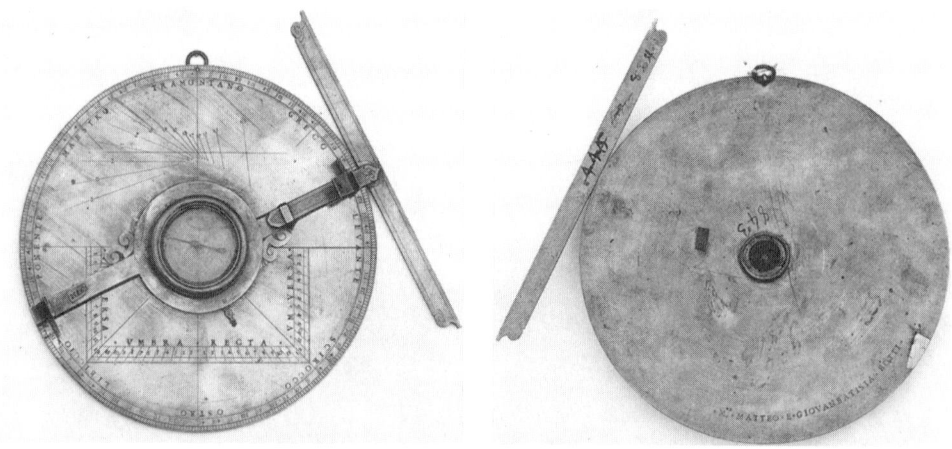

8a

8b

8. - Simple theodolite VIII, a) top, b) underside with names of owners.

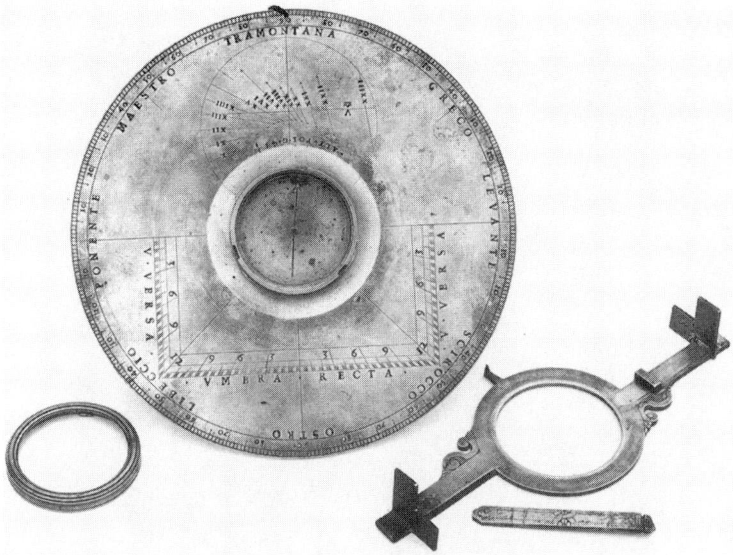

9. - Simple theodolite IXa disassembled.

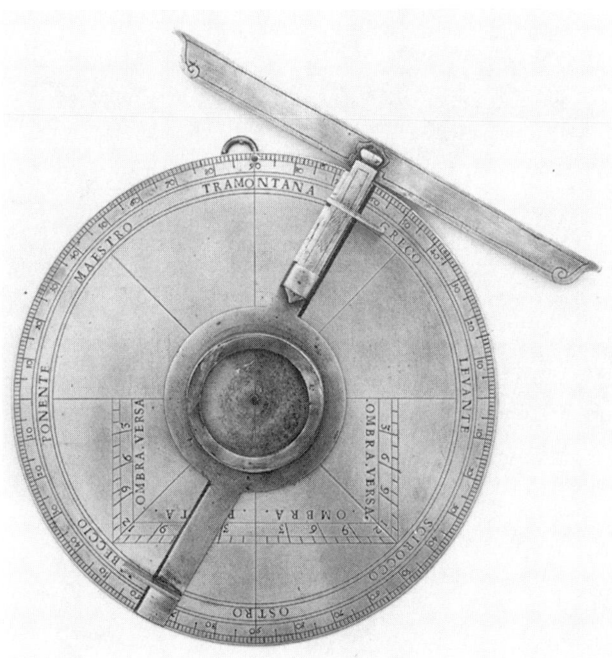

10. - Simple theodolite X.

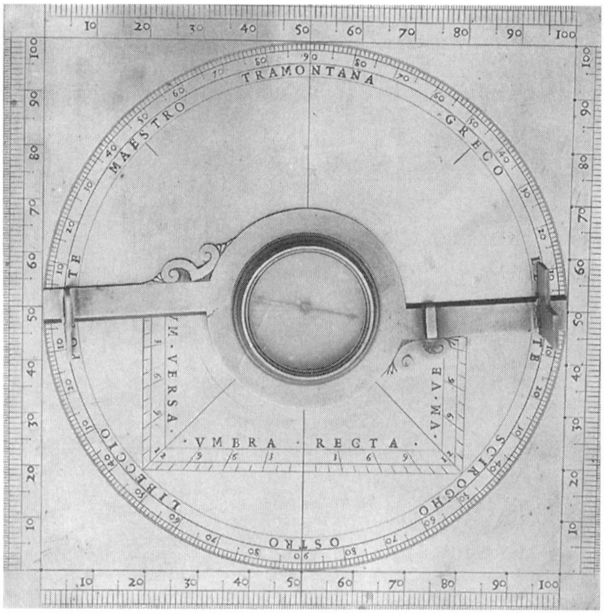

11. - Simple theodolite XI.

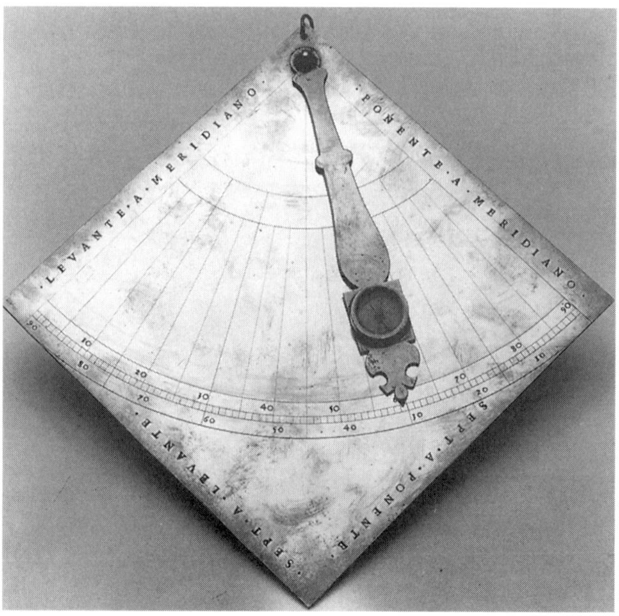

12. - Declinatory XII.

13a

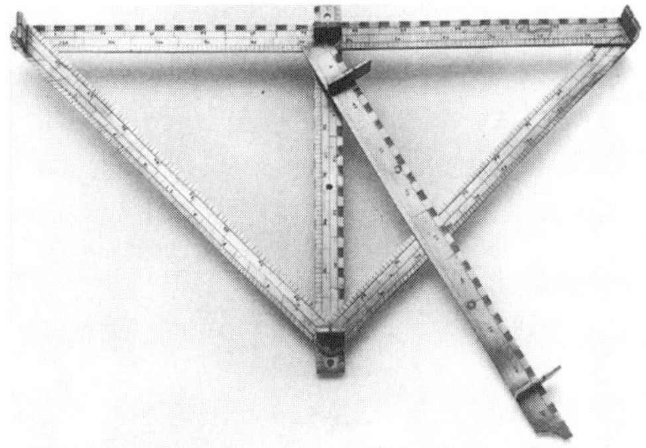

13b

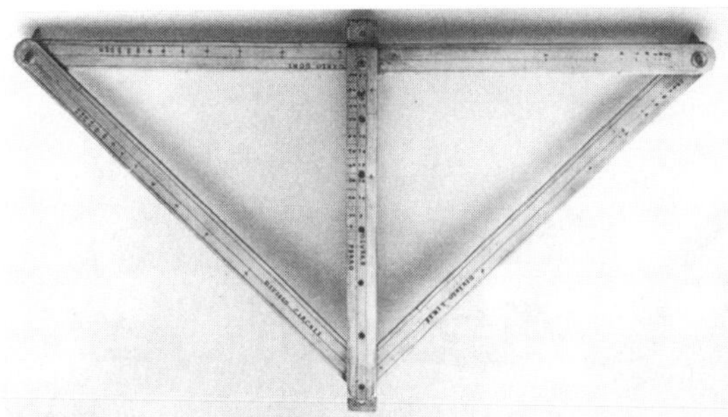

13c

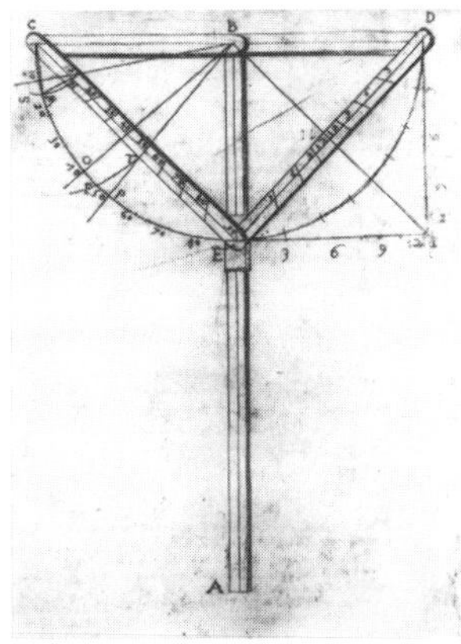

13e

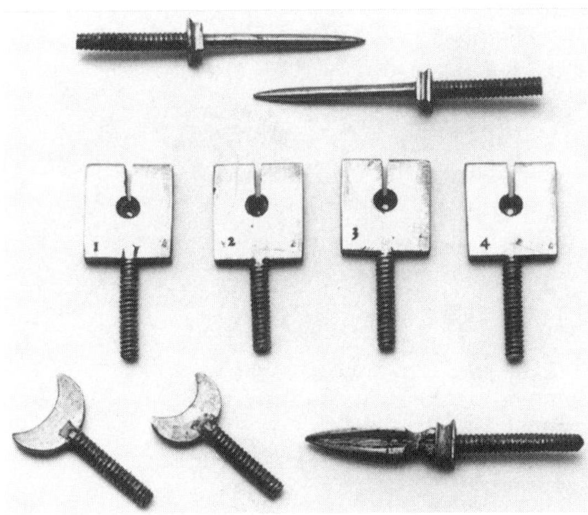

13d

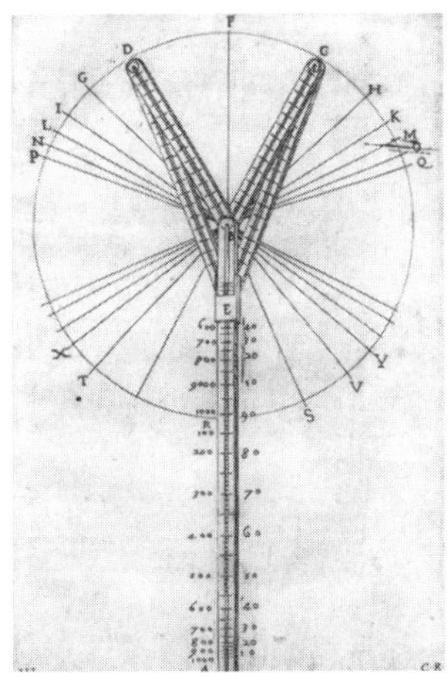

13e

13. - *Radio latino* XIII, a) the linked bars, b) arranged for angle measurement, c) reverse side, d) sights, pegs, and pen, e) two figures from Egnazio Danti's *Trattato del Radio Latino* (1583) showing the manner of use.

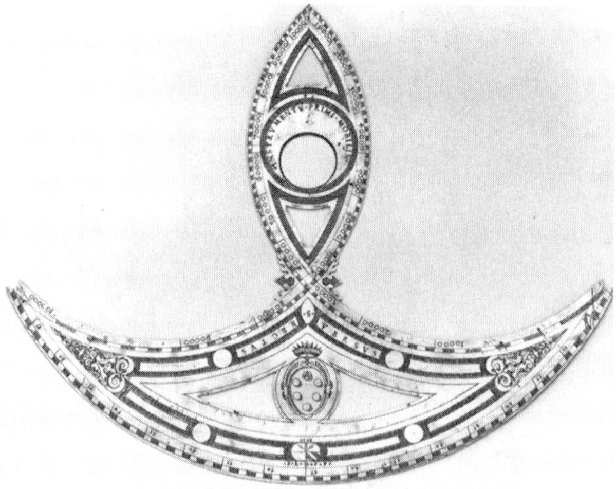

14. - *Primum mobile* XIV, 1568.

15a

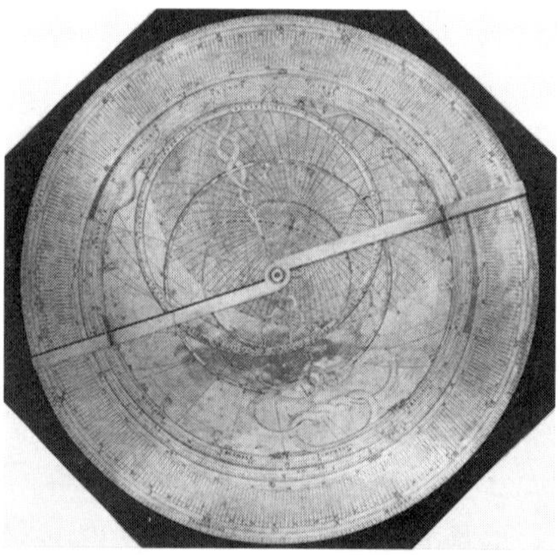

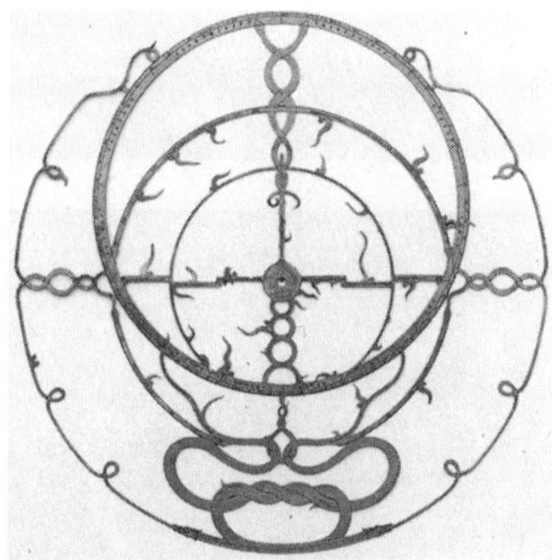

15b

15c

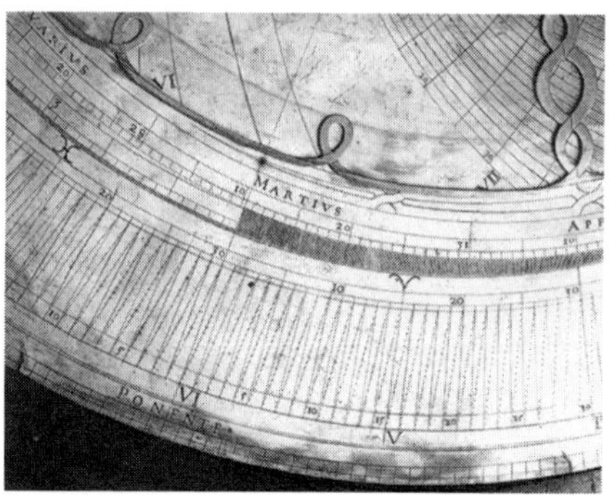

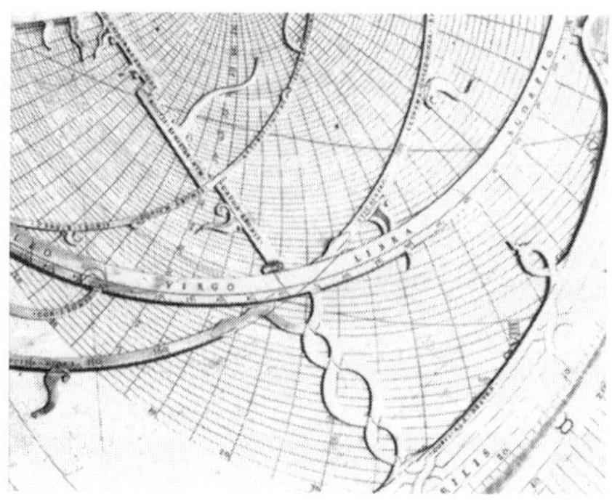

15d

15. - Very large astrolabe XV (IC 484), a) front, b) rete, c) detail of front showing beginning of Ombra Versa scale (shaded) with third division marked 3; note transversals, d) detail of front showing punches on rete: N1, N2, L1, L2.

16a

16b

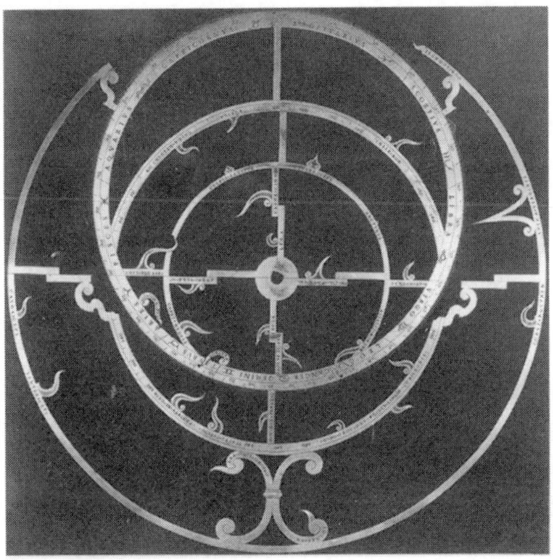
16c

16d

16. - Large astrolabe XVI (IC 180), a) front, b) back with Rojas projection, c) rete, d) detail showing coat of arms.

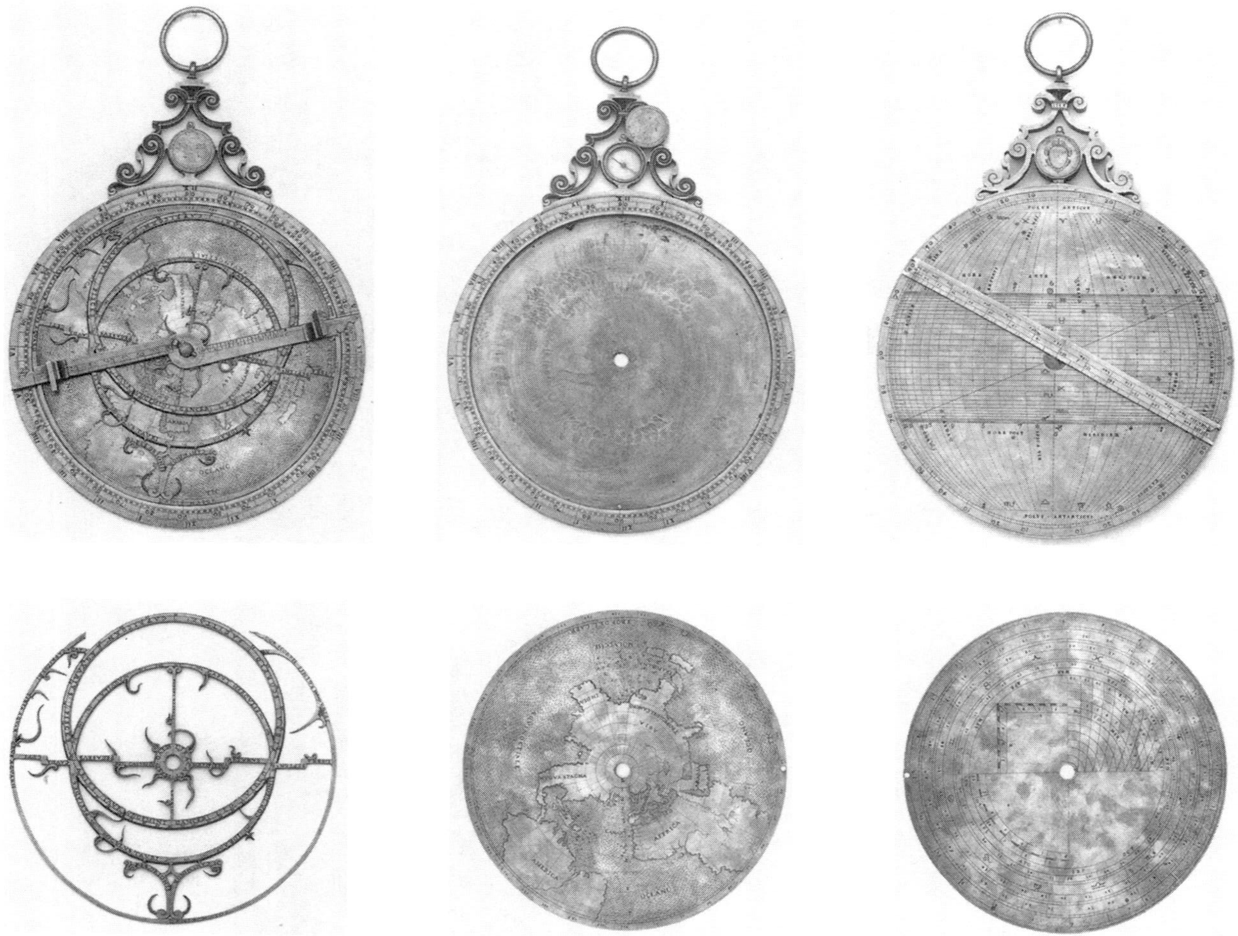

17a

17c

17d

17f

17. - Medium astrolabe XVII (IC 4006), a) front, b) mater, c) back, d) rete, e) map plate, f) multiple function plate, or «ready reckoner».

18. - Medium astrolabe XVIII (IC 2005), front.

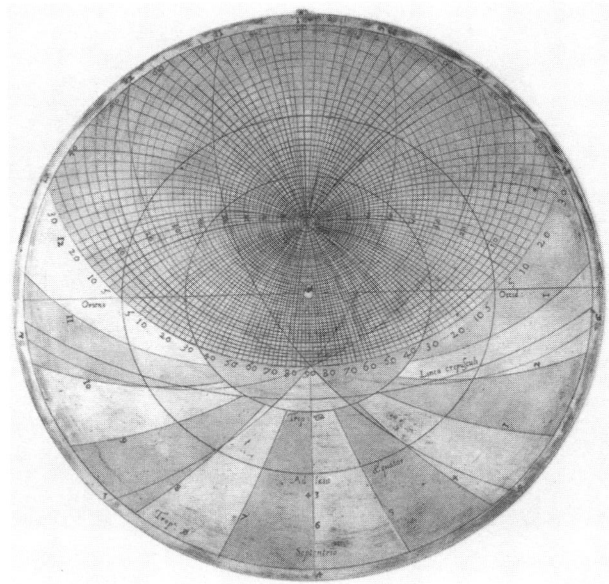

20. - Additional latitude plate XX, made by Giusti for the Mercator astrolabe (IC 490).

19. - Details of map plate XIX (IC 4005), showing layout marks for SPAGNA, and layout marks over-punched.

21. - Small astrolabe XXI (IC 458), a) front, b) back.

22. - Small astrolabe XXIII (IC 4607), a) front, b) rete, c) multiple function plate.

VII

An Astrolabe attributed to Gerard Mercator, *c.* 1570

GERARD L'E. TURNER

and ELLY DEKKER

Contents

1. Prologue

During a period of study in the Istituto e Museo di Storia della Scienza, Florence, by one of us (G.L'E.T.), several sixteenth-century astrolabes were examined, one of which included a double-sided map plate.[1] Such a plate had significance in relation to the

[1] Istituto e Museo di Storia della Scienza, Florence, inventory no. 1098. International Checklist [hereafter IC] no. 490. Sharon L. Gibbs, Janice A. Henderson, and Derek de Solla Price, *A Computerized Checklist of Astrolabes* (New Haven, Conn.: Yale University Department of the History of Science, 1973). This checklist is in course of revision and extension by David A. King, Institut für Geschichte der Naturwissenschaften, J.-W.-Goethe Universität, Frankfurt-am-Main, Germany.

astrolabes of Giovan Battista Giusti, who made quadrants, surveying instruments, and astrolabes in Florence during the period 1560–1575. Giusti's workshop is to be the subject of a forthcoming paper.[2] Five map plates by Giusti are known. They are North polar projections, and the land mass is engraved in a simple manner. The sixteenth-century astrolabe that is the subject of this paper, has a map plate with a North polar projection on one side reaching to the Tropic of Capricorn, while the other side is a South polar projection to the Tropic of Capricorn only. The engraving characteristics found on this plate are markedly similar to some products of the Mercator workshop. Not only that, but a critical examination of the rest of the instrument showed that, apart from one latitude plate, the remainder was most likely to be in the hand of Gerard Mercator (1512–1594) himself. The odd latitude plate, made for the latitude of Florence, is from the workshop of Giovan Battista Giusti. It can be shown that Giusti made the astrolabes commonly associated with Egnazio Danti (1536–1586), cosmographer to Duke Cosimo II of Florence. Danti published his *Trattato dell'uso et della fabbrica dell'Astrolabio* at Florence in 1569, the same year that Cosimo (1519–1574) was created Grand Duke of Tuscany. It is likely, therefore, that the Mercator astrolabe was acquired during a period close to 1569.

Since written information about makers of scientific instruments is rare, particularly in the early period, our knowledge of the workshops producing such artefacts, and their locations, is bound to depend largely on examination of the instruments themselves. Such examination, in particular the study of the letters and numbers engraved or stamped on these brass objects, can be the most important aid to ascribing unsigned instruments to known makers, through comparison with signed ones. This makes it possible to establish a larger corpus of instruments for a given period and location. The method has been used to identify some 110 instruments made in England during the reign of Queen Elizabeth I.[3] The same method has been used to gather the corpus of instruments attributed to the workshop of Giusti, as mentioned above.

In the case of the Mercator workshops of Louvain and Duisburg a rather different route has to be taken because there are no instruments signed by Mercator that are known.[4] What is known, however, is that he did make astrolabes and other instruments, and of course his globes and maps are available for study. These provide their evidence through printing, so the engraved copperplates for the gores and maps provide the link with the engravings on brass instruments, but in mirror image. The printing process removes the sharpness of edges and the depth of cuts by the burin, but the style of letters and numbers, and the skill, are clear. For the sixteenth century scarcely any documentary evidence remains. Although the Mercator astrolabe in Florence can be very clearly identified in a late-eighteenth-century inventory,[5] it cannot be identified in earlier inventories, even though it will have been present, because the entries are not exact, the wording being of the 'and five brass astrolabes' variety.

[2] G.L'E. Turner, 'The Reconstruction of a 16th-Century Florentine Workshop', paper read at the Eleventh Scientific Instrument Symposium, Bologna, 9–14 September, 1991. At that time, 16 instruments by Giusti were discussed; now the total is 19. See G.L'E. Turner, 'The Florentine Workshop of Giovan Battista Giusti, 1556–c 1575' *Nuncius: Annali di Storia della Scienza*, 10, pt 1 (1995), 131–72, 15 plates. (Reproduced in this volume chapter VI.)

[3] G.L'E. Turner, 'Mathematical Instrument-Making in London in the Sixteenth Century', in *English Map-Making 1500–1650*, edited by Sarah Tyacke (London: British Library, 1983), pp. 93–106, plates 47-52. The record has been greatly extended since 1983 to over 100; see G.L'E. Turner, *Elizabethan Instrument Makers: The Origins of the London Trade in Precision Instrument Making* (Oxford: Oxford University Press, 2001).

[4] Globes, but no instruments, are cited by F. Zinner, *Deutsche und niederländische astronomische Instrumente des 11.-18. Jahrhunderts*, 2nd edn (Munich, 1967; reprinted 1979), pp. 443–4

[5] A full description is given in a manuscript inventory, 'Inventario del Reale Gabinetto 1776–1779', item 669. An account of the collection is in [Felice Fontana], *Saggio del Reale Gabinetto di Fisica e di Storia Naturale di Firenze* (Rome, 1775).

We believe that the evidence supplied by the astrolabe itself, and evidence from other astrolabes and maps by various craftsmen, show that it was only the Mercator workshop of Duisburg that could have produced this astrolabe. With an example of a Mercator product postulated, it is to be expected that other instruments can be identified as Mercatorian. Indeed, part of an astrolabe in Florence,[6] where only the mater remains, could also be from Duisburg. And while writing this paper, a search through record files kept in the Museum of the History of Science, Oxford, uncovered a photograph of what may be a brother of the Florence astrolabe in the Czech Republic. The lettering and numbers are by the same hand, but the rete is to a different pattern and less embellished. Another, with an identical back to the Florence astrolabe, is in Augsburg.[7]

2. The astrolabe described

The astrolabe, which is unsigned and undated, is made of a copper-based alloy and gilded (Figures 1–5, 7, 8). Associated with it are five original latitude plates, and a further one made in Florence in about 1570. A seventh plate is engraved with projections of the Earth from the North and South Poles.

The dimensions (all measurements in millimetres) are: diameter 316; width of limb 11·4, thickness of limb 6·2; thickness of disc attached to limb 2; diameter of all plates 293; rete: diameter of ecliptic circle 202, thickness 3; rule: length 316, width 13, thickness 4·2; alidade: length 316, width 15, thickness 3·5.

The instrument is formed from a circular disc fixed to the limb by 16 rivets. The throne is formed by a bar attached to the limb and supported by two S-shaped brackets with elaborate decoration. The ring at the top swivels above a bracket that moves in a shackle left or right. The boss is foliate, and the bracket attached to the ring has a grotesque face on either side.

The face of the limb is divided in degrees, and marked, from the top, 90°–10° in each quadrant (zero degrees are not marked). The limb is also marked for 24 hours, engraved 1–12 twice.

The mater is engraved with projections for the latitudes of 90° North and 0°. Also engraved are the (un)equal hours for latitude 0° (twenty-four hours) and the astrological houses (numbering twelve).

The rete bears the names of 42 stars or groups of stars. For Gemini, two stars are marked on the ecliptic circle, there are seven pointers for Ursa Major, and Cauda Capricorni has a forked pointer. A list of the star names is given in Table 1.

On the back of the instrument is a shadow square in the lower semicircle, and in the upper left quadrant is a diagram for conversions between equal and planetary hours based on the times of sunrise and sunset; the right quadrant is blank. Around the edge are divided scales for degrees (as on the front), the zodiac, and the calendar. The calendar is eccentric with respect to the zodiac. The first point of Aries is $10\frac{1}{4}$ March.

[6] Inv. no. 1100; IC 478. The mater is blank, and on the back is a Rojas projection. No other parts remain; diameter 290 mm.

[7] Moravská galerie, Brno, Czech Republic, inv. no. 24.385; IC 4608. There is a plate for latitude 49°, and on the back lower half is a shadow square, above left a conversion graph from equal to unequal hours, above right an unequal hours diagram. The throne is the same, with the grotesque face, but lacking the foliate boss. The pointer for Cauda Capricorni is forked, and many words are engraved with the ı lacking the dot. The other is in the Städtische Kunstsammlungen, Augsburg, inv. 3537. We are grateful to Professor D. A. King for drawing attention to this astrolabe.

Figure 1. The front of the astrolabe attributed to Gerard Mercator.

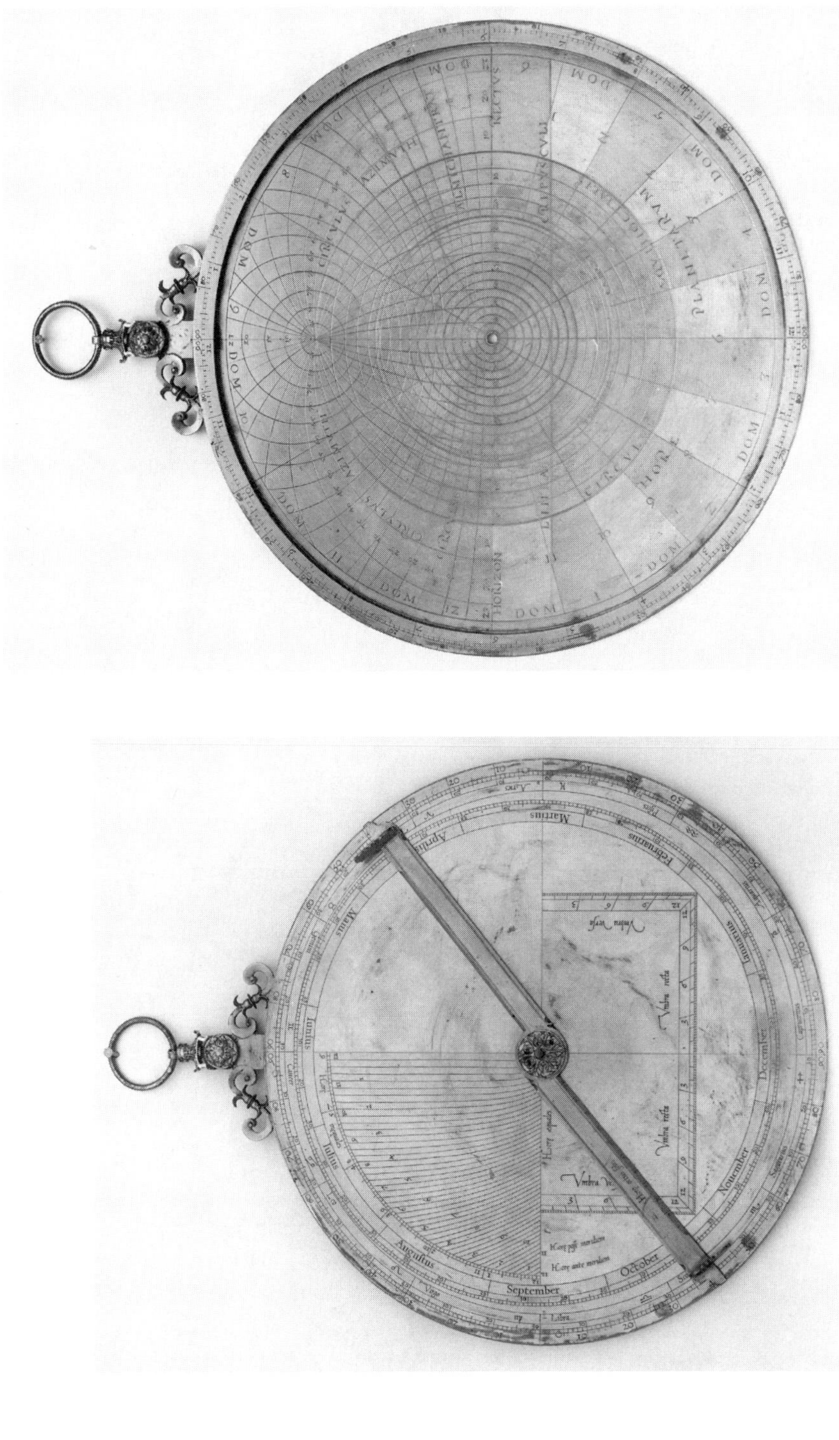

Figure 2. The back and the mater.

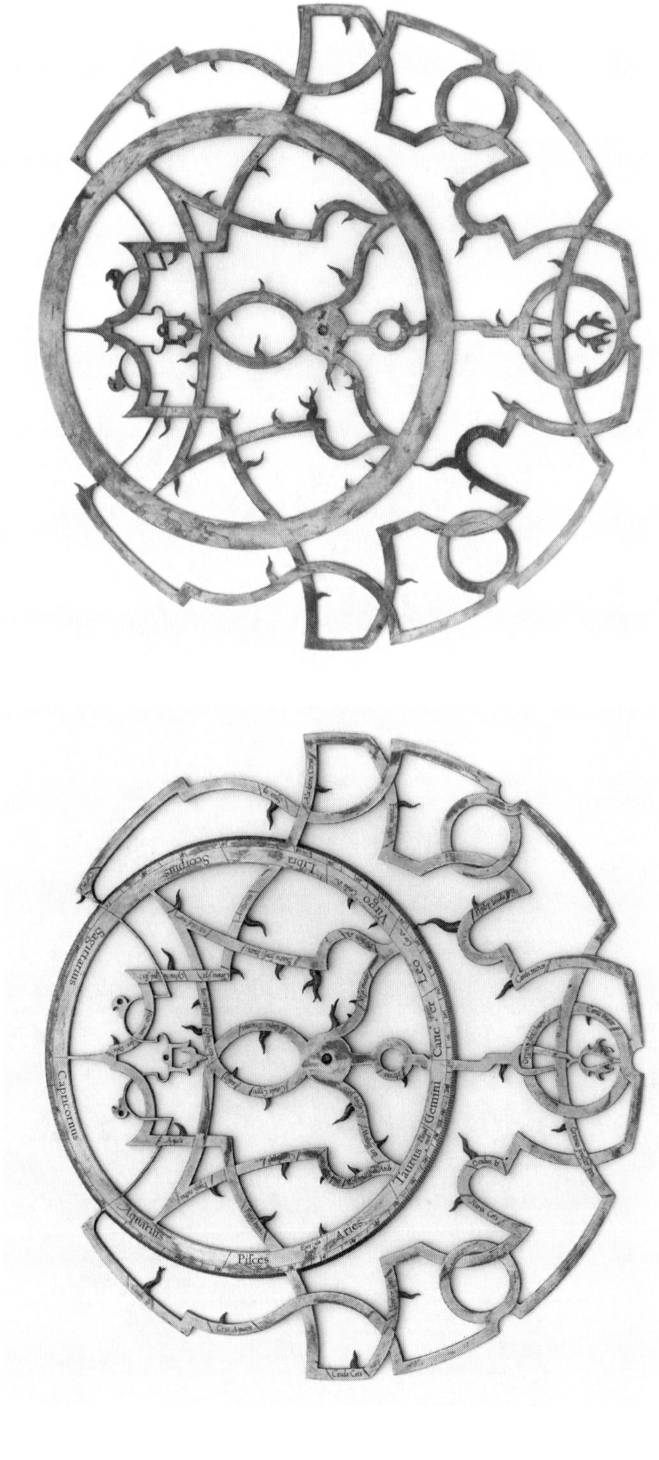

Figure 3. The rete, front and back.

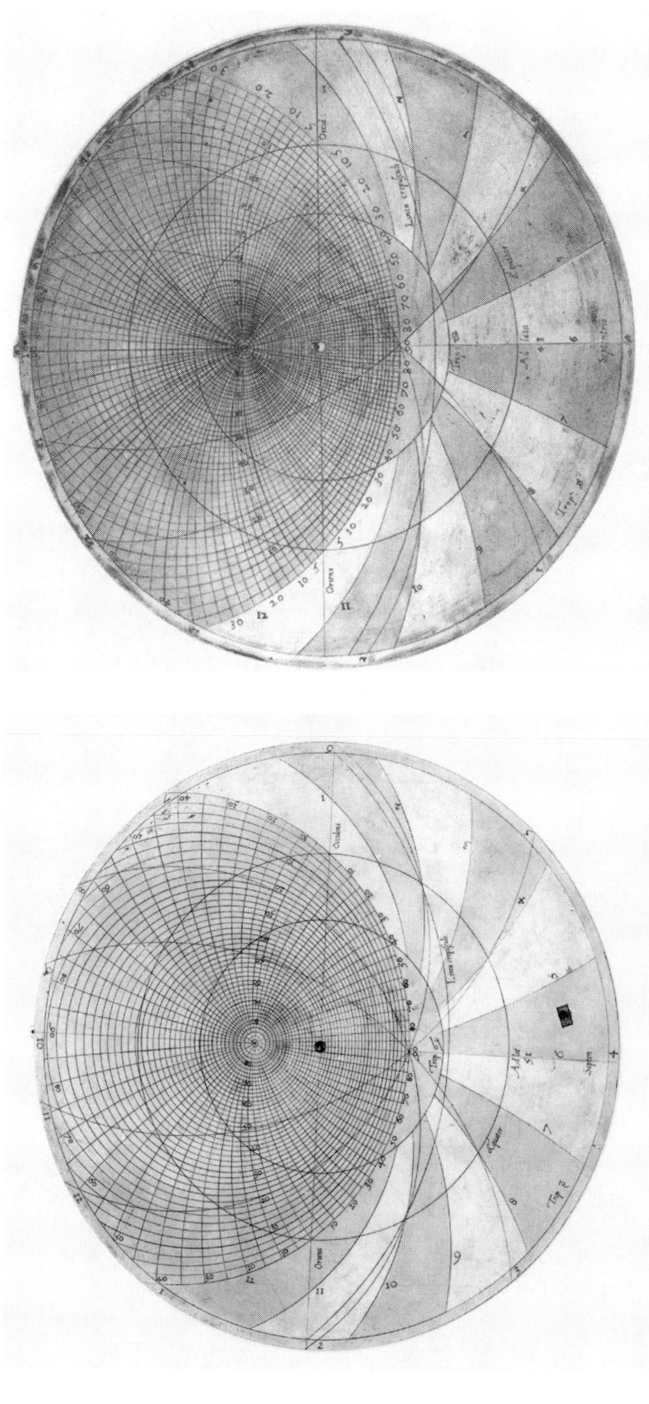

Figure 4. Plate for latitude 51°, and plate by Giusti for 43°.

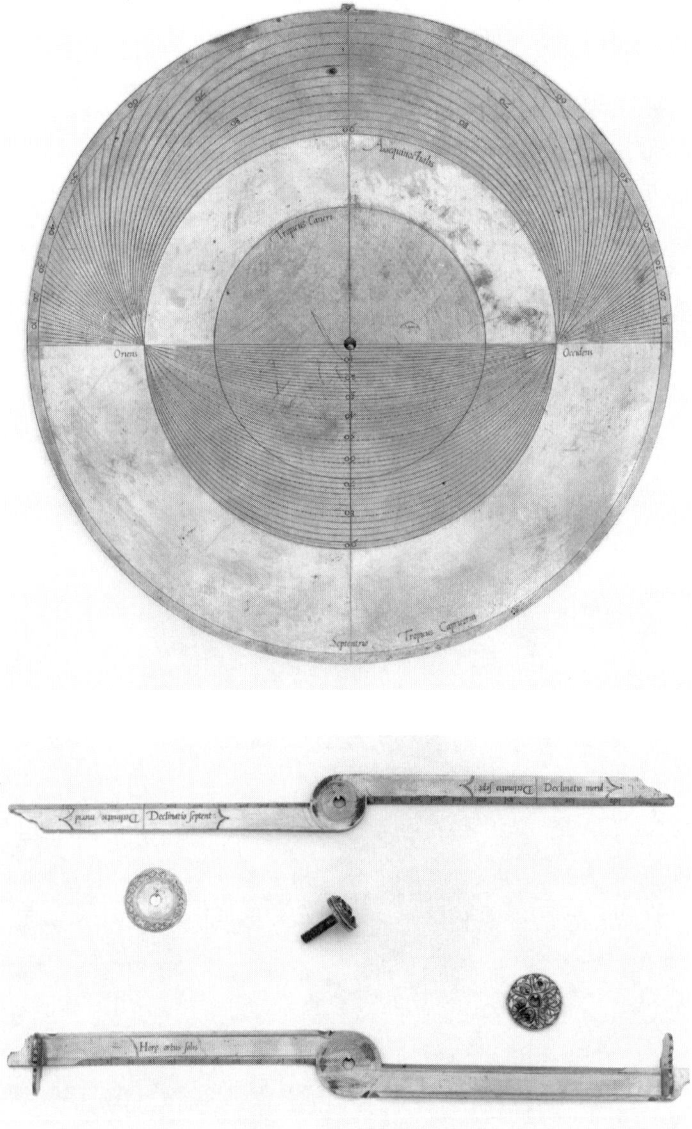

Figure 5. The tablet of horizons, and the rule, alidade, bolt, nut, and washer.

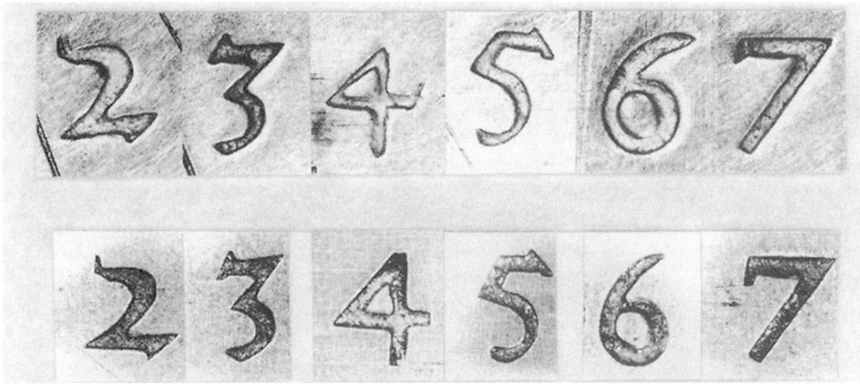

Figure 6. Number punches used by Giovan Battista Giusti *c.* 1570: top latitude plate; below, instrument in Florence, Museo di Storia della Scienza, inv. 2508.

The alidade has one of its arms divided with a scale 0–10 labelled *Horae ortus ſolıs*. The sights have a notch at the top and a pinhole below. On either side of each sight is a wing, one with a hole and the other with a dot, the disposition on the sights arranged so that a hole in one sight faces a corresponding dot on the other, for aligning on the Sun to find its altitude. The rule is divided on both arms 0°–70° N and 0°–20° S; the scales labelled *Declinatio ſeptent:/Declinatio merid:*. The pin is a bolt, with an incised triangular motif on the head and a foliate roundel on the nut. The washer has an acanthus and rose motif around its rim.

Table 1. The 42 star names engraved on the rete (letter forms reproduced).

Caſſıop: pect^{9}	Spıca [sign of Virgo]
Cauda Cetı	Arcturus
Media cıngulı Andr:	Bootıs ſiniſ: hum:
Venter Cetı	Lanx ſept: [sign of Libra]
Meduſæ cap	Coruna ſept:
Narıs Cetı	Ophıuchi pal: ſın:
Pleıad med:	Cor [sign of Scorpius]
Oculus [sign of Taurus]	Herc: cap
Hırcus	Ophıu: cap:
Orıonıs ſınıſter pes	εxtra: man: dex: Ophı:
Orıonıs dex: hum:	Caput Draconıs
Canis maior	Lyra
Gemini [two stars on ecliptic circle]	Aquıla
Canis minor	Cauda Cygnı
Hydræ lucıda	Cauda [sign of Capricornus] [two pointers]
Cor [sign of Leo]	Pegaſı rıctus
Vrſa maıor [seven pointers]	Crus Aquary
Fundus Vaſıs	Crus Pegaſı
Dorſum [sign of Leo]	Pegaſı hum:
Cauda [sign of Leo]	Peg: Vmb:
Ala detra Coruı	Extr: ala Pegaſı

Figure 7. The map plate, North Polar projection.

Figure 8. The map plate, South Polar projection.

On each of the six latitude plates protrudes a small lug, which is symmetrical to the 90° line of the projection. The lug fits into a socket in the mater below the throne. Each lug is pierced with a round hole for grouping the plates together, presumably either on a peg or on a cord. This arrangement is necessary because the mater can accommodate just one plate at a time, contrary to the normal practice with astrolabes. The plates are engraved with almucantars at 2° intervals, and azimuths at 5°, except for one side of the Florence plate which has azimuths at $2\frac{1}{2}°$ intervals. Also present are the astrological houses marked 1 to 12, the equator, tropics, crepuscular line, and planetary hours, marked 1 to 12. A remarkable feature on the plates is the burnishing of the gilding in alternate segments of the planetary hours. Such a decorative conceit has not been remarked on any other astrolabe, and suggests that the instrument came from a first-class workshop, and that it was possibly made for a notable customer.

The original plates are engraved for latitudes every 3° between 36° and 60°, and one has a tablet of horizons (Table 2). The sixth plate is made for the latitude of Florence, 43°, and it is not from the same workshop as the rest of the astrolabe. Being required to be used in Florence, a plate for 43° was made by a craftsman of the city, Gian Battista Giusti. For the attribution to this craftsman, see Section 3 below. The fabrication of this plate, although skilled, is not to the same very high standard as the rest. The brass has many fissures, and it is somewhat thicker. The gilding has not been as effective as on the original five plates. An attempt has been made to burnish alternate segments as on the originals, but with less success. The engraved lettering has been imitated as closely as possible, even to the omission of the dot over the ı,[8] so have many numerals, the exception being those labelling the unequal hour lines, where punches have been employed. These punches are characteristic of those used by Giusti on his instruments. Although both sides of Giusti's plate are labelled: *Ad latıt 43*, there is a minor difference in the azimuths. On one side they are at intervals of 5°, and on the other $2\frac{1}{2}°$; the almucantars are all at 2°.

The map plate is not provided with a lug nor with a hole for hanging. Made of copper, the land mass and decorative elements are gilded, while the seas are left as

Table 2. The plates with the astrolabe.

Latitude		Thickness (mm)
1. 36°/tablet of horizons		0·9–1·5
2. 39°/42°		0·85–1·7
3. 45°/48°		1·1–1·35
4. 51°/54°		1·2–1·9
5. 57°/60°		1·2–1·4
6. 43°/43°	(Florentine workshop)	1·3–1·8
7. Map plate		1·0–1·2

[8] The occasional use of the letter ı (without a dot) seems to be an identifying sign of Gerard Mercator's hand. See his globes of 1541 and 1551, and *Tabulae geographicae Cl: Ptolemei ad mentem autoris restitutae & emendatae Per Gerardum Mercatorem Illustriss: Ducis Cliviae etc: Cosmographū* (Duisburg, February 1578 [old style]). The Latin cartouche to Gemma Frisius's celestial globe of 1537, engraved by Gerard, shows ı eight times.

Table 3. Map texts. See Figures 9, 10, 11.

I.	AMERICA SIVE INDIA NOVA / *Anno D: 1492. a Chriftophoro Columbo nomine Regis/Caftellae primum detecta*
II.	Noua Guinea / *nuper inuenta quae an fit / infula an pars continentis / auftralis incertum est*
III.	*Pfittacorum regio fic a / Lufitanis appellata ob in / credibilem earum auiu / ibidem magnitudinem*
IV.	*Vaftifsimas hic efse regiones ex / M. Pauli Veneti & Ludouici / Vatomanni* [sic] *fcriptis peregrina: / tionibus liquido conftat*
V.	Maletur / *regnum fcatens / aromatibus*

copper. This has blackened through long contact with the atmosphere. One side is a North polar projection to the Tropic of Capricorn, while the other side is a South polar projection cut off at the Tropic of Capricorn. This side has more space for engravings of monsters and sailing ships. The prime meridian runs through the Azores, a few degrees West of the Canaries, and through the magnetic pole, *Polus Magnetis*, at longitude 180° and latitude *c.* 74°. The meridians and the parallels are engraved for every 10°. The longitude scale is on the outer rim, marked by punched numbers every five degrees from 5° to 360°. The latitude scale is along the prime meridian and is marked every ten degrees from 10° to 90°. Both sides of the plate show the polar circles, *Circulus Arcticus* and *Circulus Antarcticus*, the equator, *Circulus Aequinoctialis*, and the two tropics, *Tropicus Cancri* and *Tropicus Capricorni*. Five inscriptions are worth a special mention here (Table 3).

After the outline of each coast had been engraved, shading was added to indicate the shoreline. This has produced a confused outline after the gilding process, as the shading picked up some of the gold leaf, but could not provide a flat surface for polishing. Thus regions of estuaries, inlets, groups of small islands can be indistinct, and definition is further impaired by the blackening of the regions of plain copper. Some large and small islands in Indonesia have not received any gold leaf, and so remain dull (NW tip of New Guinea and the Moluccas).

3. The Florentine latitude plate and the Giusti-Danti connexion

The sixth plate, already described, was made for the latitude of Florence, 43° (modern figure 43° 47′), because the nearest original plates are for latitudes 42° and 45°. It is made to fit the astrolabe exactly, and what is even more telling is that the engraved letters and numbers imitate those on the other plates as well as the craftsman was able to do. Some numbers, however, those on the unequal hours, are punched, and the characteristics of the punches used readily point to other instruments made in Florence. The plate came from the workshop of Giovan Battista Giusti, of whose products nineteen items are known to exist (not counting the present plate).[9] Figure 6 shows that the numbers stamped on the additional plate are the same as those used on other Giusti instruments. Through comparing the punches used on instruments dated 1568 and 1572,[10] it is possible to see that the punches used on the latitude plate are one of the sizes to be found on the 1568 astrolabe. We have, therefore, a firm date for the use of this set of punches.

[9] See footnote 2.

[10] Istituto e Museo di Storia della Scienza, Florence, primum mobile, 1568, inv. 2643; astrolabe, 1568, inv. 1285; IC 4605; Science Museum, London, astrolabe, 1572, inv. A629481; IC 4606.

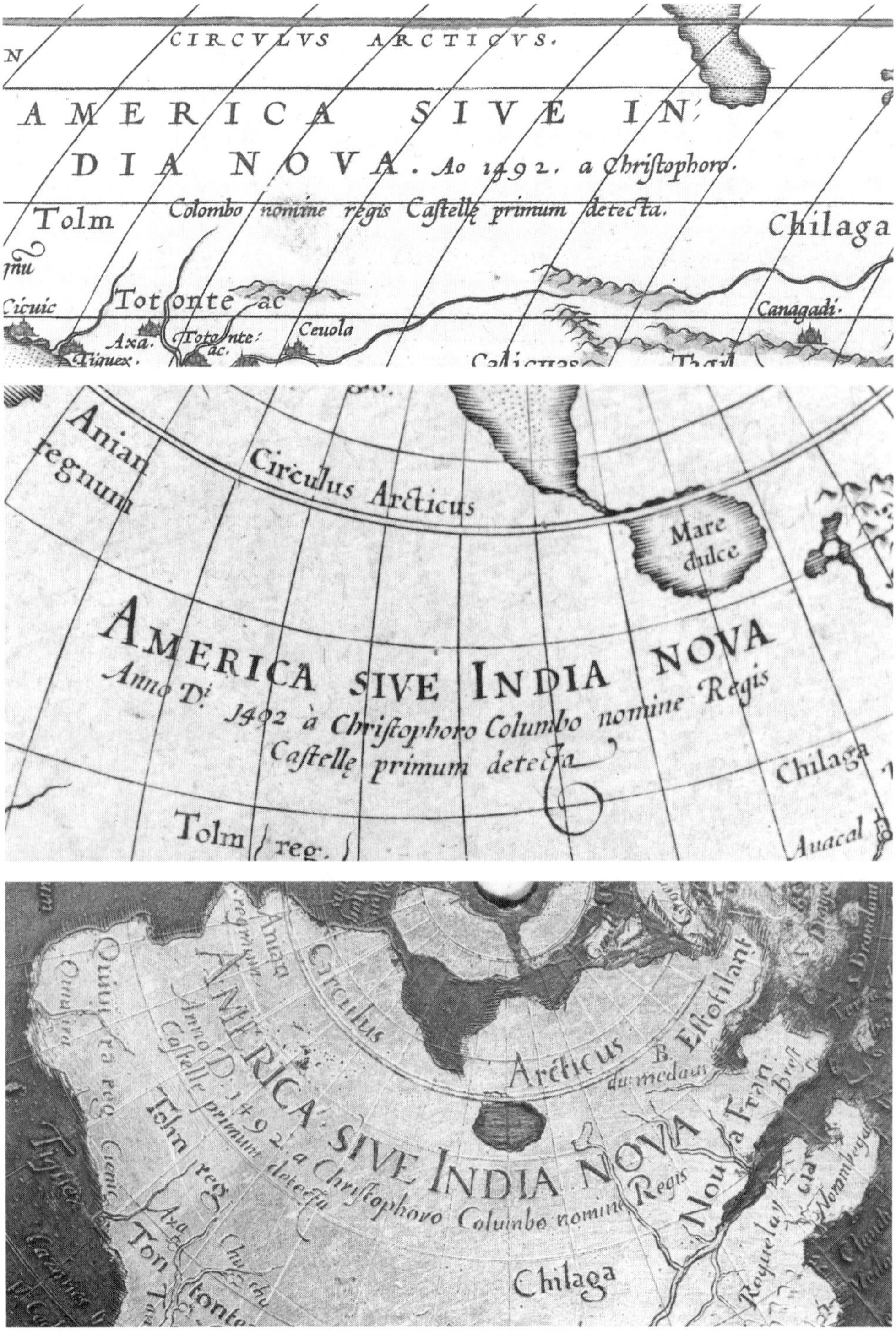

Figure 9. America: texts from Ortelius, 1570 (top); Rumold, 1587 (middle); map plate.

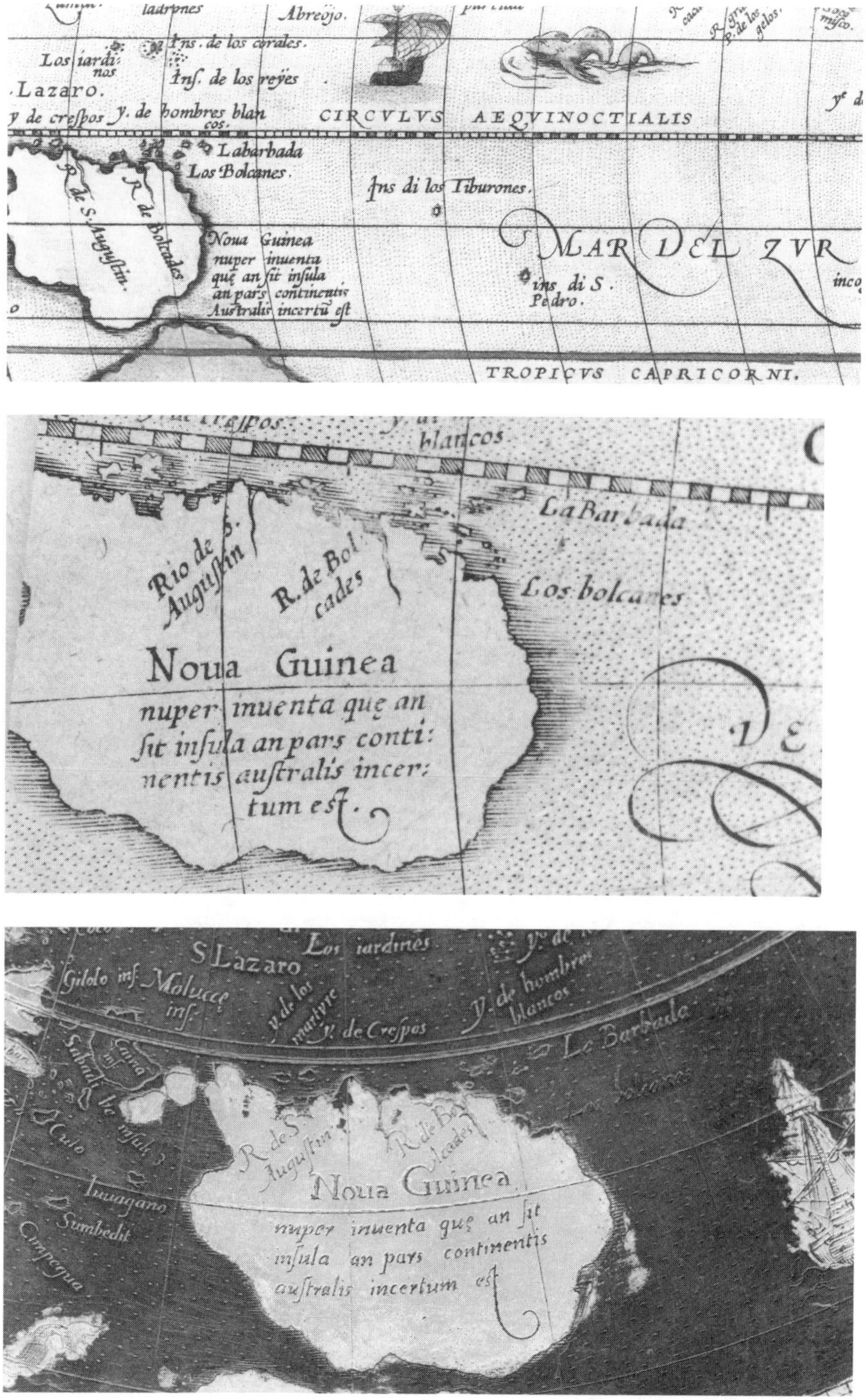

Figure 10. New Guinea: texts from Ortelius, 1570 (top); Rumold, 1587 (middle); map plate.

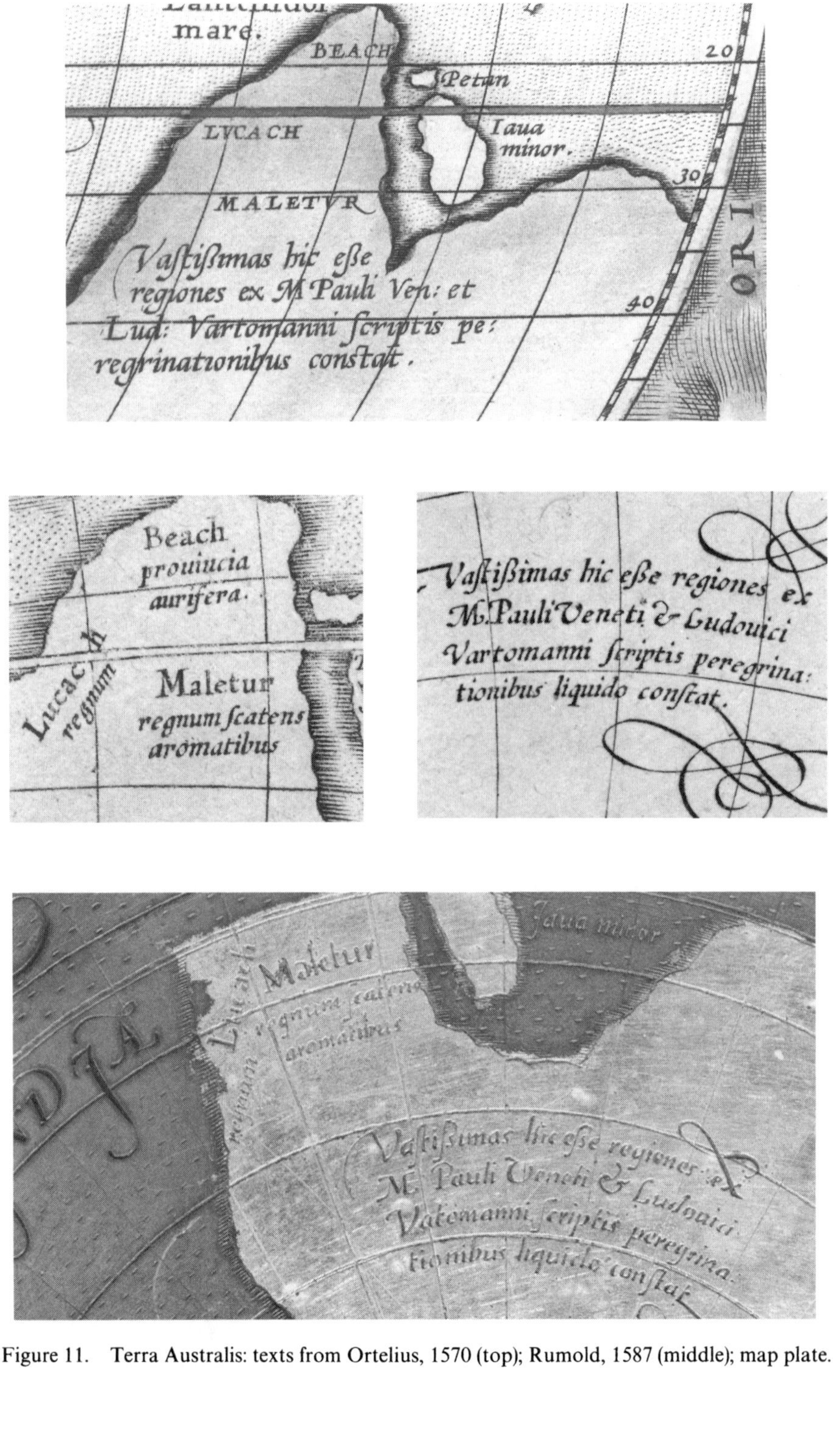

Figure 11. Terra Australis: texts from Ortelius, 1570 (top); Rumold, 1587 (middle); map plate.

Eight of the Giusti corpus are astrolabes, two of which are dated, and one large undated astrolabe is engraved with the name: '·F· EGNATIVS DANTES·', placed under a coat of arms (in chief azure an olive branch with three leaves, argent two bars azure).[11] Another instrument connects Giusti with Danti, a *Primum mobile*, which is dated 1568, and is engraved under a similar but abbreviated coat of arms: '·F·E·D·P·F·'.[12] These initials are construed as: Frater Egnatius Dantes Perusinus faciebat.[13] If the one is not marked 'fecit', why should the other be so marked? The title page of *Trattato* says that the book is 'Di F. Egnatio Danti dell'Or. di S. Domenico', while elsewhere: 'F. Egnatio Danti Perugino' (his town of birth). Giusti signed two of his common quadrants: IOANNES BAPTISTA IVSTI FLOE FACIEBAT (1565); ... FLO FAC (1568).[14] The initials might as well be construed as: Frater Egnatius Danti Perugino e Florentino. That Danti, an aristocrat and scholar, could be said to have 'made' an instrument is out of the question.

3.1. *Danti and Giusti*

Egnazio Danti was born in Perugia into a family associated with the arts, and also interested in astronomy and measurement.[15] In 1555, Egnazio entered the Dominican order, changing his baptismal name, Carlo Pellegrino, to that by which he is generally known. His education had encouraged an interest in astronomy and mathematics, and this in turn led him to study cosmography to such purpose that, in 1562, he entered the service of Duke Cosimo II of Florence (from 1569, Grand Duke Cosimo I of Tuscany). His specific task was to create the series of maps of countries of the world that the Duke wished to place on the walls of a room in the Palazzo Vecchio.[16] This room, generally described as the *Guardaroba*, but sometimes as the *Sala delle Carte geografiche*, was eventually entirely decorated with 53 maps, covering the known world. Of these, 29 were designed by Danti between 1564 and 1575 (dates from the maps) during the period when he served Cosimo as cosmographer. The room was also to contain a great terrestrial globe of over 6 metres diameter, on which the regional maps on the walls could be located; this, too, was Danti's responsibility.[17] At first he worked in the monastery of Santa Maria Novella, but in 1571, at the express request of the Duke, he was granted permission to live in the Palazzo, enabling him to work more closely with his patron. He studied to reform the Julian calendar, constructing, with the Duke's permission, an astronomical quadrant and armillary on the façade of the church of Santa Maria Novella.[18] These and other scientific projects in Florence ended with the death in 1574 of Cosimo, whose successor did not favour Danti. The map project was completed by Stefano Bonsignori from 1576 (date on map of La Francia). Danti left

[11] Museum of the History of Science, Oxford, LE 2072; IC 180. Diameter 409 mm.

[12] Museo di Storia della Scienza, Florence, inv. 2643.

[13] Giuseppe Boffito, *Gli Strumenti della Scienza e la Scienza degle Strumenti* (Rome, 1929; reprinted 1982), p. 220; but Frater Egnatius Danti Praedicatorum Fecit in M. L. Righini Bonelli, *Il Museo di Storia della Scienza a Firenze* (Milan: Electa Editrice, 1968), p. 12. Perusia is the Roman name for modern Perugia.

[14] Museo di Storia della Scienza, Florence, inv. 2524; inv. 2520.

[15] *Dizionario Biografico degli Italiani* (Rome, 1960–), Danti, Egnazio, XXXII (1986), 659–62.

[16] Ettore Allegri and Alessandro Cecchi, *Palazzo Vecchio e i Medici: Guida Storica* (Florence, 1980), pp. 303–10.

[17] Boffito (footnote 13), 212–13.

[18] M. L. Righini Bonelli and T. B. Settle, 'Egnatio Danti's Great Astronomical Quadrant', *Annali dell'Istituto e Museo di Storia della Scienza di Firenze*, 4, pt 2 (1979), 3–13.

Florence for Bologna, where he became professor of mathematics in 1576. His work on cosmography and on the calendar continued, and in 1580, after a period at home in Perugia, he moved to Rome to create more maps, this time in the Vatican.

Danti also published extensively, mainly annotated translations from Latin and Greek works of science, and commentaries. An original work, however, was his *Trattato dell'uso et della fabbrica dell'astrolabio*, of 1569.[19] This is of particular importance, both in his association with Giovan Battista Giusti, and because it suggests that both he and his patron, Cosimo, to whom the book is dedicated, were interested in the use of the astrolabe. It is significant that Danti advised, at the end of the *Trattato*, that anyone wishing to have an astrolabe made should seek out a diligent workman to do the job for him. The passage reads in the original:

> Non manchero gia di ricordare due cose à quelli, che desiderono fabbricare gl'Astrolabii: L'una che cerchino di hauere gli Artefici diligenti, perche tutte le volte, che no si userà gran diligenza, gl'instrumenti saranno poco giusti, tanto è facile à chi non è piu che accurato in tal fabbrica pigliare errore. L'altra che si ingegnino di far tali strumenti di quella maggior grandezza, che possibilesia, perche quanto saranno maggiori, tanto piu minutamente saranno diuisi, & tanto piu ageuolmente, & bene si operera con essi.

The workman he most likely had in mind was Giusti.

It is safe to assume that astrolabes were made by Giusti as a result of Danti's interest in this instrument. In view of what was being undertaken at the Medician court, it is very reasonable that an astrolabe should be acquired from the finest globe, map, and instrument maker in the world. The Mercator astrolabe can, after all, be regarded as a combined celestial and terrestrial globe. A suitable acquisition for Cosimo's elevation to the Grand Dukedom of Tuscany, which took place in 1569, would have been this magnificent instrument.

4. The map plate

In the sixteenth century, geographical map plates are a not uncommon addition to astrolabes. An early account of such a circular world map in polar stereographic projection was published by Peter Apian (1495–1552) in his *Cosmographia* of 1524.[20] Through the efforts of Gemma Frisius (1508–1555), the well-known mathematician and physician, the book was reprinted and translated into many languages. The introduction of geographical plates reflects the growing interest in geography that was first stimulated by the many printed editions of Ptolemy's *Geography* from 1472 onwards, and was further encouraged by the voyages of exploration. Until the sixteenth century the astrolabe had been a purely astronomical instrument, serving various purposes related to astronomical phenomena. The addition of a geographical map plate in fact converted the astrolabe into an image of the world as a whole, in a way comparable to a pair of globes. It is therefore not surprising to find among the plates belonging to the present astrolabe a geographical plate. What is extraordinary,

[19] *Trattato dell'Vso et della Fabbrica dell'Astrolabio. Di F. Egnatio Danti dell'Or. di S. Domenico. Con l'Aggivnta del Planisferio del Roias. All'Illvstriss. et Reveren. S. don Ferdinando Cardinal de Medici.* In Fiorenza Appresso i Giunti. M.D.LXVIIII. An engraving decorating the title page consists of the six balls of the Medici superposed over a terrestrial globe showing Africa and Asia. See Figure 18.

[20] Petrus Apianus, *Cosmographicus Liber* (Landshut, 1524).

however, is that this plate covers not only the North polar but also the South polar projection (Figures 7 and 8). In addition it provides many more geographical details than is usually the case, particularly the detailed outlining of rivers, mountain ridges, and cities, which stand out in comparison with map plates found on other astrolabes. This indicates strongly that some maker has been at work with a great interest in, and command of, contemporary cartographical knowledge.

Considering the cartographical image, it is clear that the present plate belongs to the so-called Mercatorian tradition of map making. The development of the cartographical views of Gerard Mercator is marked by his successive works: the double-cordiform map of the world of 1538, the terrestrial globe of 1541, his map of Europe of 1554 and its revised edition of 1572, the world map of 1569 in the projection that still carries his name today, the terrestrial globe of 1579 dedicated to the Sultan Murad III, and the maps included in his *Atlas*, published posthumously by his youngest son Rumold Mercator (*c.* 1546–1599).[21] From this series of maps and globes one sees that Mercator's views have developed considerably during the period 1541–1569. Thereafter no major changes are encountered.[22] Indeed, the world map in the equatorial stereographic projection produced by his son Rumold in 1587 is, apart from some minor changes, a faithful imitation of Mercator's world map of 1569.[23] The Mercatorian world picture is characterized by a number of geographical features of which a few are: the four islands centred on the North Pole; the choice of the prime meridian aligned with the magnetic pole; the shrunken Mediterranean; the western bulge of South America.

A comparison has been made between the astrolabe map, these various Mercator mappings, and the world map published by Abraham Ortelius (1527–1598) in his *Theatrum Orbis Terrarum* of 1570 (Figures 9–11). The latter is included because this is also a reproduction of Mercator's world map of 1569 in another projection. This reveals that the cartography of the map plate is closest to that of the world map of Rumold Mercator of 1587. All texts on the astrolabe map plate (see Table 3) are identical to those on Rumold's map, apart from one small error. The agreement between the texts of the map plate and those of the Ortelius map is less convincing: two texts (II and III) are again exactly the same; two texts (I and IV) show minor deviations; text V is lacking. Further, in those cases where the details of Rumold's map deviate from that of Ortelius, the astrolabe map corresponds to the former. For instance, on Rumold's map and the map plate the Caspian Sea is 'Mare de Sala', whereas on Ortelius's map it is 'Mar de Bachu'; the magnetic pole is drawn on both Rumold's map and the map plate, but it is absent on Ortelius's map. No doubt because of the scale involved, the astrolabe map is less elaborate as far as place names are concerned, but the names of countries, oceans, rivers, and cities are identical. The one small difference compared with Rumold's map of 1587 appears to be in the region around 'Groenlant'

[21] H. Averdunk and J. Müller-Reinhard, *Gerard Mercator und die Geographen unter seinen Nachkommen* (Gotha, 1914). For his calligraphy, and its enormous influence, see A. S. Osley, *Mercator: A Monograph on the Lettering of Maps, etc in the 16th Century Netherlands with a Facsimile and Translation of his Treatise on the italic hand and a Translation of Ghim's* Vita Mercatoris (London, 1969). For the Sultan's globes, see Gerard Turner, Elly Dekker, Julian Raby, and Priscilla Thomas, *The Murad III Globes* (London: Christie, Manson & Woods, 1991); 32 pp. catalogue of the sale on 30 October, 1991.

[22] A late development is the new, revised map of the North Polar region, included in Mercator's *Atlas* (Duisburg, 1595), on which Nova Zemla (Novja Zeml'a) is presented as a separate island.

[23] In a text: *DE MVNDI CREATIONE as constitutione breuis instructio*, appended to Rumold Mercator's map, Gemma Frisius is acknowledged for the projection. See Rodney W. Shirley, *The Mapping of the World: Early Printed World Maps 1472–1700* (London, 1984; reprinted with corrections, 1987), p. 179.

and the northern part of Scandinavia. However, the condition of the astrolabe map is such that it is hard to be specific.[24] Since the overall agreement is so good in all other parts, we conclude either that Rumold's map was the *source* for the geographical data presented on the astrolabe map, or that *both* derive from *one and the same source.*

In discussing the plausibility of either conclusion we note first that the astrolabe map is not a mere reproduction of an existing map. As far as we know, it is the only map in existence of Mercator's world picture in the polar stereographic projection. More importantly, the map plate is not an isolated object but a part of an astrolabe, conforming to existing traditions. If the map plate and the astrolabe were made at the same time and in the same workshop, and we see no reason why this should not be so, it appears that both were made before Egnazio Danti left Florence after the death on 21 April 1574 of his patron, Grand Duke Cosimo I. We can then rule out the possibility that Rumold's map of 1587 was the source for the map plate (see Section 3).

Let us, therefore, turn to the other hypothesis, that both the astrolabe map and Rumold's map derive from one and the same source. In discussing its implication, we first note that in fact there are now known four maps in different projections produced after 1569, which all present the world picture of Gerard Mercator. These maps are: Ortelius's map of 1570; the Murad III terrestrial globe of 1579; Rumold's map of 1587; and the present astrolabe map. At first sight the obvious source for all four maps appears to be Mercator's own world map of 1569. However, none of these maps is a mere copy of Mercator's world map in a different projection. In particular, all of them have one or more texts that were not taken from Mercator's world map.[25] Yet the texts used instead appear somehow related to each other (see Table 3). The texts I to IV engraved on the map plate were used by both Ortelius in 1570, with minor variations, and by Rumold Mercator in 1587. The text numbered V is found on the Murad III terrestrial globe of 1579 and again on Rumold's map.

The relationships between these texts on the various maps have so far not been explained. In our opinion they point to the existence of an as yet unknown common source used for all these mappings, one that is closely connected with the workshop of Mercator. Since this unknown source must have been already in existence by 1570, it might have been a manuscript map or globe by Gerard Mercator. Of course, one should be extremely hesitant in postulating such a hypothesis. Yet it seems to meet the facts. Nor is it an outrageous idea. After all, Mercator must have had a system for recording new information. It is known that Mercator kept manuscript maps—he says so explicitly of his map of Europe of 1554—for the purpose of study.[26] A manuscript map of the world would facilitate the noting of the many bits of information that in the

[24] Despite the small difference, the presentation of 'Groenlant' on the astrolabe map is far closer to that on Mercator's world map of 1569 (and related maps) than to any other contemporary presentation of this region.

[25] The texts on Mercator's world map have been edited by B. van 't Hoff, *Gerard Mercator's Map of the World (1569): In the form of an Atlas in the Maritiem Museum 'Prins Hendrik' at Rotterdam*, publication of the Maritiem Museum, No. 6, Supplement to No. 2 of *Imago Mundi* (Rotterdam, 1961).

[26] *Declaratio insigniorum utilitatum, quae sunt in globo terrestri, coelesti et annulo astronomico ad invictissimum Romanum Imperatorem Carolum Quintum: Opuscule inédit de Gérard Mercator*, publié par le Dr Van Raemdonck (St Nicolas, 1868). See also Averdunk and Müller-Reinhard (footnote 21), pp. 16, 56. A recently discovered atlas with manuscript maps by Mercator is mentioned by Osley (footnote 21), p. 74. See Sotheby's of London catalogue for 13 March, 1979: 'The Mercator Atlas of Europe. To be sold as a single lot in the sale of valuable autograph letters, ... '. See also P. Scott and J. Goss, 'Important Mercator "Discovery" under the hammer', *The Map Collector*, 6 (March 1979), 27–35.

course of time, particularly the period between 1541 and 1569, came to his notice through reports from other geographers and travellers. Such a manuscript map or globe may have contained the geographical features that he introduced for the first time on his world map of 1569, and it may have had the texts included in all four mappings listed above.

Indeed, it would have been only natural if a manuscript source of Mercator was used by his son Rumold in 1587 for making his map in the equatorial stereographic projection. It is also generally accepted that Mercator and Ortelius exchanged data and information, and that there are strong arguments that Mercator had provided the information for the Ortelius map of 1570.[27] A Gerard Mercator manuscript would surely clarify the so far unexplained overall agreement between the texts on the maps by Ortelius and Rumold of 1570 and 1587 respectively. A manuscript containing the Maletur text (V in Table 3), appears so far the only way to explain why this text appears on the Murad III terrestrial globe of 1579 and on Rumold's map of 1587, since it does not derive from Mercator's world map of 1569 nor is it encountered on any other map prior to 1587.[28]

A common origin of the astrolabe map, made before 1574, and Rumold's map can hardly be explained without the hypothesis of the use of a manuscript map by Gerard Mercator. If, indeed, the astrolabe map is based on such a manuscript source, it almost certainly was fabricated in the Duisburg workshop of Gerard Mercator. This implication is fully consistent with the conclusion reached in Section 6 below, that the astrolabe itself was made by Gerard Mercator. Telling support for our view is given by Osley:[29]

> Such was the importance that Mercator attached to the appearance of his maps that, to the end of his life, he engraved as much as possible with his own hand. But the volume of work that he undertook was so great that he had to rely increasingly on hired engravers. He must have had workshops of his own at Duisburg, though it is unlikely that he owned presses for printing. The epoch-making map of the world of 1569, which we know he engraved without any assistance (except for the margin [ref. to p. 187 for Ghim's comment]), may be regarded as a rough line of division. Everything that preceded it was either pure Mercator or contained a very high proportion of his own engraving: thereafter, we can detect other hands, working, however, to drawings made by Mercator himself.

5. Mercator as instrument maker

Gerard Mercator received his education in mathematics, astronomy, and geography in Louvain from the influential mathematician Gemma Frisius (1508–1555),[30] through whom he became involved in instrument making. Although Gemma designed scientific instruments and vigorously published his ideas, there is no evidence that he

[27] Shirley (footnote 23), pp. 144–5, and Averdunk and Müller-Reinhard (footnote 21), pp. 72–4.

[28] The best compilation of world maps of the sixteenth century is that by Shirley (footnote 23).

[29] Osley (footnote 21), p. 65. Rumold assisted: 'Dès lors aussi il reprit le burin pour travailler aux cartes que la santé de son père ne lui permettait plus d'achever et terminer l'Atlas'; *Biographie Nationale*, XIV (Brussels, 1897), col. 426.

[30] A short outline of Mercator's life, together with an English translation of Walter Ghim's *Vita Mercatoris* is found in Osley (footnote 21), pp. 19–29, and 184–94. See also Averdunk and Müller-Reinhard (footnote 21).

himself was engaged in making instruments, nor in training others to do so.[31] His reputation in this respect seems to derive from a portrait, engraved in 1557 two years after his death, in which various instruments and the tools for making them are depicted (Figure 12). It is not impossible that the portrait served to confirm the activities of his nephew, Gualterus Arsenius, who at that time was beginning successfully to produce instruments on a commercial basis, and who, from the very beginning, advertised his relationship with his famous Louvain uncle.[32] However that may be, the little evidence there is points to a less than active role for Gemma Frisius in instrument making; nor did he need to be active. From the beginning, he could and did find the required craftsmanship in Louvain in the workshop of Gaspar van der Heyden (*c.* 1496– died after 1549), otherwise known as Gaspar a Myrica.[33]

In 1524, Van der Heyden was working in Louvain as a goldsmith, and he was the first to make globes there in 1526/7, when Gemma Frisius had only just arrived (he became a student at Louvain University on 26 February 1526). Therefore, Gaspar van der Heyden was already well established as an artisan when, in 1531, Gemma used him to make his terrestrial globe, and again in 1536/7 for the production of a pair of globes.[34] The latter pair was produced in co-operation with Gerard Mercator, who had married in 1536, and who for the purpose of a comfortable income had, in 1532, given up philosophy for astronomy and mathematics. 'He applied himself with such diligence to these subjects that, within a very few years, he was teaching their elements to a number of private students, and from time to time was fashioning and constructing scientific instruments (i.e. spheres and astrolabes), astronomer's rings, and similar apparatus in bronze.'[35] Clearly, between 1532 and 1536, Mercator acquired his skills in metal working, and Van der Heyden is the most obvious artisan to have trained him. Anyway, by 1536 Mercator was already a skilled engraver, and in 1540 he was busy organizing his own workshop independently of that of Van der Heyden; in a letter to Antoine Perrenot, written in about 1540, he explains that he is trying to obtain the tools needed to make instruments.[36]

Mercator was then constructing a terrestrial globe of his own, which was published in 1541. In the preceding years Mercator had already acquired facilities for map making. His first map is of Palestine, and it was published in 1537 while he was working

[31] Ferdinand van Ortroy, *Bio-Bibliographie de Gemma Frisius* (Brussels, 1920). Osley (footnote 21), p.91, suggests that Gemma himself had a workshop at Louvain founded by Gaspar a Myrica, and developed under the directing force of Gemma Frisius. It is this workshop, supposedly, which was taken over by Gualterus Arsenius when Mercator left in 1552. This questionable point of view is found in *The Time Museum: Catalogue of the Collection*, edited by Bruce Chandler, Volume 1, 'Time Measuring Instruments', Part 1, 'Astrolabes, Astrolabe Related Instruments', by A. J. Turner (Rockford, Illinois, 1985), p.47. See pp.138–9 for the argument that Gualterus and Regnerus are one and the same, Reynier being a family name, that of Gemma in fact. See footnote 52 for a possible confirmation.

[32] For some astrolabes from the Arsenius workshop and their signatures, see R. T. Gunther, *Astrolabes of the World*, 2 vols (Oxford, 1932; reprinted London, 1976), pp.382–3. See also footnote 42. This field has now been fully covered in Koenraad Van Cleempoel, A Catalogue raisonné of *Scientific Instruments from the Louvain School*, 1530 to 1600 (Turnhout: Brepols, 2002).

[33] Antoine de Smet, 'Der Goldschmid und Graveur Gaspar van der Heyden und die Konstruktion von Globen in Löwen im ersten Drittel des XVI.', *Der Globusfreund*, 13 (1964), 32–48.

[34] P.C.J. van der Krogt, *Globi Neerlandici: The Production of Globes in the Low Countries* (Utrecht, 1993), pp. 40–57.

[35] Osley (footnote 21), p.185, Ghim's words in translation.

[36] M. van Durme, *Correspondance Mercatorienne* (Antwerp, 1959), pp. 14–17; No. 3, Mercator à Antoine Perrenot, Louvain 4 August [1540]: Mitto tandem, mi Reverendissime, calvariam, multo serius profecto quam opinari quiverim perfectam, sed dum paucis armis instructus hoc artificium absolvere vellem, nova subinde cavitatum obliqui, angusti, rotundi, acutive recessus instrumenta poposcere, quibus fabricandis plus iusto rei mora protacta est, ut fit, dum ipsi omnium indigi alienas cogimur et varias officinas petere.

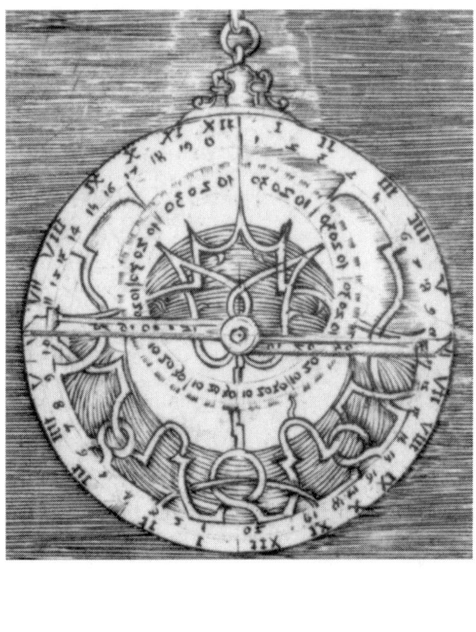

Figure 12. Gemma Frisius, posthumous portrait, 1557, engraved by J. van Stalburch. (Courtesy Rijksmuseum Stichting, Amsterdam.) *Right* Enlargement of the astrolabe, showing 'standard' rete pattern.

with Van der Heyden, and in 1538 his double-cordiform world map appeared. During the following decade, Mercator seems to have made all kinds of scientific instruments, among which was his celestial globe of 1551.[37] It is worth noting that the very fact that Mercator acquired tools in 1540 makes it highly improbable that there existed at the time a well-equipped local workshop under the 'directing force' of Gemma Frisius. As a devoted physician, the latter probably spent more time on his patients than in actual instrument making.[38] The simple fact that his name was effectively used by the Arsenius family for the promotion of their own instruments is no evidence for the direct involvement of Gemma in workshop activities.

From 1540, Mercator worked independently of both Van der Heyden and Gemma Frisius. It may well be that some of the instruments designed by Gemma, such as his astronomical ring, were made in Mercator's workshop, but he himself also designed an astronomical ring, and may well have preferred his own design in constructing such rings.[39] His skills as an instrument maker were soon recognized, and it is not surprising to learn that he made a series of scientific instruments for the Emperor in the 1540s, through the good offices of his patron, Nicolas Perrenot de Granville, Chancellor to Charles V. Because these instruments were destroyed, Mercator made a second set in about 1552, the time he left Louvain for Duisburg.[40] According to Osley, the workshop of Gemma Frisius was then taken over by Gemma's nephew, Gualterus Arsenius, but again it is questionable whether there was such a workshop to be taken over. When Mercator moved to Duisburg he took his own workshop with him, and continued there to make scientific instruments and to produce his globes (Figure 13). In Duisburg he continued to teach mathematics and the art of instrument making, witness of which is given in a letter written in 1562 by one of his apprentices, Johan van den Corput (1542–1611): 'To cut seal [matrices] and to engrave copper otherwise all belongs to our art of mathematics, which is now part of my studies, since we have to make our instruments ourselves'.[41] It is not certain, therefore, that the Arsenius workshop was a continuation of something already existing. It may have come into being after the departure of Mercator in 1552. The earliest known instrument from the Arsenius workshop dates

[37] Petrus Beausard, *Annuli Astronomici instrumenti cum certissimi tum commodissimi usus*, (Antwerp, 1553), pp. 19–20: 'Verum dictis armillis si quintus annulus accedat, qui horizontis vice fungatur (cuiusmodi nonnulos videre licuit: utque est, quem magna industria ac solerti opera his proximis annis, Carolo quinto Imperatori invictissimo et matheseos studiosissimo, confecit noster Gerardus Mercator, Rupelmundanus, vir, quo fabricae instrumentorum, ut cum omnium pace dixerim, hac tempestate nullum peritiorem noverim), non me Hercule dicere possim, huiusmodi instrumentum quot usus praestare queat: adeo ut usuum certitudine (taceo multitudinem) cum astrolabio certare facile dixeris'. See also Henri Michel, 'Mercator, Constructeur d'Instruments astronomiques', *Ciel et Terre*, 78 (1962), 191–6; Antoine de Smet, 'Mercator à Louvain', *Duisburger Forschungen*, Festschrift zum 450 Geburtstag Gerhard Mercator (Duisburg, 1962), pp. 28–90.

[38] At Louvain, Gemma Frisius graduated in medicine, practised it there, and taught in the faculty. His medical reputation is indicated in the posthumous portrait of 1557 engraved by Jan van Stalburch. See Figure 12.

[39] The astronomical ring of Mercator is described in his unpublished treatise of 1551 (footnote 26), and in Beausard (footnote 37). See also letters from Mercator to Perrenot of *c.* 1544 and 1545 in Van Durme (footnote 36), pp. 26–30, Nos 14, 15.

[40] Ibid., and Osley (footnote 21), p. 186. Idem, Ghim's words in translation: 'he made for His Majesty a large number of scientific instruments of exquisite workmanship'.

[41] Jan Postema, *Johan van den Corput 1542–1611: Kaartmaker, Vestingbouwer, Krijgsman* (Kampen, 1993), p. 42: 'Signetten te steken ende andersins in coper te graveren, behoirt al tot unser kunst van mathematica, dwelck nu een deel mynder studien is, want wy unse instrumente self moeten maecken'.

from 1554,[42] and it was only after the death of Gemma in the following year that the workshop seems to have started commercial production in full force. In contrast, the emphasis of the activities of Mercator's workshop in Duisburg shifted more and more to maps and globes, whereas instruments were probably produced on an incidental basis and only for special clients, mainly because after 1552 Mercator spent much time on scientific endeavours in geography. Against this background, it is to be expected that any instrument made by Mercator in Duisburg, although stemming from a Flemish tradition, will be distinctly different from those made in the Arsenius workshop.

6. The calligraphy

Mercator began his commercial life as a calligrapher, exercising his skill on map design and map-plate engraving. He produced his first terrestrial globe *c.* 1536, a celestial in 1537, a map of the Holy Land in 1537, and a double-cordiform map of the world in 1538. In 1540 he published his manual on calligraphy. There followed a terrestrial globe of his own design in 1541, and a celestial globe in 1551. The great world map on the Mercator projection was published in 1569.[43]

On the 1541 globe we may heed the opinion of the cartographic historian, R. W. Shirley:[44]

> The [1541] globe is the first such work intended for navigators as can be seen from the placing of compass roses in the oceans and the loxodromic lines cutting the meridians at equal angles. The location of certain stars is also marked, to help navigators at night. Mercator's engraving and cursive italic letters are of the highest standard, with the relatively large scale allowing very much more detail to be shown than on the author's smaller double-cordiform map three years earlier.

When the engravings of Mercator are compared with those by his contemporaries, one sees that he achieves the greatest accuracy and clarity. The style of an engraver may be expected to be modified over time. With Mercator, there is some development between 1536 and 1541, after which he maintains the same style for the rest of his productions. The two periods are described by Osley:[45]

> The work done by Mercator at Louvain may be thought of as his first period. The lettering is essentially that recommended in the *Literarum latinarum*, though it is performed more freely and has shed most of the gothic residue of his woodcut letters. In this version, the ascenders and descenders of the long letters (e.g. *b*, *d*, *h*, *p*, *q*), begin or end with a curved stroke. This invests the script with a look of handwriting. In his second period, Mercator gradually modified his hand. The curved extremities of the ascenders and descenders were abandoned for a neat serif. The diagonal movement is softened. The letters became small and even, and this enabled him to insert lengthy passages of descriptive matter into cartouches on the maps.

[42] An astrolabe, signed: '*Authore Gem[m]a Frisio et exaratus a Gualtero Arsenio Lovanij 1554*', is described in Fernandez Villar, *Sobre el astrolabio firmado por G. Frisius y G. Arsenius*, Museo de Historia, Castillo de Chapultepec (Mexico City, 1976). We are grateful to R. and M. Webster for providing a copy. This astrolabe may be the one acknowledged by Gemma Frisius in *De Astrolabo Catholico Liber* (Antwerp, 1556), pp. 14v–15r: 'Astrolaba quae nostro instinctu per nepotem nostru[m] Gualterum Arsenium co[n]structa sunt, ad secundu[m] annum à bissexto inscriptos habent dies'. In the same chapter the year is shown to be 1554.

[43] Osley (footnote 21), chapter 5.

[44] Shirley (footnote 23), 88–9.

[45] Osley (footnote 21), p. 65.

Figure 13. Gerard Mercator, 1574, from his *Tabulae Geographicae Cl. Ptolemaei*, 1584. (Courtesy Universiteitsbibliotheek, Leiden.)

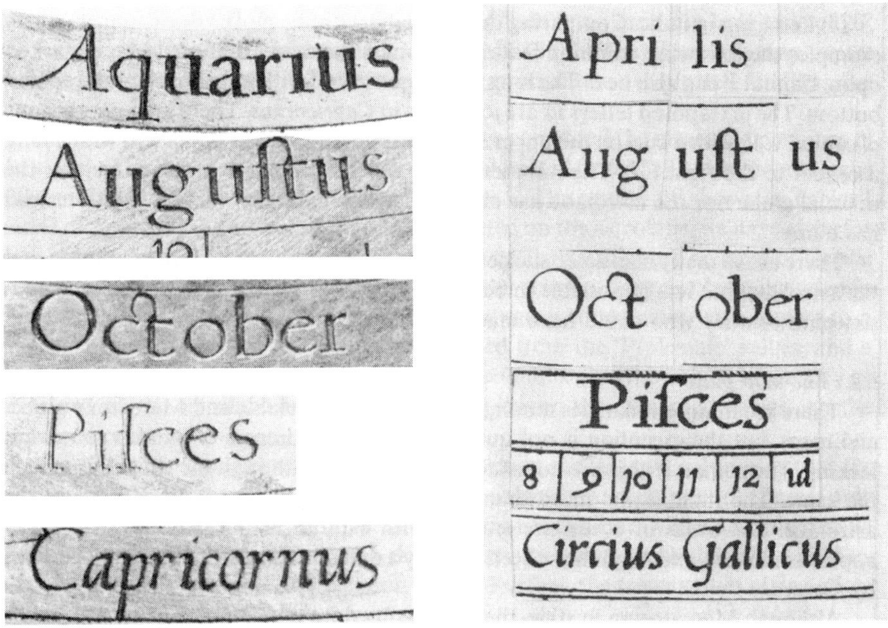

Figure 14. Comparison of lettering from the rete (left) and Mercator's globe of 1551 (right).

6.1. *The astrolabe*

A comparison has been made between the globes of 1541 and 1551 and the astrolabe, and a few examples have been selected for illustration in Figure 14.

(*a*) Lettering in roman. One can readily match like with like by taking the names of the months and signs of the zodiac from the astrolabe and globes.

astrolabe	*globe 1551*	*comments*
Aquarius	Aprilis	curl on A
Augustus	Augustus	straight A
October	October	ct ligature
Pifces	Pifces	long s

Mercator's roman is open, and has a large x height, with short ascenders and descenders. Neither capitals nor minuscules are at all condensed. The letter spacing is even and regular; the serifs on capitals and minuscules are large, but balanced by the openness of the letters. The stress is consistent with strong, but evenly executed differentiation between thick and thin. The capital G is formed by cutting the semicircle first, followed by the vertical stroke; this leaves part of the circle protruding to the right below the vertical stroke. The e has its bar close to the top, and is usually angled. The i has a flick to the right (í), not a dot, and sometimes the flick is absent (ı). Ligatures are on ct and st (with the long ʃ). The g is formed with a closed loop at the bottom. The diphthong æ is formed from an e with a loop under. The symbols used for the signs of the zodiac are the same, and this is most significant, since engravers have their own individual custom. This is borne out tellingly in the case of the extra latitude plate made by Giusti, who has copied as closely as possible the lettering, but he failed with some of the zodiac symbols.

(*b*) Lettering in italic. Comparing the astrolabe with the globes one can pick out as examples the following common features amongst the many that could be remarked upon. Capital *E* can also be Ɛ. The long s (ʃ) is squared off with long horizontals top and bottom. The juxtaposed letters *us* are joined, as in *Capricornus*. There are four versions of capital v: *V*; with a curl on the upper left; with a horizontal bar upper left; with a long ascender to the right (√). The number *5* does not have a bar at the top. Neither the celestial globe nor the astrolabe has elaborate swash capitals, whereas the terrestrial has many.

There are so many characteristic details in common between Mercator's globes and the astrolabe that it is impossible to believe there was any other craftsman of the mid-sixteenth century who could have made the astrolabe.

6.2. *The map plate*

There are many similarities among the plate, the astrolabe, and Mercator's globes and maps, but the execution is not quite the same; the sureness of hand is somewhat lacking. The roman letters are not as regular and even, although the shapes are much the same. The italic slants more than on the celestial globe and other parts of the astrolabe. The areas of ocean are lettered with capitals elaborately swashed. They appear smoother, and might have been done by a different hand from the rest—perhaps by Gerard.

Although Mercatorian in style, the calligraphy does not compare as exactly as the astrolabe does with hand of the master (see Figure 15). That the plate comes with the astrolabe and thus from Mercator's workshop can hardly be doubted. So could it have been made by one or other of his sons? There are two, Arnold (1537–1587), surveyor and engineer, who is said by Ghim to have made instruments, and Rumold (*c.* 1546–1599), who according to Ghim (1595) studied geography and helped his father by drawing maps. A map by Arnold exists: 'Thule *insula*...', signed and dated: '...Arnoldus Mercator Louaniensis dedicabat Anno Domini 1558'.[46] Although showing strongly the influence of Gerard Mercator, Arnold's hand is not nearly the equal of that of his father or younger brother. It resembles much more the typical hand of the Flemish map engravers of the period. The conventional signs for towns, mountains, etc., are virtually the same for Gerard and Rumold, but not for Arnold, to judge from this map.

The most likely engraver of the map plate is son Rumold (aged about 24 at the postulated date of the astrolabe). One can see very strong resemblances in the calligraphy with Rumold's map of the world published in 1587 (see Figures 9–11).[47] That this latter map is closely related to the map plate is discussed at length in Section 4 above.

7. The analysis of the star positions on the rete

The positional data of stars plotted on the rete of an astrolabe contain in principle valuable information. In the present case we have used this information to check whether the stellar data are consistent with our conclusion that the astrolabe was made by Mercator.

It is known that the star positions on Mercator's celestial globes of 1551 and 1579 are plotted for the epoch of 1550, which, after the Copernican theory of precession

[46] The Librarian of the Royal Geographical Society kindly provided a photocopy. For Ghim on the Mercator brothers, see Osley (footnote 21), pp. 191–2.

[47] *Orbis Terrae Compendiosa Descriptio* (Duisburg, 1587). Two hemispheres in stereographic projection.

introduced in 1543 in *De Revolutionibus*, corresponds to an excess in longitude of 20° 55′ with respect to the Ptolemaic data (epoch AD 137).[48] Therefore, one may expect that on an astrolabe made by Mercator the star positions are plotted for the same epoch 1550.

In order to verify the latter hypothesis we have compared the measured positions of the stars on the rete of the astrolabe with those calculated from the Ptolemaic catalogue after adding 20° 55′ to the longitudes of the stars.[49] The results are shown in Tables I and II of Appendix 2. In Table I, for 47 stars plotted on the astrolabe, we have given for each the modern name and the name engraved on the astrolabe. The stars of *Gemini* and the one named *Pleiades med*: engraved on the ecliptic, have not been included in our calculation. In the next two columns the x and y coordinates with respect to a chosen coordinate system are given as calculated from the 'Ptolemaic' values, and a precession correction of 20° 55′. The next four columns give the values of x and y as measured on the astrolabe and the differences between the calculated values and the measured ones.

The conclusion we draw from the values given in the last two columns is that the agreement between the calculated and measured values is extremely good. There is only one exception for *Media cinguli Andr*:. The position of this star, which according to its name should be identified with μ And, does in fact agree better with that of β And. For all the other stars the name and position match. Further, the mean values of dx and dy

Figure 15. Part of Gerard Mercator's map of Britain, from the *Atlas*, 1595.

[48] See Elly Dekker and Peter van der Krogt, 'Les Globes', pp. 243–67 in Marcel Watelet, ed., *Gérard Mercator cosmographe, le temps et l'espace* (Antwerp: Fonds Mercator Paribas, 1994). For the Copernican precession theory, see N.M. Swerdlow and O. Neugebauer, *Mathematical Astronomy in Copernicus's* De Revolutionibus. Studies in the History of Mathematics and Physical Sciences (New York, 1984), x, pp. 127–48.

[49] We used the 'Ptolemaic' catalogue edited by P. Kunitzsch in Claudius Ptolemäus, *Der Sternkatalog des Almagest, II Die lateinische Übersetzung Gerhards von Cremona* (Wiesbaden, 1990), and *III: Gesamtkonkordanz der Sternkoordinaten* (Wiesbaden, 1991).

are well within the errors indicated by their standard deviation. There is no evidence that points to systematic errors owing to precession or another source. It is, therefore, fully justified to say that the stars plotted on the astrolabe have an excess with respect to Ptolemy of 20° 55′, which is in complete agreement with the values employed by Mercator for his celestial globes.

8. Some workshops contemporary with Mercator

During the second half of the sixteenth century many beautifully made astrolabes were produced by the Arsenius workshop in Louvain.[50] The great astrolabe (diameter 590 mm) made by Gualterius Arsenius in 1566 for the Spanish monarch Philip II, now in the Archaeological Museum, Madrid, is just one example of its outstanding craftsmanship. Naturally, the question arises whether the present astrolabe could have been made in that workshop. A comparison between the products is needed, but as yet no systematic analysis of the main features of the Arsenius astrolabes has been carried out. For our present purpose four features will be considered: the selection of stars and their names, the pattern of the rete, the design of the throne, the constructional techniques. Ultimately, the study of these and other features would enable a classification scheme for Flemish astrolabes to be established. An important element in a classification scheme for Flemish astrolabes must lie in the rete pattern, and a special feature is the 'tulip' formed by the strapwork inside the zodiac circle. The 'standard pattern' is shown in the engraving (1557) of Gemma Frisius (Figure 12), who is generally thought to be the scientist behind the Louvain workshop (but see Section 5 above). Several astrolabes with standard pattern retes are known. The earliest one is from 1554, and a later example was published in 1608 by Adrian Metius (Figure 16 (*a*)). Astrolabes with standard pattern retes were made throughout the second half of the sixteenth century by a variety of makers.

Although the present rete has great similarity to the standard pattern, it is not identical to it. Several known Flemish makers, the Arsenius family included, used rete patterns that deviate in a characteristic way from the standard pattern. This is shown in Figure 16 where the standard pattern is compared with examples from the present astrolabe, Thomas Gemini (London, 1559), and Regnerus Arsenius (Louvain, 1565 and 1566). All derive from the standard, but do not conform to it exactly. More importantly, the characteristic deviations of the Gemini and Arsenius astrolabes differ from those of the present astrolabe. Since all signed examples of Arsenius astrolabes of 1559 and later, checked so far, show the same set of deviations from the standard pattern, this implies that the present astrolabe does not come from the Arsenius workshop.

Further distinctions are apparent when comparing the stars and their names. In Table 4, the result of comparing the present astrolabe with five signed by a member of the Arsenius family, and two that are neither signed nor dated, is summarized (see Appendix 1). These last two have the characteristic deviations found on the Arsenius rete pattern as well as a typical Arsenius throne. They are included to check whether the stars and their names would match those on the signed Arsenius astrolabes. The comparison shows again that the present astrolabe does not fit the tradition of the Arsenius workshop, whereas those that are unsigned but with Arsenius-type retes do fit. The distinction between the present astrolabe and the Arsenius tradition is two-fold.

[50] The International Checklist (footnote 1) lists 36 astrolabes signed or attributed to the Arsenius workshop between 1565 and 1574. Since the List was published in 1973, several more have come to light.

Figure 16. The rete patterns of six Flemish or Flemish inspired astrolabes. (*a*) Standard pattern. Adrian Metius, *De Fabrica Planispherii* (1606), in *Institutionum astronomicarum* (Franeker, 1606–1608); see Gunther (footnote 32), p. 398. (*b*) Flemish. Kings College, Cambridge, at Whipple Museum of the History of Science, Cambridge, inv. 1467; IC 235. Unsigned and not dated. Gunther (footnote 32), pp. 393–4. (*c*) Gemini. Istituto e Museo di Storia della Scienza, Florence, inv. 1093; IC 489. Signed: Thomas [sign of Gemini] 155 [last numeral omitted]. Closely similar to the rete on astrolabe signed: Thomas [sign of Gemini] 1559, at Museum of the History of Science, Oxford, inv. 1098; IC 575. (*d*) Mercator, attributed; *c.* 1570. Istituto e Museo di Storia della Scienza, Florence; IC 490. (*e*) Arsenius. Museum of the History of Science, Oxford, LE 2047; IC 229. Signed: *Regnerus Arsenius Nepos Gemmæ Frisÿ fecit Louanÿ anno 1565*. Gunther (footnote 32), pp. 384–6. (*f*) Arsenius. Museo Arqueológico Nacional, Madrid; IC 231. Signed: *Gualterus Arsenius Gemmæ Frisÿ Nepos Louanÿ fecit an: 1566·*. Known as Philip II's Great Astrolabe. The silhouette of the rete is taken from Gunther (footnote 32), p. 388.

Table 4. Summary of the analysis of the star names on the Mercator astrolabe and on astrolabes of Arsenius or Arsenius type.

	β And	α CMi	α Hya	α Crt	β Lib
Mercator	*Media cinguli Andr:	Canis minor	Hydrae lucida	Fundus Vasis	Lanx sept:
(1) A 1559	Andro vmbi	Canicula	Hydrae clara	Crateris fundus	Lanx Boreae clarior
(2) A 1561	Androm umbilic	Canicula	Lucida hydre	Crateris fundus	Lanx Boreae clarior
(3) A 1565	Andro obili	Canicula	Hydrae clara	Crateris fundus	Lanx Boreae clarior
(4) A 1571	Andro: vbi.	Canicula	Hydrae clara	Crateris fundus	Lanx Boreae clarior
(5) A type	Andrõada Ombilicus	Canicula	Hydrae clara	Crateris fundus	(absent)
(6) A type	(absent)	Canicula	Hydrae clara	Crateris fundus	Lanx Boreae clarior
(7) A Ferd.	Androm: vm:	Canicula	Hydrae clara	Crateris fundus	Lanx Boreae clar.

(1) Christie's South Kensington, sale 29 September 1988, lot 286; IC 4602. *Gualterus Arsenius Nepos Gemmæ Frisÿ fecit Louanÿ anno 1559.*

(2) Museé national de la Renaissance, Ecouen, inv. E.Cl. 2937 (previously Museé de Cluny); IC 234, *GVL Nepos Gemmæ Frisij Louanij fecit anõ 1661.* [*sic*, read 1561].

(3) Museum of the History of Science, Oxford, LE 2047; IC 229. *Regnerus Arsenius Nepos Gemmæ Frisÿ fecit Louanÿ anno 1565.*

(4) Merton College, Oxford; IC 232. *Gualterus Arsenius nepos Gemmæ Frisÿ fecit Louanÿ anno·*1571·

(5) Germanisches Nationalmuseum, Nuremberg, inv. WI 1164; IC 555. Not signed or dated.

(6) Kunsthistorisches Museum, Vienna; IC 562. Not signed or dated.

(7) Science Museum, London, inv. 1878–11; IC 233. *Ferd. Arssenius* [*sic*] *fecit Antverpiæ.* Not dated.

* Mercator used μ And instead of the usual β And.

First, the selection of stars is different. On the present astrolabe the *Pleiades*, two stars for *Gemini* and two stars for *Cauda Capricorni* are encountered, which are absent in the Arsenius tradition. It is interesting to note that these very stars are found in the star list published by Beausard in his *Annuli Astronomici* of 1553, in which Mercator is praised as an instrument maker.[51] *Persi lat⁹ dex*, *Orionis sinister humer⁹*, and *Cephei dex hume*: are typical for the Arsenius tradition, but they are lacking on the present astrolabe. Second, the naming of the stars shows characteristic differences (Arsenius second): *Gemini↔Caput* [Gem] *antecedentis; Canis minor↔Canicula; Fundus Vasis↔Crateris fundus; Lanx sept:* [Lib] *↔Lanx boreae clarior; Ophiuchi pal: sin:↔Ophiuc: sinis: manus.* In all these cases the naming of the stars on the various astrolabes in the Arsenius tradition is well in accord. One is therefore justified in concluding that the stars and their names on the present astrolabe are not in the tradition of the Arsenius workshop.

Third, we see that the throne on the present astrolabe is similar to the throne of the astrolabe in the 1557 portrait of Gemma Frisius (Figure 12), where the form of the side supports is the same (two S-shaped brackets with elaborate decoration). Astrolabes from the Arsenius workshop after 1559 all have a different suspension consisting of a curved base and two reclining figures. But remarkably, one earlier example has the S-brackets: it, too, is dated 1557. This throne is visible in an illustration in Osley's book, but the astrolabe is said to be in private possession, so it has not been possible to examine the instrument as a whole.[52]

The S-bracket will have been made from a casting, and the one mould produces a shape that is symmetrical so that two can form the supports at either side of the throne

[51] Beausard (footnote 37), p. 53.

[52] Osley (footnote 21), p. 93, figure 45. The signature is visible: *Nepos Gemmæ Frisij faciebat Louanÿ anõ 1557 GAR*. This monogram *GAR* may be read as: Gualterius Arsenius Regnerus.

Figure 17. The throne support as found on three 'Mercator'-type and one Arsenius astrolabes. By reversing and turning, the products of the one mould can produce the pair of supporters in two modes.

(Figure 17). The three 'Mercator' astrolabes at Florence, Brno, and Augsburg, and the portrait of Gemma, all show the brackets at an angle, while on the Arsenius astrolabe the brackets are lying flat and in a different sense. Castings are simple to produce once the mould has been made, and a good supply made in one day could last many years. What this observation shows is that the Mercator astrolabes are Louvain-influenced, and that Arsenius used the same brass foundry, presumably in Louvain. Since the Florence Mercator astrolabe is shown to date from about 1570, either some brackets were taken when Gerard Mercator moved from Louvain to Duisburg in 1552, or that contact was maintained. The later Arsenius thrones, from 1559 onwards, consist of two reclining figures, with the whole attached to a curved bar that is fixed to the top of the instrument by screws.

Fourth, an examination of an astrolabe signed: *Regnerus Arsenius Nepos Gemmæ Frisÿ fecit Louanÿ anno 1565* reveals details of the workshop practice.[53] Of medium size (diameter 290 mm), it is made in gilded brass, has the Saphea universal projection on the back, and contains one plate for latitudes 45° and 51°. A list of 49 stars (counting Ursa Major as one, although it has seven pointers) is published by Gunther, and is reproduced with some corrections in Appendix 1. An examination of the back of the rete shows how it was made. A sheet of brass was marked out with the pattern of banding and the position and shape of the star pointers. Lines that outline the features can still be seen in several places. On the back of the ecliptic circle the degrees are carefully scribed. By the root of each star pointer is neatly engraved a number, which will allude to a table of stars kept in the workshop. The numbers run 1–16, 53, 60, outside the ecliptic circle, while inside one finds 17–25, 33–46, 51, 57, 59, 63, 64, and 68 or 69. Three have no numbers; the familiar seven pointers for Ursa Major would not require any individual numberings and an experienced craftsman might think it pointless to engrave a number beside such an obvious group. One presumes that the workshop kept a list of stars suitable for use on astrolabes, and that the ones actually employed depended on the size of the instrument, and thus the space for placing pointers. On this astrolabe, 49 stars have been chosen from a list of some 70 stars. The

[53] Astrolabe in the Museum of the History of Science, Oxford, LE 2047; IC 229. See Gunther (footnote 32), pp. 384–6, Pls XCII, XCIII.

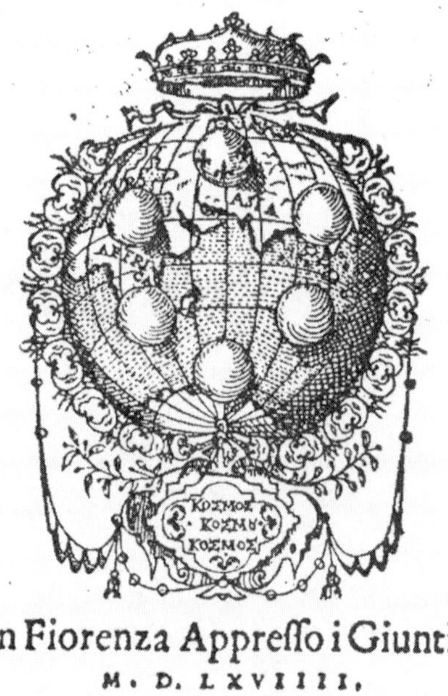

Figure 18. The cartouche from the title page of Egnazio Danti, *Trattato*, 1569 (footnote 19).

rete was laid out mathematically, the star pointers numbered, and then handed over to a craftsman who formed the front by cutting, rounding, and polishing. Then the engraver added the star names as indicated by the numbers selected, and the signs of the zodiac, which were standard. To finish, the front of the rete was gilded. This workshop practice shows that the making of the astrolabe was passed from specialist to specialist. This implies a considerable order book, borne out by the large number of Arsenius astrolabes still existing today.

The rete on the astrolabe attributed to Gerard Mercator shows much less working on the back. The positions of the ecliptic circle and the strapwork are scribed lightly. The star pointers are not numbered or marked in any way.

The conclusion we derive from the various comparisons is clear. The *similarities* between the rete pattern and the stars and their names point to a common Flemish origin. The *differences* show that the present astrolabe was made *neither* in the Arsenius workshop *nor* by a maker who adhered to the tradition of this workshop.

9. When was the astrolabe made?

There are several independent pieces of evidence that, taken together, focus a date for the astrolabe to quite a narrow span of years. We will summarize the evidence that has been presented in this paper under three main headings.

9.1. *Calligraphy*

The sheer controlled skill of the engraver's hand, and so many characteristic letter forms, point to Gerard Mercator's second period, which was at Duisburg. The style has settled down by the time of the celestial globe (1551), and found its great expression with the world map of 1569.

The map plate does not appear to be in the hand of Gerard, and a case has been put forward that this was engraved by the second son, Rumold. He was born *c.*1546, and by 1565, at the beginning of his twenties, he would have been mature enough to produce the map plate. This is the period in the run-up to the publication of the world map, when Gerard would have needed additional help because of the volume of work. From the evidence of his own signed map of the world published in 1587, Rumold was highly skilled, and his hand seems to match that of the map plate's engraver.

9.2. *Egnazio Danti*

The role of Danti in Florence is almost certainly an important reason for the presence there of this astrolabe. He was appointed cosmographer to Duke Cosimo de'Medici in 1562 at the age of 26. He designed wall maps for the Palazzo Vecchio and a globe. In 1569 he published a treatise on the astrolabe. A Florentine craftsman made for Danti a large astrolabe that bears Danti's name. This astrolabe has a map plate, on a North polar projection similar to the Mercator astrolabe, but is extremely simple by comparison. The craftsman who made Danti's astrolabe, and seven others, most with map plates, has been identified as Giovan Battista Giusti of Florence. Giusti labelled his instruments by punches, not generally by engraving. The Mercator astrolabe possesses a latitude plate for Florence, 43°, that was made by Giusti. The wear on some punches points towards a date near 1570 for its inclusion amongst the Mercator fabricated plates.

9.3. *The Grand Duke of Tuscany*

Cosimo de'Medici began his rule in 1537. With skill, luck, and force of arms he was created the first Grand Duke of Tuscany on 27 August 1569. He was now 50, and his cosmographer, whose *Trattato* on the astrolabe had just been published with a dedication to Cosimo showing the six balls of the Medici over a globe of the Earth, was 33 years old. Clearly in great favour, Danti was, in 1571, granted permission to live in the Palazzo Vecchio. Unfortunately for Danti, Cosimo died on 21 April 1574, and was succeeded as Grand Duke by his son Francesco (1541–1587). He was, therefore, five years younger than Danti, and had acquired a dislike for the man. Danti was summarily dismissed on 28 September 1575.

Danti's presence in Florence between 1562 and 1575 sensitized the court to cosmography, and one would expect to find here the reason for a copy of Mercator's world map to have been purchased in November 1569 for delivery to Florence.[54]

An astrolabe with a map plate is a portable amalgamation of the celestial and terrestrial spheres, as well as an astronomical observing instrument and calculator. Although the local craftsmen were reasonably competent, they were no match for the renowned cartographic and craft centre: Flanders. At the time of the crowning of the first Grand Duke of Tuscany in 1569, what would be more appropriate than to acquire

[54] On 1 November 1569 a world map was sold by Plantin to 'D. Balthasar Rovelascha p[our] Rafael Bruenello de fratelli Friscobaldi p[ar] commission de Claude de Lorneau'; see Léon Voet, 'Les relations commerciales entre Gérard Mercator et la maison Plantinienne à Anvers', *Duisburger Forschungen*, 6 (1962), pp. 171–229, expecially p. 205 table 4, p. 211 note 71. Rafael Brunelli will have been the agent for the Florentine bankers Frescobaldi. The production of the world map was already in full swing, for in September 1569 the Plantin firm in Antwerp bought 56 maps from Mercator (idem). For the European banking empire of the Fuggers, and its influence, see Götz Freiherr von Pölnitz, *Die Fugger* (Frankurt-am-Main, 1960). With headquarters in Augsburg, and foreign trading post at Antwerp, the Fuggers had *Grossfaktorei* also in Venice, Genoa, and Rome during the sixteenth century.

from the world's finest astrolabe maker an example of his art? After all, Gerard Mercator had made mathematical instruments for the Holy Roman Emperor himself. And Cosimo was rising in the courtly circles of Europe.

All things considered, the date of *c.*1570 seems to fit this astrolabe of Gerard Mercator and his Duisburg workshop.

Appendix 1: Comparison of astrolabe star names, Mercator *v.* Arsenius

In this table is summarized the result of a comparison of the star names on the Mercator astrolabe with those on a number of astrolabes from the Arsenius workshop. In comparing the star names we have ignored differences in capitals, abbreviations and order. For example, *Cap: Medusæ*, *Caput Meduse*, and *Medusæ caput* are essentially the same names, whereas *Lucida hydræ* and *Hydræ clara* are not.

Mercator name	Modern name	Arsenius 3 name	1	2	4	5	6	7
Cassiop: pect	α Cas	Cassio: pec	●	●	●	●	×	●
Cauda Ceti	β Cet	Cauda ceti	●	●	●	●	●	●
Media cinguli Andr:	μ And	×	×	×	×	×	×	×
×	β And	Andro: ōbili:	●	●	●	●	×	●
Venter Ceti	ζ Cet	Venter ceti	●	●	●	●	●	●
Medusae cap	β Per	Meduse caput	●	●	●	●	●	●
×	α Per	Persei lat' dex:	●	●	●	●	●	●
Naris ceti	α Cet	Nares ceti	●	●	●	●	●	●
Pleiad med:	Tau	×	×	×	×	×	×	×
Oculus [Tau]	α Tau	Oculus [Tau]	●	●	●	●	●	●
Hircus	α Aur	Hircus	●	●	●	●	●	●
Orionis sinister pes	β Ori	Orio: sinis: pes	×	●	●	●	●	●
×	γ Ori	Orionis sinister humer'	●	●	●	●	●	●
×	ε Ori	Cinguli orio: media	●	×	×	×	×	×
Orionis dex: hum:	α Ori	Orio: dex: hume:	●	●	●	●	●	●
Canis maior	α CMa	Canis maior	●	●	●	●	●	●
Gemini	α Gem	Caput [Gem] antecedentis	●	●	●	●	×	●
Gemini	β Gem	×	×	×	×	×	×	×
Canis minor	α CMi	Canicula	●	●	●	●	●	●
Hydrae lucida	α Hya	Hÿdrae clara	●	□	●	●	●	●
Cor [Leo]	α Leo	Cor [Leo]	●	●	●	●	●	●
Vrsa maior/1	α UMa	Vrsa maior/1	●	●	●	●	●	●
Fundus Vasis	α Crt	Crateris fundus	●	●	●	●	●	●
Vrsa maior/2	β UMa	Vrsa maior/2	●	●	●	●	●	●
Dorsum [Leo]	δ Leo	Dorsum leonis	●	●	●	●	●	●
Vrsa maior/3	γ UMa	Vrsa maior/3	●	●	●	●	●	●
Cauda [Leo]	β Leo	Cauda leonis	●	●	●	●	●	●
Vrsa maior/4	δ UMa	Vrsa maior/4	●	●	●	●	●	●
Ala detra Corui	γ Crv	Corui ala dextra	●	●	●	●	●	●
Vrsa maior/5	ε UMa	Vrsa maior/5	●	●	●	●	●	●
Spica [Vir]	α Vir	Spica virginis	●	●	●	●	●	●
Vrsa maior/6	ζ UMa	Vrsa maior/6	●	●	●	●	●	●
Vrsa maior/7	η UMa	Vrsa maior/7	●	●	●	●	●	●
Arcturus	α Boo	Arcturus	●	●	●	●	●	●
Bootis sinis: hum:	γ Boo	Bootis sinister humerus	●	●	●	●	●	●
Lanx sept: [Lib]	β Lib	Lanx boreae clarior	●	●	●	×	●	●

Mercator name	Modern name	Arsenius 3 name	1	2	4	5	6	7
Coruna sept:	α CrB	Corona septentri:	●	●	●	●	●	●
Ophiuchi pal: sin:	δ Oph	Ophiuc: sinis: manis	●	●	●	●	×	●
Cor [Sco]	α Sco	Cor [Sco]	●	●	●	●	●	●
×	ζ Oph	Ophiuc:genu sinis:	×	×	×	×	×	×
Herc: cap	α Her	Caput herculis	●	●	●	●	●	●
×	η Oph	Ophiuchi genu dex:	●	●	●	×	×	●
Ophiu: cap:	α Oph	Ophiuchi caput	●	●	●	●	●	●
Extra: man: dex: Ophi:	τ Oph	Ophiuchi ma: dex:	●	●	●	●	×	●
Caput Draconis	γ Dra	Caput draconis	●	●	●	●	●	●
Lyra	α Lyr	Lÿra	●	●	●	●	●	●
Aquila	α Aql	Aquila	●	●	●	●	●	●
Cauda Cygni	α Cyg	Cauda cÿgni	●	●	●	●	●	●
×	β Aqr	Sinister hume [Aqu]	●	×	×	●	×	×
×	α Cep	Cephei dex hume:	●	●	●	×	●	●
Cauda [Cap]	γ Cap	Cauda [Cap]	●	●	●	●	●	●
Cauda [Cap]	δ Cap	×	×	×	×	×	×	×
Pegasi rictus	ε Peg	Pegasi rictus	●	●	●	●	●	●
×	α Aqr	Dexter hume: [Aqu]	●	×	●	×	●	●
Crus Aquary	δ Aqr	Crus [Aqu]	●	●	●	●	●	●
Crus Pegasi	β Peg	Crus pegasi	●	●	●	●	●	●
Pegasi hum:	α Peg	Pegasi humerus	●	●	●	●	●	○
Peg: Vmb:	α And	Pegasi vmbili:	●	●	●	●	●	●
Extr: ala Pegasi	γ Peg	Ala Pegasi	●	*	●	*	*	*

● star present, name essentially as on 3;
Variant names: □ Lucida hydra, ○ Pegasi ala dextra, * Pegasi extrema ala;
× star absent

Key:

1. Christie's South Kensington, sale 29 September 1988, lot 286; IC 4602. *Gualterus Arsenius Nepos Gemmæ Frisÿ fecit Louanÿ anno 1559.*
2. Museé National de la Renaissance, Ecouen, inv. E.Cl. 2937 (previously Museé de Cluny); IC 234. G*VL Nepos Gemmæ Frisij Louanij fecit anõ 1661*. [*sic*, read 1561]
3. Museum of the History of Science, Oxford, LE 2047; IC 229. *Regnerus Arsenius Nepos Gemmæ Frisÿ fecit Lounaÿ anno 1565.*
4. Merton College, Oxford; IC 232. *Gualterus Arsenius nepos Gemmæ Frisÿ fecit Louanÿ anno·1571·*
5. Germanisches Nationalmuseum, Nuremberg, inv. WI 1164; IC 555. Not signed or dated.
6. Kunsthistorisches Museum, Vienna; IC 562. Not signed or dated.
7. Science Museum, London, inv. 1878–11; IC 233. *Ferd. Arssenius* [sic] *fecit Antverpiæ*. Not dated.

Appendix 2: Analysis of the star positions: some technical notes

In order to measure the positions of the stars on the rete of the astrolabe, we used a one-to-one photocopy of the rete. For our coordinate system, we chose initially as x-axis the line connecting the two points on the ecliptic corresponding to longitude 0° and 180°; our y-axis was chosen as the line connecting the two points on the ecliptic

Table I. Analysis in x-y coordinates (units in cm).

Modern name	Astrolabe name	Calculated Ptolemy + 20° 55′		Measured Astrolabe		Difference Calculated-Astrolabe	
		x	y	x	y	dx	dy
α Cas	Cassiop: pect	2·98	0·21	2·92	0·21	0·06	0·00
β Cet	Cauda Ceti	13·15	1·23	13·18	1·16	−0·03	0·07
μ And	Media cinguli Andr:	4·65	0·63	4·80	0·89	−0·15	−0·26
ζ Cet	Venter Ceti	10·60	4·37	10·54	4·28	0·06	0·09
β Per	Medusae cap	3·30	2·80	3·44	2·81	−0·14	−0·01
α Cet	Naris Ceti	6·75	5·69	6·74	5·67	0·01	0·02
α Tau	Oculus [Tau]	3·22	6·21	3·27	6·16	−0·05	0·05
α Aur	Hircus	1·22	3·63	1·21	3·63	0·01	0·00
β Ori	Orionis sinister pes	3·09	10·41	3·08	10·41	0·01	0·00
α Ori	Orionis dex: hum:	0·98	8·23	0·95	8·16	0·03	0·07
α CMa	Canis maior	−1·47	12·17	−1·61	12·08	0·14	0·09
α CMi	Canis minor	−2·77	7·87	−2·71	7·82	−0·06	0·05
α Hya	Hydrae lucida	−7·36	6·90	−7·29	6·83	−0·07	0·07
α Leo	Cor [Leo]	−5·98	4·08	−6·03	4·16	0·05	−0·08
α UMa	Vrsa maior/1	−2·00	0·83	−2·06	0·91	0·06	−0·08
α Crt	Fundus Vasis	−11·50	4·38	−11·38	4·33	−0·12	0·05
β UMa	Vrsa maior/2	−2·44	0·99	−2·58	0·98	0·14	0·01
δ Leo	Dorsum [Leo]	−5·91	1·89	−5·87	1·91	−0·04	−0·02
γ UMa	Vrsa maior/3	−2·82	0·42	−2·72	0·51	−0·10	−0·09
β Leo	Cauda [Leo]	−6·82	1·04	−6·74	1·03	−0·08	0·01
δ UMa	Vrsa maior/4	−2·54	0·14	−2·51	0·16	−0·03	−0·02
γ Crv	Ala detra Corvi	−12·13	0·43	−12·11	0·21	−0·02	0·22
ε UMa	Vrsa maior/5	−2·66	−0·37	−2·61	−0·35	−0·05	−0·02
α Vir	Spica [Vir]	−10·41	−2·87	−10·34	−2·79	−0·07	−0·08
ζ UMa	Vrsa maior/6	−2·65	−0·72	−2·56	−0·69	−0·09	−0·03
η UMa	Vrsa maior/7	−3·02	−1·23	−3·01	−1·19	−0·01	−0·04
α Boo	Arcturus	−5·47	−3·02	−5·51	−2·99	0·04	−0·03
γ Boo	Bootis sinis: hum:	−3·63	−2·32	−3·60	−2·31	−0·03	−0·01
β Lib	Lanx sept: [Lib]	−7·72	−7·25	−7·71	−7·19	−0·01	−0·06
α CrB	Coruna sept:	−3·64	−4·14	−3·61	−4·09	−0·03	−0·05
δ Oph	Ophiuchi pal:sin:	−5·21	−8·20	−4·99	−8·11	−0·22	−0·09
α Sco	Cor [Sco]	−7·09	−12·64	−7·16	−12·49	0·07	−0·15
α Her	Herc: cap	−2·12	−6·74	−2·26	−6·62	0·14	−0·12
α Oph	Ophiu: cap:	−1·50	−7·20	−1·42	−7·13	−0·08	−0·07
τ Oph	Extra: man: dex: Ophi:	−0·88	−10·82	−0·96	−10·67	0·08	−0·15
γ Dra	Caput Draconis	−0·21	−3·17	−0·26	−3·12	0·05	−0·05
α Lyr	Lyra	0·38	−4·44	0·37	−4·44	0·01	0·00
α Aql	Aquila	2·99	−7·55	3·02	−7·52	−0·03	−0·03
α Cyg	Cauda Cygni	2·37	−3·14	2·33	−3·09	0·04	−0·05
γ Cap	Cauda [Cap]	9·64	−8·41	9·68	−8·37	−0·04	−0·04
δ Cap	Cauda [Cap]	9·74	−8·07	9·72	−8·14	0·02	0·07
ε Peg	Pegasi rictus	6·34	−5·08	6·30	−5·10	0·04	0·02
δ Aqr	Crus Aquary	11·67	−4·86	11·66	−5·02	0·01	0·16
β Peg	Crus Pegasi	5·51	−1·94	5·45	−2·02	0·06	0·08
α Peg	Pegasi hum:	6·94	−2·41	6·97	−2·52	−0·03	0·11
α And	Peg: Vmb:	5·65	−0·31	5·60	−0·39	0·05	0·08
γ Peg	Extr: ala Pegasi	7·40	−0·29	7·47	−0·39	−0·07	0·10
Mean						−0·010	−0·005
Standard deviation						0·076	0·086

Table II. Analysis in ecliptic coordinates (units in degrees).

Modern name	Astrolabe name	Calculated Ptolemy + 20° 55′ Longitude	Calculated Ptolemy + 20° 55′ Latitude	Measured Astrolabe Longitude	Measured Astrolabe Latitude	Difference Calculated-Astrolabe *d*Lon	Difference Calculated-Astrolabe *d*Lat
α Cas	Cassiop: pect	31·75	46·75	32·47	47·26	−0·72	−0·51
β Cet	Cauda Ceti	356·58	−20·33	356·28	−20·28	0·30	−0·05
μ And	Media cinguli Andr:	22·75	30·00	24·15	27·31	−1·40	2·69
ζ Cet	Venter Ceti	15·92	−20·00	15·83	−19·43	0·09	−0·57
β Per	Medusae cap	50·58	23·00	49·25	22·14	1·34	0·86
α Cet	Naris Ceti	38·58	−12·33	38·60	−12·13	−0·02	−0·20
α Tau	Oculus [Tau]	63·58	−5·17	63·15	−4·88	0·44	−0·29
α Aur	Hircus	75·92	22·50	76·08	22·50	−0·16	0·00
β Ori	Orionis sinister pes	70·75	−31·50	70·85	−31·48	−0·10	−0·02
α Ori	Orionis dex: hum:	82·92	−17·00	83·16	−16·50	−0·24	−0·50
α CMa	Canis maior	98·58	−39·17	99·37	−38·79	−0·79	−0·37
α CMi	Canis minor	110·08	−16·17	109·73	−15·77	0·36	−0·40
α Hya	Hydrae lucida	140·92	−20·50	140·70	−19·94	0·22	−0·56
α Leo	Cor [Leo]	143·25	0·17	143·20	−0·51	0·05	0·68
α UMa	Vrsa maior/1	128·58	49·00	128·62	47·82	−0·04	1·18
α Crt	Fundus Vasis	167·25	−23·00	166·99	−22·42	0·26	−0·58
β UMa	Vrsa maior/2	133·08	44·50	134·93	43·66	−1·84	0·84
δ Leo	Dorsum [Leo]	155·08	13·67	154·65	13·85	0·43	−0·19
γ UMa	Vrsa maior/3	143·92	46·50	141·63	46·62	2·29	−0·12
β Leo	Cauda [Leo]	165·42	11·83	165·11	12·43	0·31	−0·60
δ UMa	Vrsa maior/4	144·08	51·00	143·45	51·14	0·64	−0·14
γ Crv	Ala detra Corvi	184·42	−14·83	185·27	−14·29	−0·86	−0·55
ε UMa	Vrsa maior/5	153·08	53·50	152·15	53·92	0·94	−0·42
α Vir	Spica [Vir]	197·58	−2·00	197·10	−1·67	0·48	−0·33
ζ IMa	Vrsa maior/6	158·92	55·67	157·42	56·46	1·50	−0·79
η UMa	Vrsa maior/7	170·75	54·00	170·00	54·02	0·75	−0·02
α Boo	Arcturus	197·92	31·50	197·55	31·19	0·37	0·31
γ Boo	Bootis sinis: hum:	190·58	49·00	190·49	49·38	0·10	−0·38
β Lib	Lanx sept: [Lib]	223·08	8·50	222·82	8·73	0·27	−0·23
α CrB	Coruna sept:	215·58	44·50	215·22	44·95	0·37	−0·45
δ Oph	Ophiuchi pal:sin:	235·92	17·00	236·54	18·31	−0·62	−1·31
α Sco	Cor [Sco]	243·58	−4·00	243·04	−3·75	0·54	−0·25
α Her	Herc: cap	248·58	37·50	246·86	37·87	1·72	−0·37
α Oph	Ophiu: cap:	255·75	36·00	256·36	36·71	−0·61	−0·71
τ Oph	Extra: man: dex: Ophi:	265·25	14·33	264·76	15·09	0·49	−0·75
γ Dra	Caput Draconis	260·58	75·50	258·25	75·95	2·33	−0·45
α Lyr	Lyra	278·25	62·00	277·83	61·99	0·42	0·01
α Aql	Aquila	294·75	29·17	295·01	29·25	−0·26	−0·08
α Cyg	Cauda Cygni	330·08	60·00	330·61	60·61	−0·53	−0·61
γ Cap	Cauda [Cap]	315·75	−2·17	315·97	−2·22	−0·22	0·05
δ Cap	Cauda [Cap]	317·25	−2·00	316·94	−1·99	0·31	−0·01
ε Peg	Pegasi rictus	326·25	21·50	326·01	21·75	0·24	−0·25
δ Aqr	Crus Aquary	332·58	−7·50	331·89	−7·44	0·69	−0·06
β Peg	Crus Pegasi	353·08	31·00	352·34	31·61	0·74	−0·61
α Peg	Pegasi hum:	347·58	19·67	346·70	19·50	0·88	0·17
α And	Peg: Vmb:	8·75	26·00	8·25	26·70	0·50	−0·70
γ Peg	Extr: ala Pegasi	3·08	12·50	2·19	12·34	0·89	0·16
Mean						0·27	−0·16
Standard deviation						0·80	0·63

corresponding to longitude 90° and 270°. The intersection of these two axes was taken as our origin. The radius of the equator was obtained by measuring the distance between the points of intersection of the *x*-axis with the ecliptic, and was found to be 93 mm. We then determined for each star the *x* and *y* coordinates in this system of reference. Next we identified the stars by the names inscribed on the astrolabe, and, using the value of 93 mm for the radius and the value of 23° 30′ for the inclination of the ecliptic plane, calculated for these stars their *x* and *y* coordinates from their longitude and latitude for the epoch AD 1550. The latter coordinates were found from the star catalogue of Ptolemy by adding 20° 55′ to the values of the longitudes of the stars (see Section 7 above, where the reason for doing so is explained). These values of *x* and *y*, calculated for the epoch AD 1550, were subsequently compared with those obtained by measuring the positions of the astrolabe stars themselves with respect to our coordinate system.

Our preliminary procedure produced agreement between the calculated and measured positions. But the sensitivity of the results with regard to small changes in the radius of the equator made it clear that the latter, as well as the position of the north pole, should be fixed more reliably. The following method was used to this end. We measured the coordinates of a series of points along the ecliptic, using the coordinate system defined already. We then determined the origin and radius of the circle that, by means of the method of least squares, gave the best fit to our measured points. Having found in this way the most reliable parameters for the ecliptic, we derived, by a simple calculation, the most reliable position for the north pole and the radius of the equator. This procedure resulted in some minor corrections of our initial position for the north pole (along the *x*-axis it shifted by 0·05 mm; along the *y*-axis by 0·4 mm), and in a correction of our initial radius for the equator from 93 to 92·6 mm. With these new and now reliable parameters, we subsequently repeated the measurements and calculations described above. The result presented in Table I is discussed in Section 7.

For the sake of completeness, we have calculated in Table II the longitude and latitude of the corresponding values of the *x* and *y* coordinates. If the errors encountered in the stellar data are the result either of plotting the stars according to the makers, or from our own measurements, these errors unavoidably represent displacements in the *x*–*y* plane in arbitrary directions. In order to control these errors, comparisons should always be made in terms of *x* and *y* coordinates rather than longitude and latitude, because, as is clear from a comparison between our Tables I and II, a perfectly acceptable pattern of errors in *x* and *y* can cause very strange results when converted into another coordinate system. Leaving the case of *Media cinguli Andr*: aside (since another type of error seems to be involved), we see that the excellent agreement found in Table I is not found in Table II. In both series *d* Lon and *d* Lat there are now two cases for which the difference found is more than twice the standard deviation (*d* Lon of γ UMa and γ Dra, and *d* Lat of α UMa and δ Oph), and the mean values of *d* Lon and *d* Lat are twice the error to be expected in the mean from the standard deviation. Clearly, both mean values of *d* Lon and *d* Lat represent a systematic effect, and one could argue that the effect in longitude indicates that the precession should be adjusted. However, equal displacements in *x* and *y* coordinates of a star can cause different quantities in longitude and latitude, the effects of which add up to some false systematic errors, as is seen from the present analysis.

Acknowledgements

Particular thanks are due to Professor Paolo Galluzzi, Director, and Dr Mara Miniati, Curator, of the Istituto e Museo di Storia della Scienza, Florence, for their skilful promotion of the history of scientific instruments. They make working at the Istituto, with its library and extraordinary collection, one of the pleasures in life. To Senora Franca Principe, also of the Museo, we are indebted for her photography, which is scarcely to be matched elsewhere (Figure 1–5, 7, 8). Ready access has been granted to the reference collections at the Science Museum, London, thanks to Dr D. A. Robinson, Keeper of Physical Sciences, and Mr K. Johnson; at the Museum of the History of Science, Oxford University, thanks to Dr W. D. Hackmann, Acting Curator, and A. V. Simcock, Librarian; and at Leiden University, thanks to Drs Dirk de Vries, Librarian of the Map Room. Assistance with information and references was kindly provided by Professor G. Schilder and Dr P. van der Krogt of Utrecht University; Dr R. Distelberger of the Kunsthistorisches Museum, Vienna; Mr R. Schmidt of Vienna; Dr Liba Taub, and Mr and Mrs R. S. Webster of the Adler Planetarium, Chicago. For helpful discussions, we are indebted to Mrs J. A. Bigos, Dr Jon Darius, Professor D. A. King, Frankfurt University, and Dr W. F. Ryan, Warburg Institute, London University.

VIII

The Three Astrolabes of Gerard Mercator

Contents

1. Introduction

It is both remarkable and appropriate that the 400th anniversary of the death of Gerard Mercator (1512–1594) should be the occasion of the first publication of three astrolabes made by him, one bearing his monogram. No instruments by Mercator were known to exist until the first identification was made by the author in early 1992, though it is recorded that he did make astrolabes and other instruments. Two further astrolabes were discovered as a result of work on the first. It must be rare to establish an attribution for an instrument and then later to have the attribution confirmed by the maker himself. It shows in a startling manner that the methodology employed on the Florence instrument is sound.

The methodology used to attribute the astrolabes to Mercator involved detailed examination of all aspects of an instrument—the brass, the engraved letters and numbers, the mathematical features, the star names and positions, and the decorative elements. In the case of Mercator, only his maps and globes were available for comparative study, and their evidence comes through the medium of printing. The engraved copperplates provided the link with the engravings on the brass instruments, but, of course, in mirror image. The printing process removes the sharpness of the edges and the depth of cut produced by the burin, but the style and skill are fully apparent.

Very little documentary evidence exists to establish the provenance of instruments. The Mercator astrolabe in Florence can be identified in a late-eighteenth-century museum inventory, but cannot be distinguished in earlier ones, where the descriptions are not sufficiently detailed. For the other two, there is no provenance before their acquisition by the museums. The instruments themselves therefore must, and do, provide the only clues to their maker.

The first Mercator astrolabe to be identified is in the Istituto e Museo di Storia della Scienza, Florence.[1] It is unusual because it has a double-sided map plate.[2] This has a North polar projection on one side reaching to the Tropic of Capricorn, while the other side is a South polar projection to the Tropic of Capricorn only. The engraving characteristics found on this plate are markedly similar to some products of the Mercator workshop. Not only that, but a critical examination of the rest of the instrument showed that, apart from one latitude plate, the remainder is most likely to be in the hand of Gerard Mercator himself. The odd latitude plate, made for the latitude of Florence, is from the workshop of Giovan Battista Giusti. It can be shown that Giusti made the astrolabes commonly associated with Egnazio Danti (1536–1586), cosmographer to Duke Cosimo II of Florence.[3] Danti published his *Trattato dell'uso et della fabbrica dell'Astrolabio* at Florence in 1569, the same year that Cosimo (1519–1574) was created Grand Duke of Tuscany. It is likely, therefore, that the Florence Mercator astrolabe was acquired close to 1569.

While the Florence instrument was being studied, two more astrolabes were identified as closely similar. On the evidence supplied by examination of all three astrolabes, and by comparison with astrolabes and maps of contemporary craftsmen, only the Mercator workshop could have produced these instruments. The astrolabe in Augsburg is of exactly the same size as that in Florence, and the engraved information on the back is identical.[4] The Brno astrolabe is slightly smaller,[5] and it is signed with a monogram. This is located on the bottom edge below the hour symbol 12 on the limb, and reads 'GMR', standing for: Gerardus Mercator Rupelmundanus (see Figure 1).[6] Gerard Mercator was born at Rupelmonde, and referred to himself by this style on his productions from 1536.[7]

2. The astrolabes described

The most highly finished of the three Mercator astrolabes is the one at Florence. This has already been discussed in detail in the previous paper. All three will now be described as a group, with remarks on points of difference. The key information is given in Table 1, and their main parts are shown in Figures 2–11.

[1] Istituto e Museo di Storia della Scienza, Florence, inventory no. 1098. International Checklist [hereafter IC] no. 490. Sharon L. Gibbs, Janice A. Henderson, and Derek de Solla Price, *A Computerized Checklist of Astrolabes* (New Haven, Conn.: Yale University Department of the History of Science, 1973).

[2] This astrolabe is fully described in Gerard L'E. Turner and Elly Dekker, 'An Astrolabe attributed to Gerard Mercator, *c.* 1570', *Annals of Science*, 50 (1993), 403–43. This paper will be referred to subsequently as Turner and Dekker.

[3] Turner and Dekker, Section 3.

[4] Städtische Kunstsammlungen, Augsburg, Germany, inv. no. 3537 (IC 4609).

[5] Moravská Galerie, Brno, Czech Republic, inv. no. 24–385 (IC 4608).

[6] For examples of his signature, and information on his maps and calligraphy, see A. S. Osley, *Mercator: A Monograph on the Lettering of Maps, etc in the 16th Century Netherlands with a Facsimile and Translation of his Treatise on the italic hand and a Translation of Ghim's* Vita Mercatoris (London, 1969).

[7] Gerard Kremer had Latinized his name to Mercator by the time he enrolled at the University of Louvain on 29 August 1530. See R. W. Karrow, Jr, *Mapmakers of the Sixteenth Century and their Maps* (Chicago, 1993), p. 276. Karrow provides a full listing of Mercator's cartographic output, starting with the terrestrial globe he engraved for Gemma Frisius *c.* 1536, pp. 376–406.

Figure 1. Brno astrolabe IC 4608. Monogram GMR: Gerardus Mercator Rupelmundanus.

Table 1. The key information on the three astrolabes. Measurements in millimetres.

	Florence	Augsburg	Brno
IC number	490	4609	4608
Inventory no.	1098	3537	24–385
Signature	no	no	GMR
Diameter	316	317	278
Plates, latitude	5	1	1
Plate, latitude, extra	1	–	–
Plate, map	1	–	–
Plate diameter	293	293	258
Star pointers	50	50	48
Stars named	42	42	40
Ecliptic diameter	202	202	177
1st point Aries	10·3	10·2	10·3
Limb width	11·4	12	9·5
thickness	6·2	5·3	6
Alidade length	316	317	276
width	15	13	5 to 7
thickness	3·5	4	3·5
Rule length	316	317	278
width	13	15	12·4
thickness	4·2	3·4	3·5

The astrolabes are made of brass, with the Florence example gilded. Associated with this are five original latitude plates, and a further one made in Florence *c.* 1570. A seventh plate is engraved with projections of the Earth from the North and South Poles. The other two astrolabes each have a single latitude plate. In all three of the astrolabes the mater can accept only one plate at a time. The latitudes for which the plates are made are listed in Table 2.

Figure 2. Florence astrolabe IC 490. Front and back.

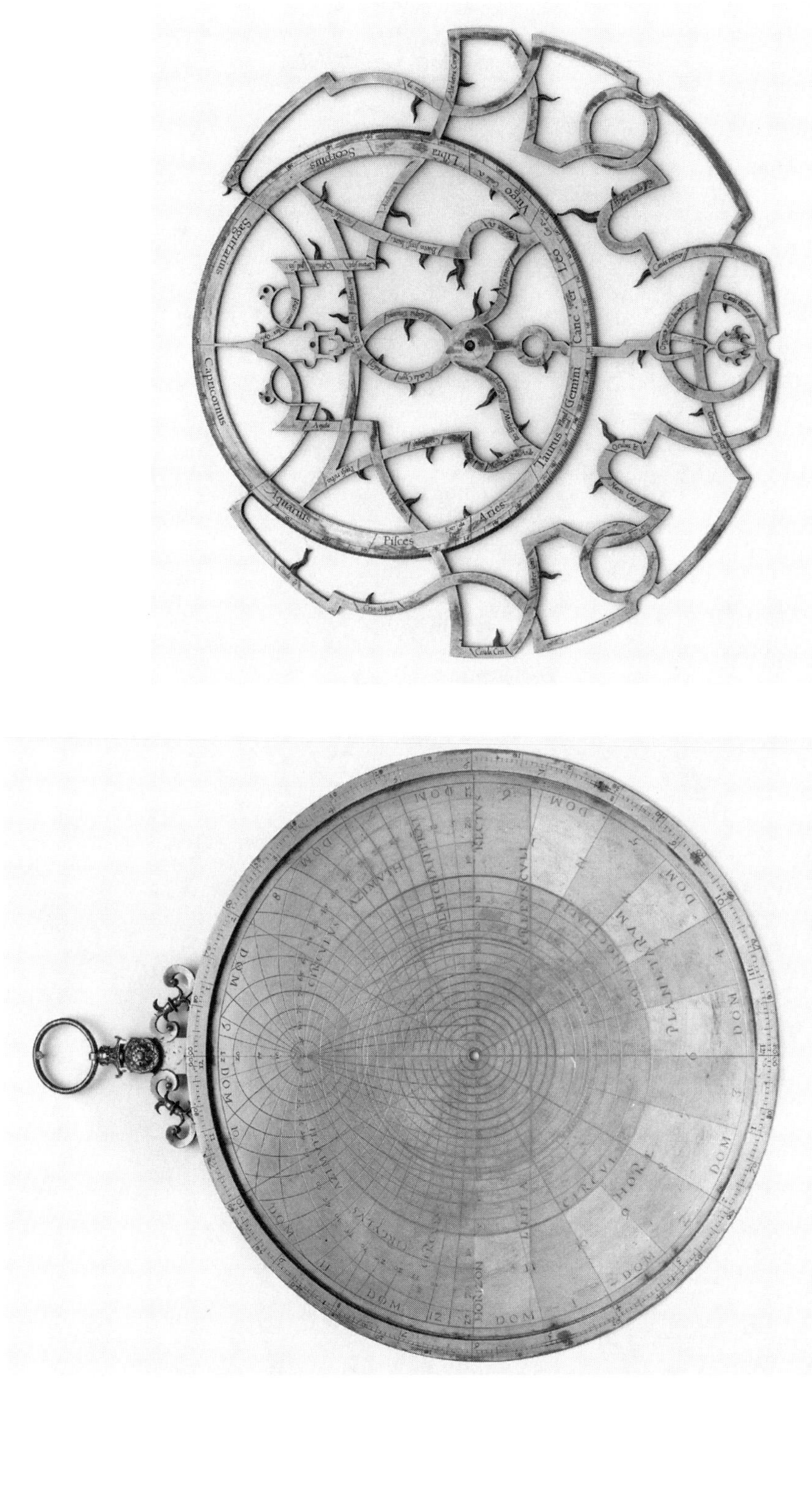

Figure 3. Florence astrolabe IC 490. Mater and rete.

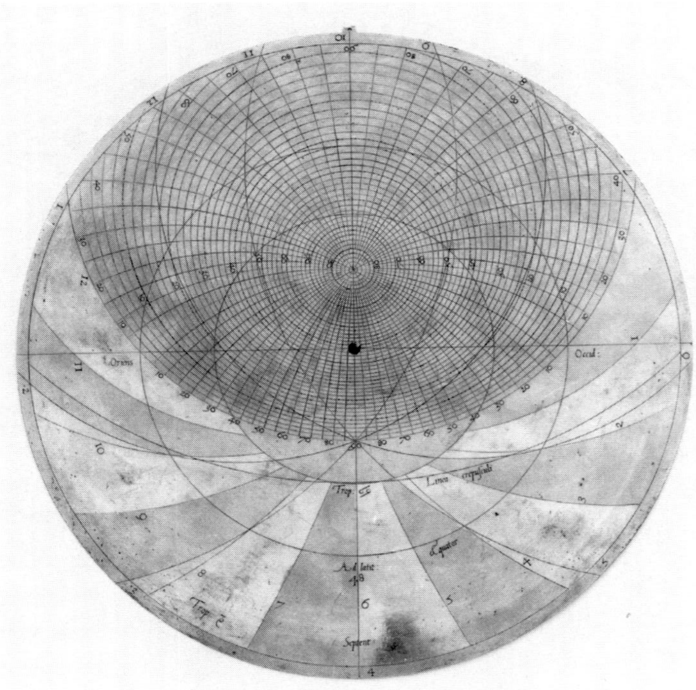

Figure 4. Florence astrolabe IC 490. Plate for latitude 48° and map plate, North Polar projection.

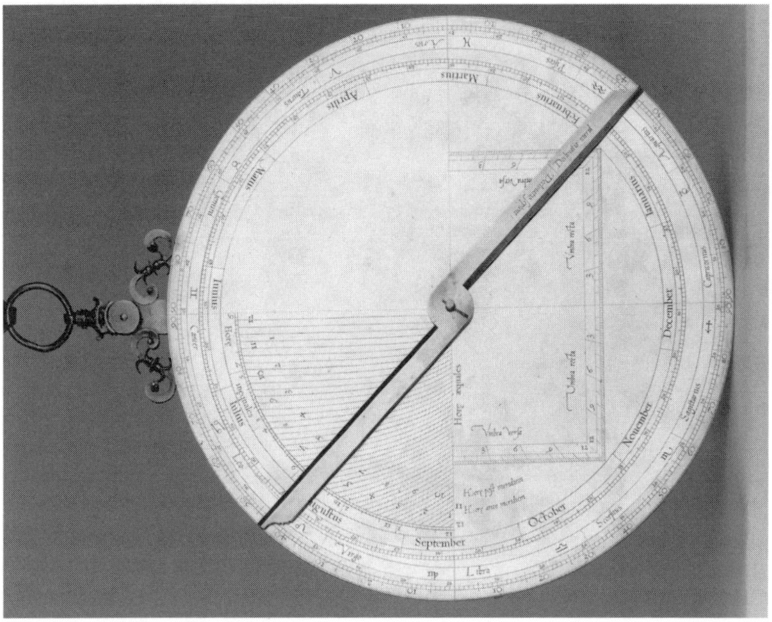

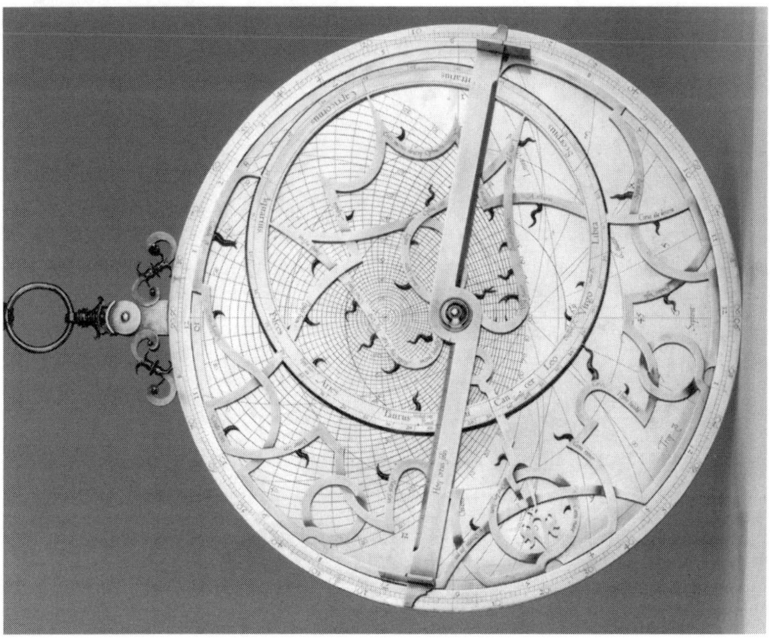

Figure 5. Augsburg astrolabe IC 4609. Front and back. The alidade and rule have been interchanged.

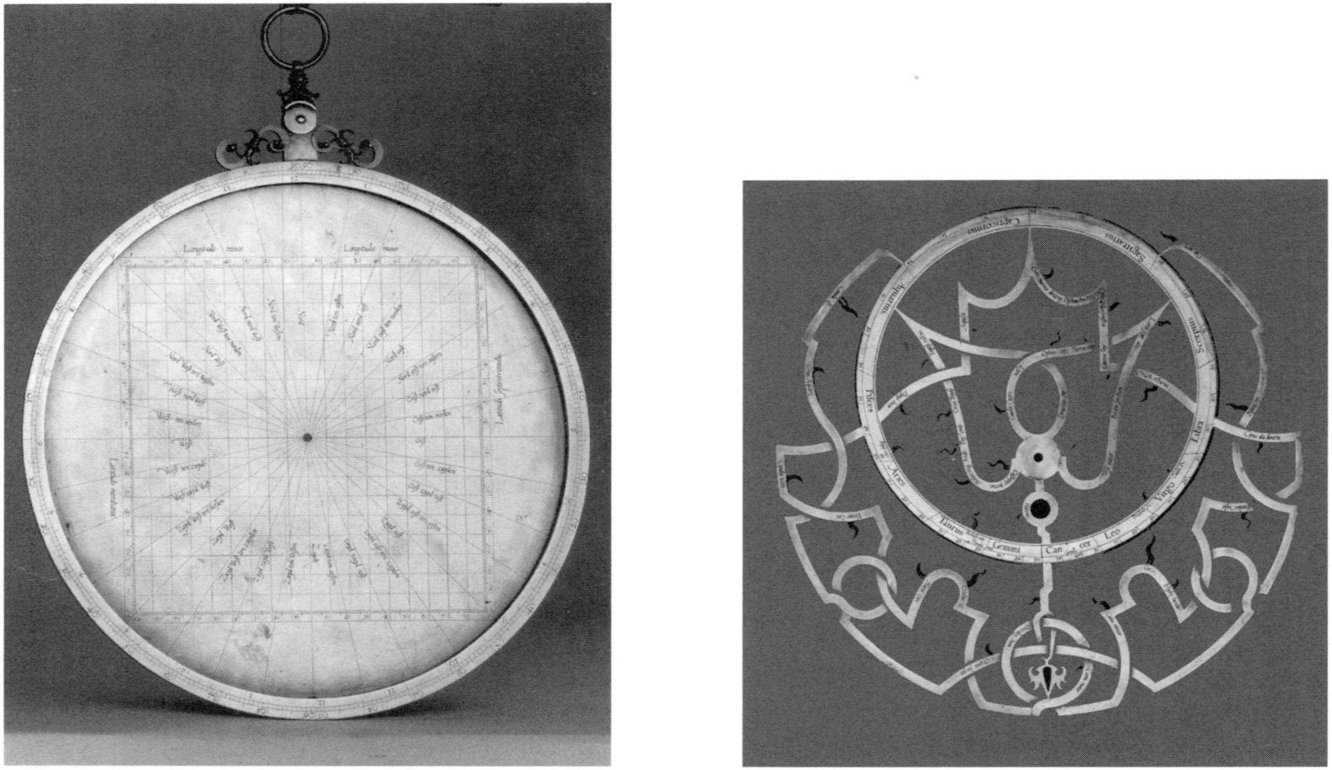

Figure 6. Augsburg astrolabe IC 4609. Mater and rete. The star pointers are in shadow.

Limb. The limbs on all three astrolabes are divided in degrees, marked, from the top, 90°–10° in each quadrant (zero degrees are not marked). Inside the degree scale is a 24-hour scale, engraved 1–12 twice. The Brno limb has the degree scale inside the hour scale.

Mater. The mater of the Florence astrolabe is engraved with projections for the latitudes of 90° North and 0°. Also engraved are the (un)equal hours for latitude 0° (twenty-four hours) and the astrological houses (numbering twelve). The maters of the other two astrolabes have engraved on them a *quadratum nauticum*, or diagram of the wind directions. On the Augsburg instrument, the sides are labelled *Longitudo minor; Longitudo maior; Latitudo meridiana; Latitudo ſeptentrionalis.* Superimposed are thirty-two radial lines each labelled in Dutch with the wind directions. The Brno mater is similar to Augsburg, but rotated through 180°. The winds are not named except for the cardinal points: Septentrio, *Meridies, Oriens, Occidens.* Two sides of the square are labelled: *Latitudo merid:, Latitudo ſeptent:.*

Back. The backs of both the Florence and Augsburg instruments are identical, with a shadow square in the lower semicircle, and in the upper left quadrant a diagram for conversions between equal and unequal hours (also called planetary or temporary hours) based on the times of sunrise and sunset; the right quadrant is blank. On the

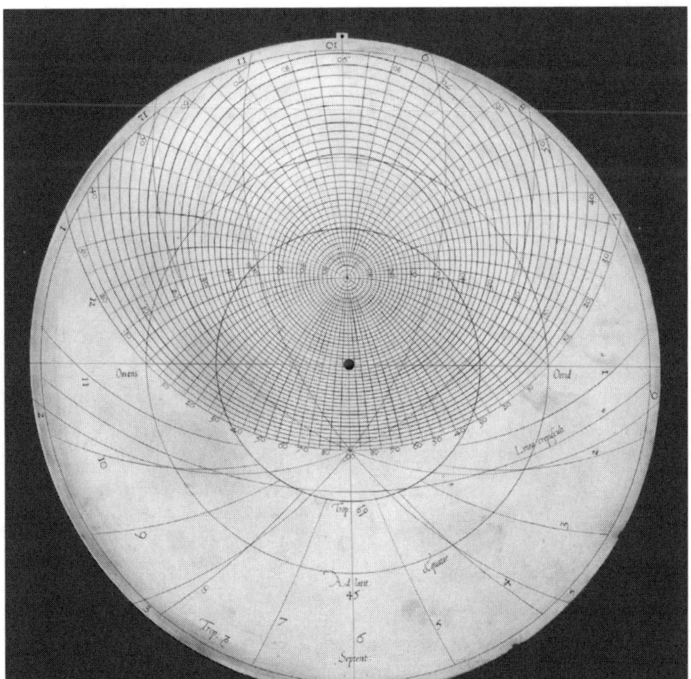

Figure 7. Augsburg astrolabe IC4609. Plate for latitude 45°.

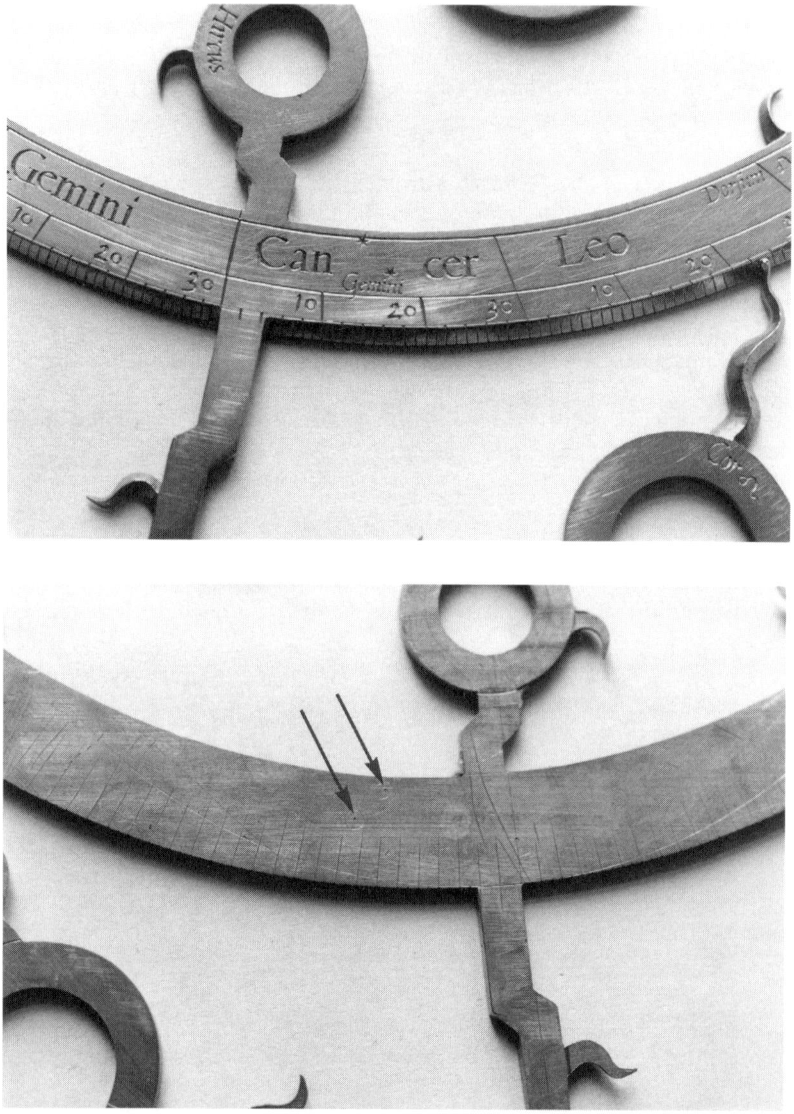

Figure 8. Augsburg astrolabe IC 4609. Detail of front and back of rete showing the pair of layout dots for the Gemini stars.

Brno astrolabe this quadrant is not blank, being filled with a diagram for measuring time in unequal hours based on the Sun's altitude; the left quadrant has its hour conversion diagram like the other two, but it is a mirror image and turned through 90°. In other respects it is the same as the Florence and Augsburg astrolabes. Around the edges of all three are divided scales for degrees, the zodiac, and the calendar. The calendar is eccentric with respect to the zodiac.

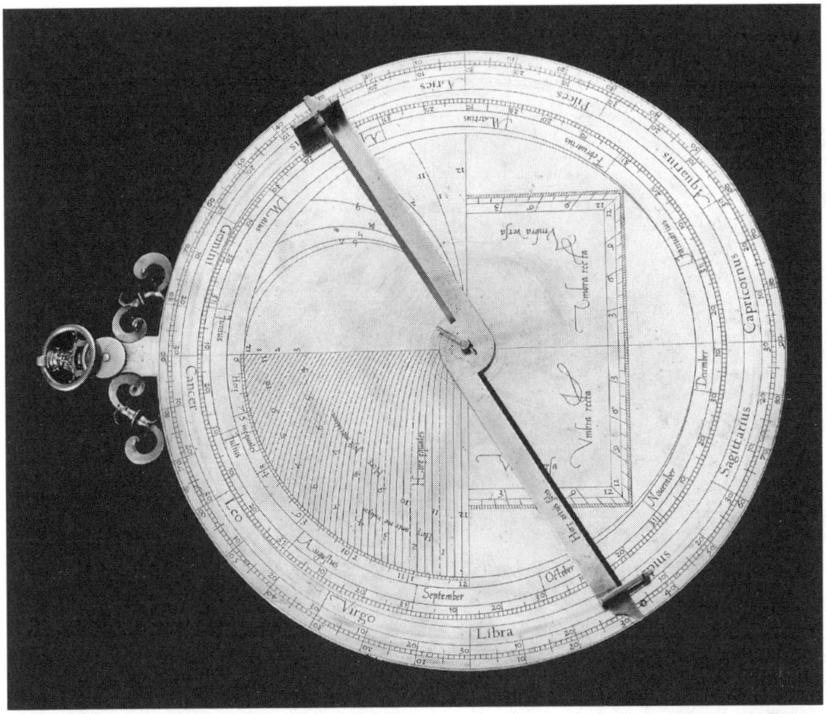

Figure 9. Brno astrolabe IC 4608. Front and back.

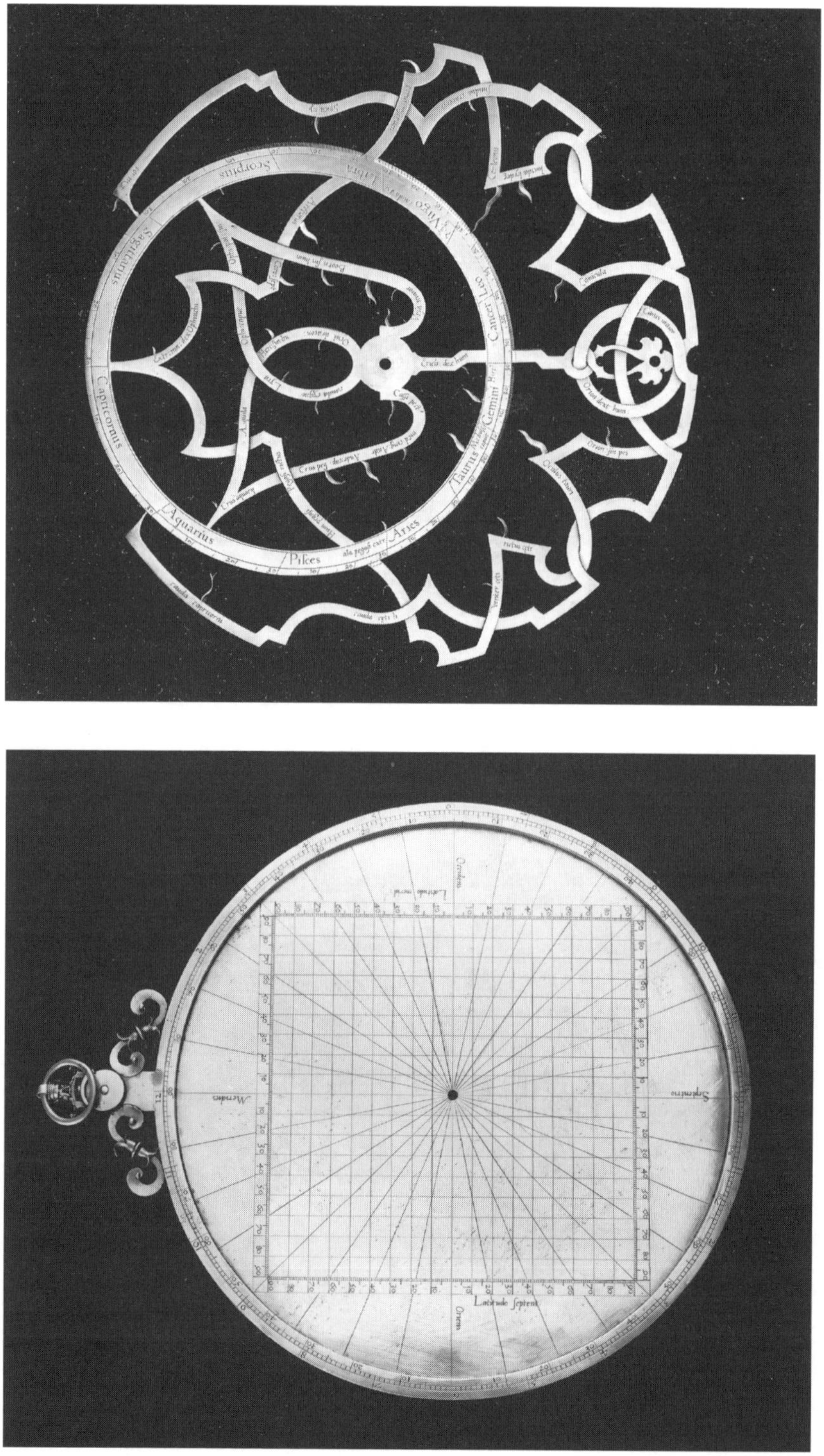

Figure 10. Brno astrolabe IC 4608. Mater and rete.

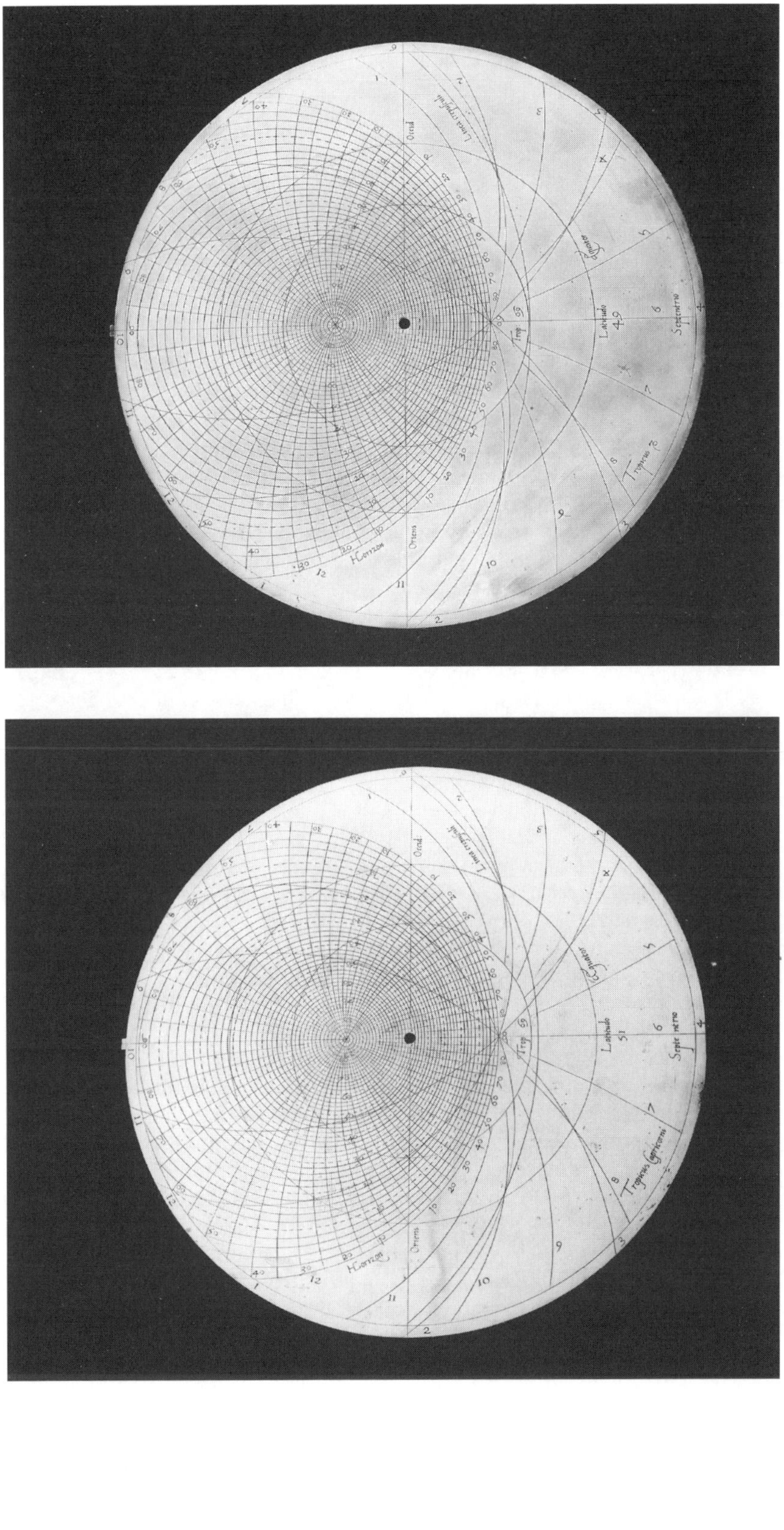

Figure 11. Brno astrolabe IC 4608. Both sides of plate for latitudes 49° and 51°.

Table 2. The plates with the astrolabes.

Florence	Augsburg	Brno
1. 36°/tablet of horizons	45°/48°	49°/51°
2. 39°/42°		
3. 45°/48°		
4. 51°/54°		
5. 57°/60°		
6. Map plate		
7. 43°/43° made by Giusti		

Plates. The original plates with the Florence instrument are engraved for latitudes every 3° between 36° and 60°, and one has a tablet of horizons. These, and the latitude plates on the other two astrolabes, are each engraved with almucantars at 2° intervals, and azimuths at 5°. The sixth plate is made for the latitude of Florence, 43°, and it is not from the same workshop as the rest of the astrolabe. As the instrument was to be used in Florence, a plate for 43° was made by a craftsman of the city, Giovan Battista Giusti.[8] The fabrication of this plate, although skilled, is not to the same very high standard as the rest. The maters of all three astrolabes can accommodate only one plate at a time, contrary to the usual practice. This makes for a better fit, and it may explain why the other two instruments have just the one plate with them; any others could have been lost over the centuries. The Brno plate has a projection for the latitude of 49°, unlike the others. This is backed by a projection for 51°, so giving a deliberate 2° difference; the others are at 3° intervals, normal for the climates and their extensions. The Augsburg astrolabe has one plate for 45°/48°, which matches one of the Florence plates. The Brno plate does not match any in the Florence set, nor, by inference, that of the Augsburg instrument. The 2° latitude range covers the region from Antwerp (modern value 51° 13′) to Paris (48° 50′). The form of words indicating the latitude on each projection is: *Ad latit* (Florence and Augsburg)/*Latitudo* (Brno); *Trop* [Cap] (Florence and Augsburg/*Tropicus Capricorni* (Brno). This supports the presumption that the Florence and Augsburg astrolabes were made at about the same time.

A remarkable feature on the Florence plates is the burnishing of the gilding in alternate segments of the unequal hours. Such a decorative conceit has not been remarked on any other astrolabe, and means that the instrument came from a first-class workshop, and that it was probably made for a notable customer.

The Florence astrolabe's maps are engraved on a copper disc, the land mass and decorative elements are gilded, while the seas are left as copper. This has blackened through long contact with the atmosphere. One side is a North polar projection to the Tropic of Capricorn, while the other side is a South polar projection cut off at the Tropic of Capricorn. This side has more space for engravings of monsters and sailing ships. On both sides, the outer rim is divided into degrees, marked every five from 5° to 360°. The prime meridian runs through the Azores, a few degrees West of the Canaries, and through the magnetic pole, *Polus Magnetis*, at longitude 180° and latitude *c.* 74°. The meridians and the parallels are engraved for every 10°. The longitude scale is on the outer rim, marked by punched numbers every five degrees from 5° to 360°. The latitude

[8] For the attribution to Giusti, see Turner and Dekker, Section 3.

scale is along the prime meridian and is marked every ten degrees from 10° to 90°. Both sides of the plate show the polar circles, *Circulus Arcticus* and *Circulus Antarcticus*, the equator, *Circulus Aequinoctialis*, and the two tropics, *Tropicus Cancri* and *Tropicus Capricorni.*

Rete. The retes are in the 'tulip' strapwork pattern typical of the Flemish astrolabes of the sixteenth century. The Florence rete has some extra flourishes within the ecliptic circle, otherwise it is the same as the Augsburg. Both have pointers for fifty stars; the Brno rete has pointers for forty-eight stars. The names are listed in Table 3. The Brno rete is not identical to the other two; it bears a closer resemblance to the retes of Thomas Gemini,[9] a Flemish engraver working in London *c.* 1540–1562 (see Figure 12 for comparative silhouettes of the retes).

The retes on both the Florence and Augsburg astrolabes bear the names of the same forty-two stars or groups of stars. The Brno instrument has thirty-one stars in common with those engraved on the Florence and Augsburg instruments with the same names. Of the remaining, one is not named, but is in the correct position for *Lanx septemtrionalis* (β Lib); three stars are common, but have different names; one has the same name, but refers to a different star (ι Cet). Additionally, the Brno rete has five stars that are completely lacking on the others. On all three retes there are seven pointers for Ursa Major, and Cauda Capricorni has a bifurcated pointer. For the constellation Gemini, two stars are marked on the ecliptic circle, but these are not included on the Brno instrument. Two stars on the Brno astrolabe show careless, but all too human error in their positions. 'Orıon: dext: hum:' (α Ori) has been given a Declination of −6° instead of +6°. 'Crus aquary' (δ Aqr) is displaced in Right Ascension by 30°, and in Declination by 10° (RA 307·38°, Dec −7·02° not RA 337·39°, Dec −17·56°).

The Brno rete provides the evidence that the laying out of the strapwork and star pointers was performed on the back. For an engraver not used to making astrolabes, the reversal of the star positions on to the back of the rete is bound to demand extra concentration. The mistakes in positioning these two stars are very simple ones. Six degrees to the wrong side of the ecliptic circle; 30°, starting from the wrong sign of the zodiac, and missing the ten, so measuring −7° instead of −17°. Could this have been a result of converting from the ecliptic co-ordinates of Copernicus to the equatorial co-ordinates used to mark out the astrolabe?

Alidade. On the Florence alidade one of the arms is divided with a scale 0–10 labelled *Horae ortus ſolıs.* The sights have a notch at the top and a pinhole below. On either side of each sight is a wing, one with a hole and the other with a dot; the hole in one sight faces a corresponding dot on the other for aligning on the Sun. The Augsburg alidade is similar, but with no decorative lines. The sights are rectangular with both pinhole and dot at each end. For the Brno instrument, the alidade is tapered in width; it is marked as before on the other two. The sights have a slit and pinhole at each end.

Rule. The Florence rule has both arms divided 0°–70°N and 0°–20°S. Scales are labelled *Declinatio ſeptent:/Declinatio merid:.* The rules for the Augsburg and the Brno astrolabes are the same as for the Florence instrument, but with no decoration. Measurements of the alidades and the rules on the Florence and Augsburg instruments raise the possibility that they were made at the same time; see Section 3.3.

[9] For Thomas Gemini, see C. D. O'Malley in *Dictionary of Scientific Biography* (New York, 1972), v, 347–9.

Table 3. The stars: their names and identifications.

Modern name	Florence IC 490	Augsburg IC 4609	Brno IC 4608
α Cas	Caʃʃıop: pect^9	Caʃʃıopæ pectus	Caʃʃı pect^9
β Cet	Cauda Cetı	Cauda Cetı	×
ι Cet	×	×	cauda cætı
μ And	Media cıngulı Andr:	Medıa cıng: Andr:	med: cıng: Andr
ζ Cet	Venter Cetı	Venter Cetı	Venter cætı
β Per	Meduʃæ cap	Meduʃæ cap:	Meduʃæ caput
α Cet	Narıs Cetı	Narıs Cetı	×
γ Cet	×	×	rictus cætı
Tau	Pleıad med:	Pleıadũ med:	×
α Tau	Oculus [Tau]	Oculus [Tau]	Oculus taurı
α Aur	Hırcus	Hırcus	Hirc^9
β Ori	Orıonıs ʃınıʃter pes	Orıonıs ʃın: pes	Orıon: ʃın: pes
β Aur	×	×	εrıch: dex hum
α Ori	Orıonıs dex: hum:	Orıo: dex: hum	Orıon: dext: hum: ★
α CMa	Canis maior	Canıs maior	Canıs maıor
α, β Gem	Gemini {2}	Gemini {2}	×
α CMi	Canis minor	Canıs minor	Canicula
α Hya	Hydræ lucıda	Hydræ lucıda	lucıda hydræ
α Leo	Cor [Leo]	Cor [Leo]	Cor leonıs
UMa	Vrʃa maıor {7}	Vrʃa maıor {7}	Vrsa maıor {7}
α Crt	Fundus Vaʃıs	Fundus Vaʃıs	fundus craterıs
γ Leo	Dorʃum [Leo]	Dorʃum [Leo]	dorsum [Leo]
β Leo	Cauda [Leo]	Cauda [Leo]	cauda [Leo]
α Cvr	×	×	coruı roʃtrum
γ Crv	Ala detra Coruı	Coruı ala dextra	×
α Vir	Spıca [Vir]	Spıca [Vir]	Spıca [Vir]
α Boo	Arcturus	Arcturus	Arcturus
γ Boo	Bootıs ʃiniʃ: hum:	Bootıs ʃin: hum:	Bootıs ʃin hvm
β Lib	Lanx ʃept: [Lib]	Lanx ʃept: [Lib]	*not named = Lanx sept.*
α CrB	Coruna ʃept:	Coruna ʃept:	Coro: ʃept
δ Oph	Ophıuchi pal: ʃın	Ophı: pal: ʃınıʃtra	Ophı: pal: ʃın:
α Sco	Cor [Sco]	Cor [Sco]	Cor [Sco]
α Her	Herc: cap	Hercu: cap	×
δ Her	×	×	Herc: ʃın:hu
α Oph	Ophıu: cap:	Ophıu: cap:	Ophı: caput
β Dra	×	×	Ocul dracon:
τ Oph	εxtra: man: dex: Ophı:	εxtr: manus dextræ Ophıuchı	εxtr: ma: dex Ophıuchı
γ Dra	Caput Draconıs	Cap: Draconıs	×
α Lyr	Lyra	Lyra	Lyra
α Aql	Aquıla	Aquıla	Aquıla
α Cyg	Cauda Cygnı	Cauda Cygnı	cauda cygni
γ, δ Cap	Cauda [Cap] {2}	Cauda [Cap] {2}	cauda capricornı {2}
ε Peg	Pegaʃı rıctus	Pegaʃı rıctus	Pegaʃı rıctus
δ Aqr	Crus Aquary	Crus Aquary	Crus aquary ★
β Peg	Crus Pegaʃı	Crus Pegaʃı	Crus peg
α Peg	Pegaʃı hum:	Pegaʃı hum:	Hum pegaʃı
α And	Peg: Vmb:	Peg: vmb:	Andr: cap:
γ Peg	Extr: ala Pegaʃı	Pegaʃı extr: ala	ala pegaʃı extr

	The star names have been printed to resemble the original letter forms.
Column one	The modern name identifies the stars on all three astrolabes.
×	No star pointer.
[Tau]	The constellation name in brackets means the sign (symbol) was engraved.
{2}	The number of star pointers under the single name of the constellation or part of constellation. *Cauda capricorni* has a single pointer with a bifurcated tip; *Ursa maior* has seven pointers; *Gemini* has two dots marked on the ecliptic circle (see Figure 8).
★	Star pointer in wrong position; for explanation, see text.

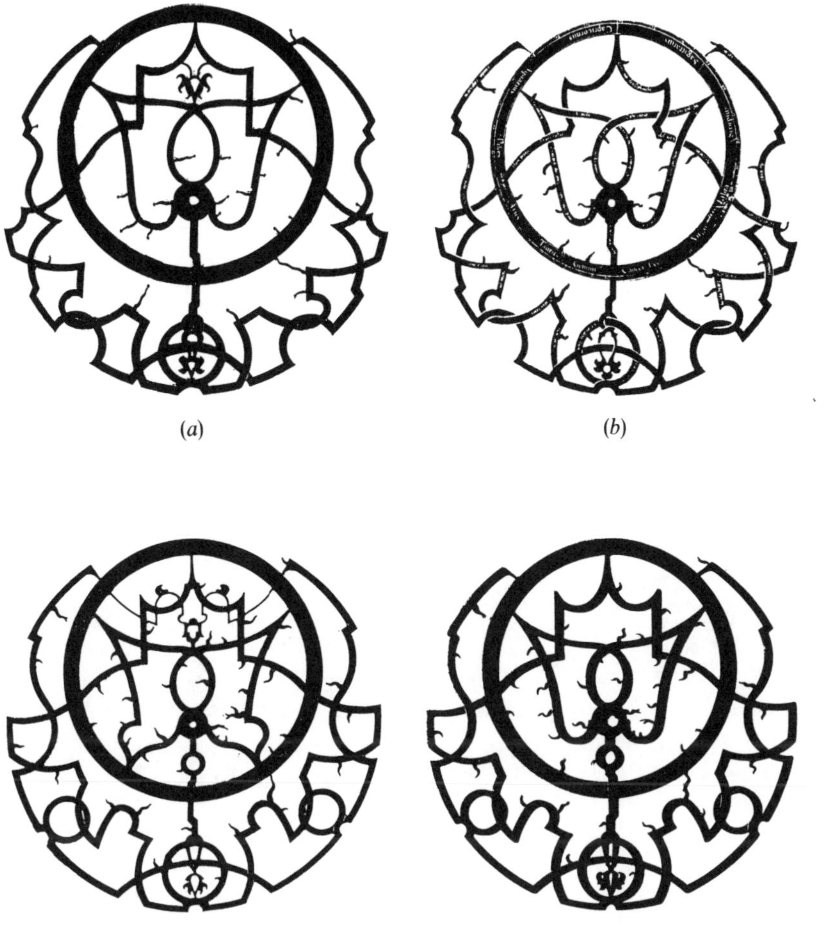

Figure 12. The rete patterns of the three Mercator astrolabes and one Gemini astrolabe. (*a*) Thomas Gemini, 1559, Florence (IC 489); (*b*) Gerard Mercator, *c.* 1545? Brno (IC 4608); (*c*) Gerard Mercator, *c.* 1570, Florence (IC 490); (*d*) Gerard Mercator, *c.* 1570, Augsburg (IC 4609).

Pin. For the Florence astrolabe, the central fastener is a bolt, with incised triangular motif on the head and a foliate roundel on the nut. The washer has an acanthus and rose motif around its rim. For the Augsburg astrolabe, it is a rod and wedge, with the head turned in concentric circles; the same for the Brno instrument, with a plainer circular head.

Throne. The thrones on all three astrolabes are the same: they are formed by a bar attached to the limb and supported by two S-shaped brackets with elaborate decoration. The ring at the top swivels above a bracket that moves in a shackle left or right. The boss is foliate on the Florence instrument and plain on the others. The pendant attached to the ring has a grotesque face on either side. The face and brackets

Figure 13. The throne on a Flemish astrolabe of the sixteenth century showing S-shaped supports in the form of dolphins. King's College astrolabe at the Whipple Museum, Cambridge (IC 235).

are cast from the same mould. The brackets have already been discussed.[10] They seem to be stylized dolphins, a sea mammal whose form has had a perennial appeal in all manner of decoration. The dolphin is found as a support in architectural features such as the corbel or bracket. A sixteenth-century Flemish astrolabe with a standard rete belonging to King's College, Cambridge, has a throne that is mechanically the same as those on the Mercator instruments, but with a pair of dolphins that support the central bar in the same manner as the S-brackets (see Figure 13).[11]

Monogram. On the Brno astrolabe, the monogram 'GMR' has the G engraved in the way that is very typical of Gerard Mercator. The capital G is formed by cutting the semicircle first, followed by the vertical stroke, which leaves part of the circle protruding to the right below the vertical stroke. If this is an early piece, one of the first astrolabes made by Gerard, then it might well have contained errors that he was not proud of. Since it was usable to one who was aware of its faults, he put his initials on it in order to mark it as not for sale: purely personal to him.

[10] Turner and Dekker, 435.

[11] Not signed or dated; at the Whipple Museum of the History of Science, Cambridge, inv. no. 1467 (IC 235). Described and illustrated in R. T. Gunther, *The Astrolabes of the World* (Oxford, 1932), pp. 393–4, plate XCVIII. Here dated to 1570–1580.

Epoch. A careful measurement of the mean value and standard deviation in Right Ascension point to an epoch consistent with 1550 for the Florence astrolabe, and, because of its very close similarity, for the Augsburg astrolabe as well.[12] It is also known that Mercator's celestial globe of 1551 was plotted for the epoch of 1550.[13] It is the Copernican theory of precession explained in *De Revolutionibus* (1543) that was applied to produce the star positions, through an excess in longitude of 20° 55′ with respect to the Ptolemaic date (epoch A.D. 137).[14] Taking this value, a table has been made comparing the values of Right Ascension and Declination calculated from Ptolemy with measured values from the Brno astrolabe. The table is in the Appendix below, from which it can be seen that the agreement is close. The values for the epoch of 1550 could not have been applied *before* about 1545, which is therefore the theoretical lower limit. If Mercator made the astrolabe between 1545 and 1550, he could have chosen an epoch of 1550. Dr Dekker has pointed out that Mercator's terrestrial globe of 1541 has forty-eight stars marked on it calculated on the pre-Copernican theory, which is 1° *less* than the figure for the precession used for the 1551 globe and the three astrolabes. The leading mathematician and astronomer of the period in the Low Countries was Gemma Frisius (1508–1555) of Louvain, with whom Mercator was associated. Dr De Smet has shown that Gemma Frisius had a close link with Poland through that country's ambassador to The Netherlands, Johannes Dantiscus (1485–1548), who was patron to Gemma. In Gemma's book, *De radio astronomico et geometrico* (Antwerp, 1545), there is enthusiasm for *De Revolutionibus:* 'il se réfè plusieurs fois à Copernic et de manière fort élogieuse'.[15] Naturally, the latest information obtained by Gemma would have been available to Mercator.

3. When were the astrolabes made?

3.1. *Evidence from the rete patterns*

As has been pointed out, the rete patterns of the Florence and Augsburg astrolabes are the same. The Brno rete, however, has a closer kinship with the two astrolabe retes made by Thomas Gemini of London that are dated 1559.[16] Thomas Gemini (originally Lambert, Lambrit, or Lambrechts), was born in eastern Flanders at Lixhe (Lieze) *c.* 1510, which is really an educated guess. He died in London in June 1562. He had migrated to Blackfriars, London in around 1540, and set up as an engraver, printer, and instrument maker. Gemini is known chiefly for his edition of Versalius' *Fabrica* that he published in London in 1545 with the title *Compendiosa totius anatomie delineatio.* For this he received a pension from King Henry VIII. In 1555, he published maps of Spain and of England, and in the same year published Leonard Digges's *Prognostication of Right Good Effect.* What is important about Thomas Gemini (the name he assumed during his London period) in the present context are the similarities between his style and that of Gerard Mercator. Gemini was a highly skilled engraver, and his calligraphy owes everything to that expounded and taught by Mercator.

[12] Data computed by Dr Dekker and published in Turner and Dekker, Appendix 2.

[13] Turner and Dekker, Section 7, and footnote 48.

[14] *Nicholas Copernicus Complete Works: On the Revolutions*, translation and commentary by Edward Rosen (Baltimore and London, 1978), Book III, chapters 11 and 12 (pp. 141–4).

[15] A. De Smet, 'Gemma Frisius et Nicolas Copernic', *Der Globusfreund*, Nr. 21–23, October 1973 (for 1972/73/74), pp. 72–80 (p. 75). Karrow (footnote 7), 205–15, lists Gemma's production of globes and books.

[16] Museum of the History of Science, Oxford, dated 1559, and dedicated to Queen Elizabeth of England, inv. no. 1098 (IC 575); Istituto e Museo di Storia della Scienza, Florence, dated 155 [last numeral omitted], inv. no. 1093 (IC 489). For Gemini, see footnote 9.

Thomas and Gerard, being much the same age, may have trained together in the workshop of Gaspar van der Heyden, a goldsmith of Louvain, who produced in 1536, with the co-operation of Mercator, a globe for Gemma Frisius.[17] Mercator acquired tools in 1540, and from then on worked at Louvain independently of Van der Heyden and Gemma Frisius. This is also about the time that Thomas Gemini is thought to have left Flanders for London. During the 1540s, Mercator is known to have made mathematical instruments, and the group of instruments made for the Holy Roman Emperor Charles V points to his high degree of skill. It must be stressed that the style and quality of the work by Thomas Gemini is such that he could have learned it only at Louvain before his London period. One would expect that he travelled with ideas for instruments (of which seven are known, four being astrolabes), so that his rete pattern, examples of which date from the end of the 1550s, could have originated from 1540. Mercator could have kept in mind, over a few years, the same rete pattern.

For Mercator, the decade began with his terrestrial globe (1541) and ended with his celestial globe (1551). During this period his calligraphy settled down into the style he kept through the rest of his life, which was spent in Duisburg from 1552. A careful scrutiny of the gores that form these two globes prompts the impression that the engraving on the Brno astrolabe resembles more the globe of 1541 than it does that of 1551.[18]

3.2. *The horary quadrant*

The back of the Brno astrolabe has two horary diagrams, the upper left is for conversion between equal and unequal hours, and the upper right (blank on the Florence and Augsburg astrolabes) is for finding the time in unequal hours from the Sun's altitude. For equal hours the day/night period is divided into twenty-four equal hours. Unequal hours are formed by dividing the day and night periods into twelve hours each, which become equal to each other only at the equinoxes. Clocks keep equal hours, but many sundials, from Roman times into the seventeenth century, are constructed to read unequal hours. The diagram of unequal hours and its approximate readings, except at sunrise, noon, and sunset, is fully analysed mathematically by Dr Archinard.[19]

The horary quadrant that appears on the upper right of the Brno astrolabe occurs in the medieval period, but an earlier Islamic source is likely, and it has been pointed out that a ninth-century Baghdad text, recently discovered in Cairo, describes the horary quadrant.[20] Western astrolabes are not infrequently provided with this quadrant, possibly through inertia or as decoration: it had become expected, traditional even, or looked nice in an otherwise blank space. On the astrolabe it was always small, and was inherently inaccurate, getting much worse as the user moves to higher latitudes (e.g., above 30°). Professor North has shown that it is found more often on western than eastern astrolabes, and points out that on 132 astrolabes that he had examined, 25 out of 57 European had a diagram, but only 16 out of 75 eastern astrolabes. North concludes: '41 instruments have the unequal-hour lines, and yet only four could have been used in at best a rough and ready way to find unaided the unequal

[17] Turner and Dekker, Section 5.

[18] Turner and Dekker, Section 6.

[19] Margarida Archinard, 'The Diagram of Unequal Hours', *Annals of Science*, 47 (1990), 173–90.

[20] D. A. King, *Islamic Astronomical Instruments* (Variorum Reprint CS253) (London, 1987), chapter 1, p. 9. A useful review of quadrants is still that of R. T. Gunther, *Early Science in Oxford*, II (Oxford, 1923), 154–81.

hour'.[21] Its presence on an instrument of the accuracy that Mercator intended his to have is anomalous. The other diagram, for the interconversion of equal and unequal hours, is useful because the astrolabe measures the equal hours and many monumental sundials the unequal, and so it is these that people would be used to. Mercator chose to leave a blank space on his masterpiece and its close kin, so the presence of the diagram on the Brno astrolabe points to an earlier period in Mercator's life.

3.3. *The astrolabes at Florence and Augsburg*

There are several independent pieces of evidence that, taken together, focus a date for two of the astrolabes to quite a narrow span of years. The sheer controlled skill of the engraver's hand, and so many characteristic letter forms, point to Gerard Mercator's second period, which was at Duisburg. The style has settled down by the time of the celestial globe (1551), and found its great expression with the world map of 1569.

The map plate does not appear to be in the hand of Gerard, and a case has been put forward that this was engraved by the second son, Rumold.[22] He was born *c.* 1546, and by 1565, at the beginning of his twenties, he would have been mature enough to produce the map plate. This was the period in the run-up to the publication of the world map, when Gerard would have needed additional help because of the volume of work. From the evidence of his own signed map of the world published in 1587, Rumold was highly skilled, and his hand seems to match that of the map plate's engraver.

The role of Egnazio Danti in Florence is almost certainly an important reason for the presence there of a Mercator astrolabe. He was appointed Cosmographer to Duke Cosimo de'Medici in 1562, and designed a globe and twenty-nine wall maps between 1564 and 1575 for the *Guardaroba* in the Palazzo Vecchio.[23] In 1569 he published a treatise on the astrolabe. Cosimo de'Medici came to power in 1537; with skill, luck, and force of arms he was created the first Grand Duke of Tuscany on 27 August 1569. He was now 50, and his Cosmographer, whose *Trattato* on the astrolabe had just been published with a dedication to Cosimo showing the six balls of the Medici over a globe of the Earth, was 33 years old. Clearly in great favour, Danti was, in 1571, granted permission to live in the Palazzo Vecchio. Unfortunately for Danti, Cosimo died on 21 April 1574, and was succeeded as Grand Duke by his son Francesco (1541–1587), who disliked Danti and summarily dismissed him on 28 September 1575.

Danti's presence in Florence between 1562 and 1575 sensitized the court to cosmography, and one would expect to find here the reason for a copy of Mercator's world map to have been purchased at Antwerp on 1 November 1569 for delivery to Florence.[24] Although the local craftsmen were reasonably competent, they were no match for the renowned cartographic and craft centre: Flanders. At the time of the crowning of the first Grand Duke of Tuscany in 1569, what would be more appropriate than to acquire from the world's finest astrolabe maker an example of his art? After all, Gerard Mercator had made mathematical instruments for the Holy Roman Emperor himself, and Cosimo was rising in the courtly circles of Europe.

All things considered, the date of *c.* 1570 seems to fit this astrolabe of Gerard Mercator and his Duisburg workshop. The Augsburg astrolabe is the same size and has

[21] J. D. North, *Stars, Minds and Fate: Essays in Ancient and Medieval Cosmology* (London and Ronceverte, West Virginia: Hambledon Press, 1989), pp. 221–2 (p. 222).

[22] Turner and Dekker, Section 6.2, 430.

[23] Ettore and Alessandro Cecchi, *Palazzo Vecchio e i Medici: Guida Storica* (Florence, 1980), pp. 303–10.

[24] See Turner and Dekker, 437.

the same back as the Florence instrument, has the same star names, but differs in having a less elaborate rete and it is not gilded. Another common feature is in the sizes of the alidades and rules. The Florence alidade and the Augsburg rule have the same width (15 mm) and the same thickness (3·5/3·4 mm); the Florence rule and the Augsburg alidade have the same width (13 mm) and the same thickness (4·0/4·2 mm).[25] It is tempting to believe that these parts were made at the same time. The weight of evidence means that a date near 1570 is also reasonable for the Augsburg astrolabe.

3.4. *The astrolabe at Brno*

The Brno astrolabe was most probably made earlier than the others. The calligraphy is slightly different, which is suggestive of an earlier date. If one examines the style of the engraved letters and numbers of Mercator's globes of 1541 and 1551, one readily sees many small variations in the layout of the letters, and the decorative flourishes, which are more pronounced in the earlier globe and on the Brno astrolabe (see Figure 14). The pattern of the Brno rete has a greater similarity to the retes cut by Thomas Gemini, who left Louvain for London *c.* 1540, than it has to the other two Mercator astrolabes. Then there is the choice of stars and star names on the rete, which has a number of layout mistakes in its construction. This mispositioning of two star pointers, and forgetting to engrave the name of the star on one pointer, suggest a certain unfamiliarity with the details of the construction of astrolabes. The unequal hour diagram in a quadrant left blank on the other two instruments is yet another pointer to an earlier date. It is not an accurate device, and if the unequal hour is required, then it is much better to take the reading on the astrolabe of the equal hour and then convert by means of the diagram engraved on the left-hand quadrant that is to be found on all three of Mercator's astrolabes. The use of the epoch of 1550 for the laying out of the rete stars shows that the Copernican theory had already reached Louvain, which it did by 1545.

The calligraphy and other features point strongly to a date during the 1540s for the Brno astrolabe. This would mean that it was made in Mercator's Louvain workshop, while the astrolabes now at Florence and Augsburg were made in the Duisburg workshop. The dimensions of the instruments support this surmise. Any craftsman wishing to make an independent instrument like an astrolabe, sundial, or globe, would select a suitable size in a measure familiar to his workshop. Thus in England, Humfrey Cole's large astrolabe of 1575 is 24 English inches in diameter, and globes are customarily 3, 9, 12, 18 or 24 inches in diameter. Mercator's astrolabes are about one foot in diameter, and his globes are over one foot in diameter. But which foot? In the sixteenth century, Mechelen (Malines; only 22 km from Louvain) was the most influential city of the Low Countries, and its foot is equivalent to 278 mm. Van der Heyden came from this city, and Mercator lived there after his university years at Louvain. Van der Heyden's globes of 1536 and 1537 have diameters of 370 mm, exactly $1\frac{1}{3}$ Mechelen feet. Mercator's 1541 and 1551 globes have diameters of 420 mm, equivalent to $1\frac{1}{2}$ Mechelen feet.[26] The Brno astrolabe has a diameter of 278 mm, exactly one Mechelen foot. The other two astrolabes, made in Duisburg on the Rhine, have diameters of 316 mm and 317 mm, and the Rhineland foot is equivalent to 315 mm.

[25] See Table 3. The thickness measurements were made in a few places to provide a guide. The alidades and rules do not have a completely uniform thickness.

[26] Globe diameters from Karrow (footnote 7), 377, 382, 384. Measures from Horace Doursthr, *Dictionnaire Universel des Poids et Mesures anciens et modernes* (Brussels, 1840; 3rd edn, 1976).

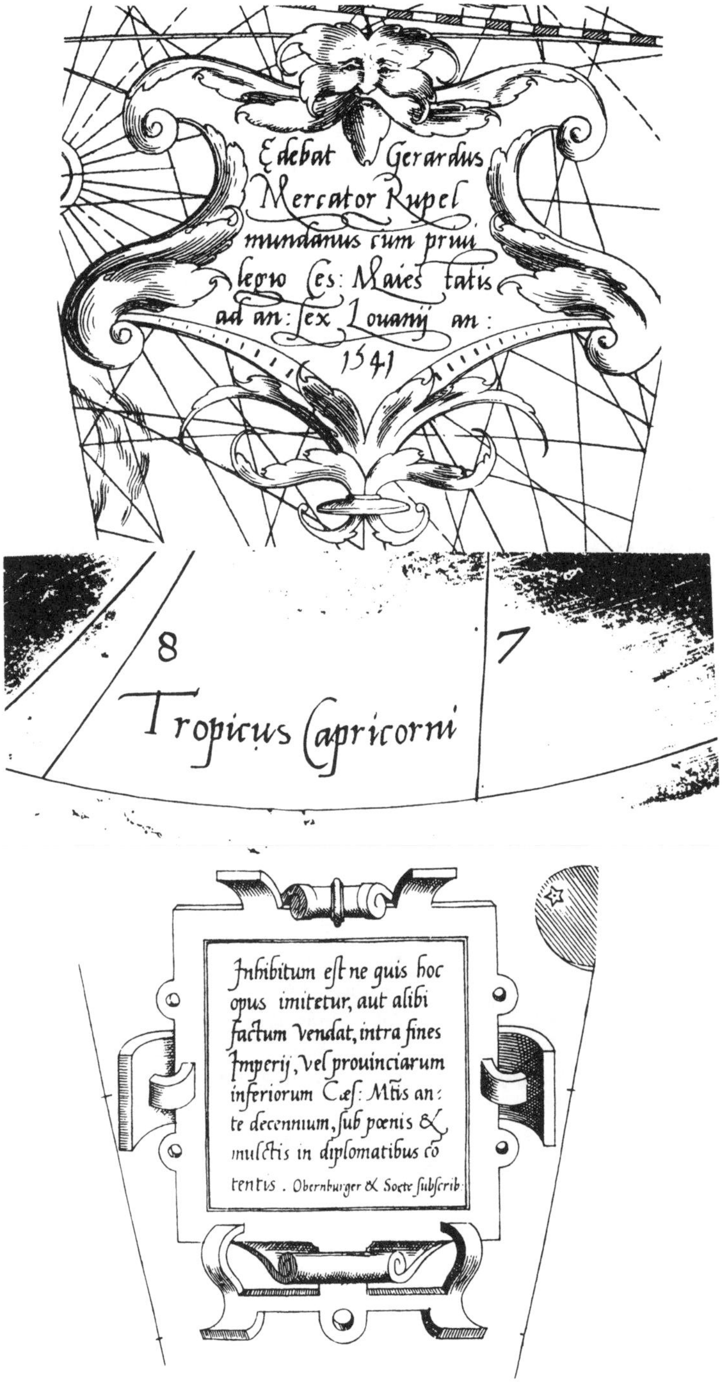

Figure 14. Comparison of lettering on two globes and the Brno astrolabe. (*a*) Cartouche on terrestrial globe of 1541; (*b*) plate for 51°, Brno (IC 4608); (*c*) cartouche on celestial globe of 1551. Compare the contraction 'Ces:' on the globes with 'Capricorni' on the astrolabe. The embracing C on the latter resembles that on the 1541 globe. Notice the greater restraint by 1551.

Appendix: Analysis of the star positions on the Brno astrolabe IC 4608

The positional data of the stars on the Brno astrolabe have been analysed in the same way as those on the Florence astrolabe.[27] From the analysis on equatorial co-ordinates (see Table I), it follows that the positions of two stars, δ Aqr and α Ori, deviate seriously from what can be expected. The nature of these deviations, a change in sign and a multiple of 10, implies that the star list used for the astrolabe must have described the positions of the stars in Right Ascension and Declination, and not in the more traditional mediation and declination, because only in terms of the former co-ordinates do we find a 'logical' explanation for the positional errors of these stars.

The other important conclusion to be drawn from the analysis is that the data on the Brno astrolabe, like those on the Florence astrolabe, are plotted for the epoch of 1550 according to the trepidation theory published by Copernicus in *De Revolutionibus* (1543). According to this theory the ecliptic longitudes of the stars for the epoch of 1550 exceed those in the Ptolemaic catalogue for the epoch A.D. 137 by 20° 55′. On the other hand, should the Alphonsine theory of trepidation be used, the figure computed is 20° 6′ for the same epoch of 1550.[28] In order to decide between these two theories we have first converted the equatorial data shown in Table I to ecliptic co-ordinates, and then calculated the *mean difference* between the longitudes calculated for Ptolemy +20° 55′ and those on the astrolabe. The resulting mean difference is −0·18°, or −11′, whereas a standard deviation of 0·65° was found. The minus sign implies that the excess of the astrolabe stars is on average somewhat larger than 20° 55′. Since the analysis is based on the data for forty stars, the error in the *mean difference* is about ±0·11°, or ±7′. These latter limits show that the astrolabe stars have an excess with respect to Ptolemy that is in accord with the Copernican values of 20° 55′ for the epoch 1550, and absolutely exclude the possibility that a pre-Copernican value of 20° 6′ was used for the longitudes of the astrolabe stars. This in turn implies that the astrolabe must have been made after 1543, when the new theory of Copenicus became public.[29]

Acknowledgements

I have pleasure in acknowledging my indebtedness to my co-author of the previous paper, Dr Elly Dekker; her analysis of the star positions is given in the Appendix. I am also grateful for the co-operation of the directors and curators of the museums holding the Mercator astrolabes. Professor Paolo Galluzzi, and Dr Mara Miniati, of the Istituto e Museo di Storia della Scienza, Florence; Dr Renate Eikelmann, Maximilian Museum, Städtische Kunstsammlungen, Augsburg; Dr K. Holešovský, Moravaká Galerie v Brně, Brno, Czech Republic. Dr J. A. Bennett, Whipple Museum of the History of Science, Cambridge, and A. V. Simcock, Museum of the History of Science, Oxford, gave their customary support.

[27] See Turner and Dekker, Section 7 and Appendix 2.

[28] See R. Mercier, 'Studies in the Medieval conception of Precession', *Archives International d'Histoire des Sciences*, 26 (1976), 197–220, and idem, 27 (1977), 33–71, especially pp. 58–9.

[29] *De Revolutionibus* was known in Louvain almost directly after its publication; see De Smet, footnote 15.

Table I. Analysis of the star positions on the Brno astrolabe in equatorial co-ordinates (units in degrees).

Modern name	Astrolabe name	Calculated Ptolemy + 20°55′ RA	DEC	Measured Astrolabe RA	DEC	Difference Cal-Measured DRA	DDEC
α Cas	Cassi pect	3·95	54·27	4·07	55·10	−0·12	−0·84
μ And	med: cing: Andr	7·68	36·30	7·13	36·91	0·56	−0·61
ζ Cet	Venter caeti	22·41	−12·17	22·93	−12·25	−0·51	0·08
γ Cet	rictus caeti	35·23	1·91	35·83	1·53	−0·60	0·38
β Per	Medusae caput	40·34	39·93	40·54	39·39	−0·20	0·55
α Tau	Oculus tauri	62·57	15·85	62·64	15·50	−0·07	0·35
α Aur	Hircus	71·43	45·09	71·19	44·54	0·24	0·55
β Ori	Orion: sin: pes	73·46	−9·10	73·47	−9·26	−0·01	0·16
β Aur	Erich: dex hum	81·92	43·32	82·46	43·95	−0·54	−0·63
α Ori★	Orion: dex: hum:	83·19	6·33	83·41	−6·04	−0·22	12·37
α CMa	Canis maior	96·91	−15·87	97·50	−16·06	−0·59	0·19
α CMi	Canicula	109·37	5·99	109·64	5·37	−0·28	0·62
α Hya	lucida hydrae	136·87	−4·92	137·39	−5·16	−0·52	0·25
α Leo	Cor [Leo]	145·65	13·96	146·43	12·99	−0·78	0·97
β UMa	Vrsa maior	157·86	58·27	157·09	58·02	0·76	0·24
α Crt	Fundus crateris	159·14	−16·10	159·63	−16·33	−0·49	0·23
α UMa	Vrsa maior	157·51	63·72	159·78	61·83	−2·26	1·89
δ Leo	dorsum [Leo]	162·30	22·33	161·86	21·21	0·44	1·12
β Leo	cauda [Leo]	171·34	16·64	170·64	15·36	0·70	1·28
γ UMa	Vrsa maior	171·56	55·78	172·79	54·93	−1·23	0·85
δ UMa	Vrsa maior	176·78	59·30	176·12	59·30	0·66	0·00
α Crv	corui rostrum	176·63	−22·27	176·44	−22·48	0·19	0·21
ε UMa	Vrsa maior	187·90	57·62	187·75	57·24	0·15	0·38
ζ UMa	Vrsa maior	195·22	56·95	194·04	55·91	1·19	1·03
α Vir	Spica [Vir]	195·43	−8·77	195·43	−9·81	0·00	1·05
η UMa	Vrsa maior	202·13	51·22	200·91	50·99	1·22	0·23
α Boo	Arcturus	208·96	22·00	208·75	21·36	0·20	0·64
γ Boo	Bootis sin hum	212·54	40·10	213·06	40·06	−0·52	0·04
β Lib	not named	223·21	−7·69	223·04	−8·12	0·17	0·43
α CrB	Coro: sept	228·69	28·51	228·37	28·34	0·33	0·17
δ Oph	Ophi: pal: sin:	237·55	−2·73	236·88	−2·42	0·68	−0·32
α Sco	Cor [Sco]	240·72	−24·85	240·32	−24·40	0·40	−0·45
δ Her	Herc: sin: hu	253·30	25·81	253·64	26·73	−0·35	−0·92
α Oph	Ophi: caput	258·20	13·08	258·16	13·74	0·04	−0·65
β Dra	Ocul dracon:	259·46	52·78	258·52	53·47	0·94	−0·69
τ Oph	Extr: ma: dex Ophiuchi	265·34	−9·09	265·26	−8·90	0·08	−0·19
α Lyr	Lyra	274·95	38·64	275·05	38·82	−0·10	−0·18
α Aql	Aquila	291·64	7·51	291·79	7·86	−0·15	−0·35
α Cyg	cauda cygni	307·05	44·01	306·53	44·79	0·52	−0·78
δ Aqr★	Crus aquary	337·39	−17·56	307·38	−7·02	30·00	−10·54
γ Cap	Cauda capricorni	318·90	−18·22	319·18	−18·29	−0·27	0·07
δ Cap	Cauda capricorni	320·35	−17·61	320·61	−17·78	−0·26	0·17
ε Peg	Pegasi rictus	321·28	7·47	321·37	7·48	−0·08	−0·01
β Peg	Crus Peg	340·58	25·54	340·05	25·13	0·53	0·41
α Peg	Hum pegasi	340·82	13·17	340·53	12·74	0·30	0·43
α And	And: cap:	356·82	27·16	356·47	27·68	0·34	−0·52
γ Peg	ala pegasi extr	357·76	12·68	357·97	12·27	−0·21	0·41
ι Cet	cauda caeti	0·60	−10·45	359·47	−10·37	1·13	−0·08
Mean value without Orion dex hum and Crus aquary						0·03	0·18
Standard deviation without Orion dex hum and Crus aquary						0·63	0·60

★ Star pointer in wrong position, see Section 2.

IX

A TUDOR ASTROLABE BY THOMAS GEMINI AND ITS RELATIONSHIP TO AN ASTROLOGICAL DISC BY GERARD MERCATOR OF 1551

By Gerard L'E Turner, F S A and Koenraad Van Cleempoel

The collection of planispheric astrolabes at the National Maritime Museum, Greenwich, is currently being catalogued. The collection, which is well balanced between Eastern and Western examples, comprises almost sixty instruments ranging in dates from the fourteenth to the seventeenth century, a time when the astrolabe lost its importance in the Latin West. One of the treasures at Greenwich is an astrolabe made in London c 1550 *by Thomas Gemini. It has not come down to us complete; only the throne, limb, and back survive, the rete, plates, alidade, and rule having been discarded. Its importance, however, lies in the engraved astrological information on the back, which is apparently copied from a disc made by Gerard Mercator that is dated* 1551. *This disc has only recently received scholarly attention.*

THE ASTROLABE

The Greenwich astrolabe by Thomas Gemini is no stranger to this Society.[1] Its former owner, George Gabb who sold it to the Museum in 1938, read a brief paper to the Society on what he called 'The Astrological Astrolabe of Queen Elizabeth'.[2] Astrological, because of the unusual set of concentric scales found engraved on the back (fig 1). In his paper Gabb presents Queen Elizabeth as the first possessor of the instrument, through the evidence of the initials E R set either side of the royal arms. If one makes a comparison with three other astrolabes by Gemini, it is soon evident that Gabb's opinion is untenable, and that the Greenwich astrolabe was made for King Edward VI, who ruled from January 1547 to July 1553. The Mercator disc confirms this view.

Gemini made a signed astrolabe in 1552 for Edward VI.[3] Like the one in Greenwich, this astrolabe bears the initials E R set at the sides of the royal arms. But since it is clearly dated 1552, we know it was meant for Edward, and not Elizabeth, who became Queen in November 1558. Gemini also made a signed astrolabe for Queen Elizabeth in 1559, now at Oxford.[4] Elizabeth's ownership is recorded on the round bracket of the throne: *Elizabeth Dei Gratia Angliae, Franciœ et Hibernœ Regina* on one side, with the royal coat of arms and initials E R on the other. In style and overall appearance the Greenwich astrolabe stands closer to the 1552

Fig 1. Back of the astrolabe attributed to Thomas Gemini, *c* 1550. National Maritime Museum, Greenwich, AST0567. IC425. *Photograph*: Copyright National Maritime Museum, Greenwich

Fig 2. Detail of the astrolabe, showing the royal arms of Edward VI and the astrological bands. *Photograph*: Copyright National Maritime Museum, Greenwich

astrolabe than to the one of 1559. This is especially so in the shape of the throne, which consists of two very flat S-shaped brackets set in juxtaposition, completed with contemporary ornamental designs. The 1559 throne is much more elevated and includes a round bracket.

What now refutes Gabb's dating of the Greenwich astrolabe is the information on the back, contained in seventeen concentric circles (fig 2).[5] They can be divided into two groups: for degrees and mundane information, and for astrological purposes. (The scales are numbered starting from the outer edge.)

Scales 1–7: (1) degree scale numbered 4×0–$90°$; (2) degree scale units for numbered scales no 1 and 3; (3) degree scale numbered 12×0–$30°$, for each Sign of the Zodiac; (4) name and sigil of the Sign; (5) degree scale units for numbered scale no. 6; (6) numbering of the days of the calendar month; (7) names of the calendar months.

Scales 8–17: (8) The lunar mansions, numbered from 1–28 and accompanied by information on their properties, eg, moist (*humida*), fortunate (*fortunata*), etc (9) The properties of particular degrees, with the following abbreviations: *p* = pitted (the matter stands in doubt), *a* = lame (some party in the affair is maimed), *f* = increasing fortune. Also given are the degrees of exaltation (*gaudium*) and fall (*casus*) of the planets (including the head and tail of the dragon). (10) The division into tenebrous (*teneb.*), light (*lucida*) and void degree arcs of a Sign. (11) The division into male and female degree arcs of a Sign. (12) The division into decans, or every tenth sector of the Zodiac, each ruled by one of the planets. (13) The unequal division of each Sign into five sectors, each ruled by a planet (excluding the Sun and the Moon). (14) The triangle to which one particular Sign belongs, and which planet rules each part of that triangle (the triangles are four groupings of three Signs). (15) The planet that is in exaltation in a Sign, including the head and tail of the dragon. (16) The planet which is lord over a Sign or 'house' (*domus*) as well as the planet that is in detriment in that Sign (*occasus*). (17) This scale is marked *Termini secundum Cardani*, ie, the terms according to Girolamo Cardano (1501–76). It consists of an unequal division of each sign into seven parts, each of which is ruled by one of the planets.

As yet no thorough study on these scales has been published, and no suggestion has ever been made towards a possible source of inspiration. Hitherto unnoticed is that the back of the Greenwich astrolabe bears material almost identical to that on the front plate of an astrological disc published by Gerard Mercator in 1551, now preserved in Basel.[6] The similarity between the two instruments is striking: the arrangement and setting of the scales are identical, as are the spelling and the italic letter font. The written information is copied almost verbatim. Because of this strong similarity, a dating of the Greenwich astrolabe near 1551 seems justified.

THE ASTROLOGICAL DISC

Mercator's astrological disc is made of wood, with paper pasted on each side. On the front are the astrological scales printed from a copperplate engraving (figs 3 and 4), and on the back are the printed instructions for the use of the instrument (fig 5). Passing over the front are nine rules, and at the centre is a small volvelle showing the aspects. This text has recently been edited by Steven Vanden Broecke.[7] He explains how the instrument acts as an astrological complement to Mercator's celestial globe, which he issued in April 1551, one month before the publication of the astrological disc.[8] The celestial globe carries information on the nature of the

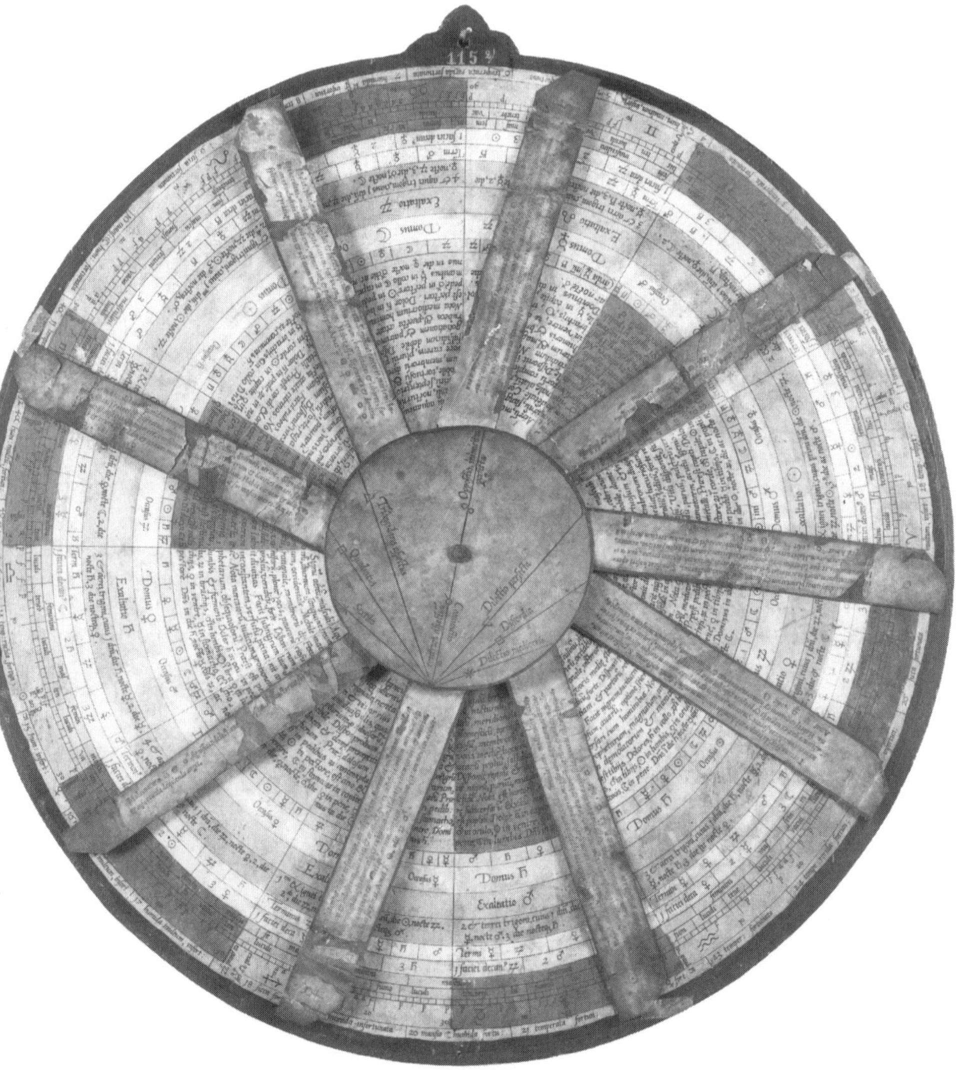

Fig 3. The front of the astrological disc, signed and dated by Gerard Mercator, 1551. Historisches Museum, Basel, inv 1876–20. IC3201. *Photograph*: M Babey. Copyright Historisches Museum Basel

fixed stars, while the horizon circle is divided into the twelve mundane houses, and within each house a brief description of its qualities. These elements constitute only part of the common astrological parameters, the rest of which can be found on the disc. It is, therefore, likely that it was published simultaneously with the globe, as the instruction sheet on the back constantly refers to a combined use of globe and disc.

Fig 4. Detail of the astrological disc, showing the astrological bands. *Photograph*: P Portner. Copyright Historisches Museum Basel

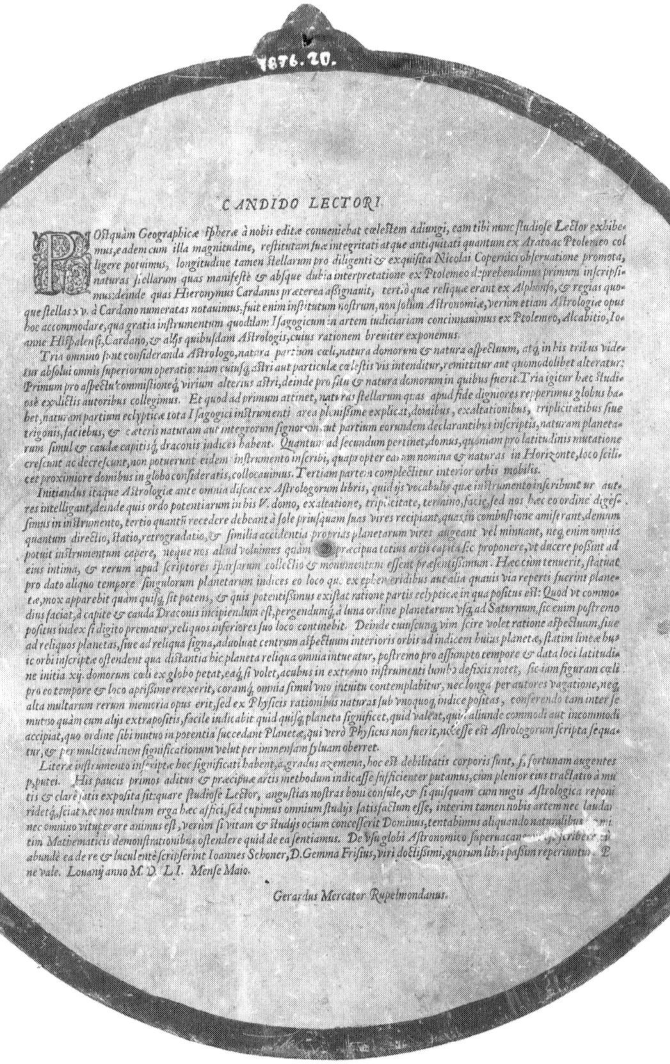

Fig 5. Back of the astrological disc, showing the printed instructions for using the disc. Signed: 'Louanÿ anno MDLI mense Maio. Gerardus Mercator Rupelmondanus'. *Photograph*: M Babey. Copyright Historisches Museum Basel

The design of the disc stems from Mercator's own research in cosmology. Prior to his celestial globe and astrological disc of 1551, Mercator had already established himself as a fine instrument maker of brass instruments.[9] This aspect of Mercator's life deserves more attention, but there are indications that he was very much engaged in making and writing about instruments from *c* 1540 onwards until his departure to Duisburg in 1552. Some of the instruments he made were also accompanied by manuals, usually handwritten. The astrological disc is no exception. The back presents practical information on use, as well as a brief account of the consulted sources. Vanden Broecke observed some discrepancy between the sources mentioned on the back, and the ones that were actually used to design the front of the disc. On the back Mercator puts an emphasis on cosmological astrology, whereas the front presents the sources for a judicially orientated astrology. That is to say: the back is more critical and concerned with providing astrology with a solid physical basis, whereas the front presents a more traditional scheme for setting and interpreting horoscopes that are intended for predictions.

THE ROLE OF JOHN DEE

There are good reasons to believe that John Dee is the 'intellectual co-author' of the astrological disc. The English polymath visited Louvain twice: from May 1547 for some months, and from midsummer 1548 until 15 July 1550, then moving on to Paris.[10] It was during the second stay that he spent much time with Mercator, a period that coincides with the creation of Mercator's celestial globe and the astrological disc. There is no direct evidence of the contents of their debates, but an important indirect source is John Dee, *Propaedeumata Aphoristica* (1558), which he dedicated to Mercator, ending: 'and by far my dearest friend'.[11] Dee announces in the preface how he will address the problems they discussed in Louvain: providing astrology with a firm physical and epistomological basis. The contents of the treatise, together with other elements, strongly show that the interests and ideas of Dee and Mercator were developing along parallel lines.

Dee writes how Mercator made him an armillary sphere incorporating a mechanism for demonstrating trepidation, an erroneous concept that is associated with the precession of the equinoxes.[12] Instruments of this type are extremely rare. Only six examples have been recorded, but none are by Mercator.[13] However, the six surviving examples all come from the Louvain workshops, and the Mercator sphere was possibly their model as it would easily predate them by at least ten years. A slow oscillation of the ecliptic was introduced by the ninth-century Islamic astronomer, Thabit ibn Qurrah, to explain a supposed variation in the precession of the equinoxes. This effect was called trepidation, resulting from an error in Ptolemy's determination of precession. Mercator, however, must have found the theory interesting enough to construct a three-dimensional model of it. That this happened around 1550 is no coincidence. The fact that Mercator made such a sphere for Dee echoes their mutual interest in the history of astronomy and astrology.[14]

John Dee may well form the link between the astrolabe of Thomas Gemini at Greenwich and the astrological disc by Mercator. That the astrolabe was copied from the disc and not vice versa, is most likely. There are at least two other examples of Mercator's influence on Gemini. His astrolabe of 1559, for example, has a rete shape which is close to that of an earlier astrolabe by Mercator.[15] A sinical quadrant by Gemini, dated 1551 and inscribed *Edwardus Rex*, in the British Museum (perhaps a companion piece to the Greenwich astrolabe), was probably copied from a lost instrument by Mercator of which the handwritten manual is still preserved.[16] But Gemini not only imitated scientific instruments, he also published in 1545 a copy of Andreas Vesalius, *De humani corporis fabrica* (1543), illustrated with his own engravings copied from the woodcuts in the book of the Fleming, Vesalius.

The fact that Gemini modelled the back of the Greenwich astrolabe after Mercator's disc is thus perfectly in line with the rest of his work, and method of working. The question then arises: how did Gemini get access to the design of the disc? A possibility is that it was passed on to him through Dee, who could have brought it back with him after his extended stay on the Continent, where he was in Louvain from midsummer 1548 for two years. The two polymaths also conducted a lively correspondence after Dee's departure.

We do not know whether Gemini was in direct contact with Mercator. The latter was a great cosmographer with unsurpassed talents as an instrument maker, the former a fine engraver and draughtsman, but a less inventive instrument maker, who copied original work from prominent academics in Louvain.

ACKNOWLEDGEMENTS

We are grateful to the Director of the National Maritime Museum, Greenwich, for access to the Gemini astrolabe, and to the Director of the Historisches Museum, Basel, for providing photographs of the Mercator disc.

NOTES

1. Brass; diameter 352mm. Inv no AST0567; old no A.36/37–92C. IC 425.
2. Gabb 1937, 101–3. Gabb acquired the instrument in 1927 and published a brief note in *The Illustrated London News* on 14 May 1927. He thought, without foundation, that Humfrey Cole was the maker, and this caused R T Gunther to insert a note dated 17 June 1927 into the proofs of his paper on Cole presented to the Society a year earlier. See Gunther 1927, 302.
3. Brussels, Observatoire Royale de Belgique, IC 450. All seven instruments of Gemini are discussed in Turner 2000, 95–111.
4. Found in 1936 at the University Observatory, Oxford. Paper read to the Society on 12 November 1936 by R T Gunther (1937, 65–72).
5. The scales were briefly described by Gunther (1927, 302). Their significance for an astrologer is explained in Eade 1984, 59–88.
6. Wood and paper; diameter 380mm. Historisches Museum, Basel, inv no 1876.20. IC 3201.
7. Broecke 2001. We thank Vanden Broecke for useful discussions.
8. The signature is on the back: *Louanij anno m.d.li. Mense Maio. Gerardus Mercator Rupelmondanus.*
9. An astrolabe of *c* 1550 with the monogram GMR was first recognized as by Mercator by G L'E Turner, who had already attributed two unsigned astrolabe to Mercator. There are now three known astrolabes by Mercator. These instruments, as well as Mercator's influence on other Louvain makers, are discussed by K Van Cleempoel. See Turner and Dekker 1993; Turner 1994; Cleempoel 2000.
10. Crossley 1851, 5–7. After the first visit, in 1547, Dee brought back an astronomer's staff and astronomical ring both devised by Gemma Frisius, and two great globes made by Mercator. Initially left at Trinity College, Cambridge, they were damaged by the rioters who entered Dee's house at Mortlake in 1588 (ibid 28). Dee assessed the loss of his instruments at £30, comprising a 5ft quadrant, 10ft astronomer's staff, geographical and terrestrial globes by Mercator, and an armillary sphere made for Dee by Mercator that showed the theory of the eighth, ninth, and tenth spheres (idem).
11. Shumaker 1978.
12. The system represented by the ninth sphere, trepidation, and its historical context is dealt with in Dekker 1999, 'The trepidation armillary sphere of Gualterus Arsenius', 96–9.
13. One is at the National Maritime Museum, Grenwich (inv no AST 0618), made by Arsenius, 1568. For a complete listing, see Dekker 1999, 149.
14. For the link between instruments, astronomy, and astrology, see Dunn 1994.
15. The rete of the 1559 astrolabe is illustrated in Gunther 1937, pl XII. The astrolabe of Mercator in question is now preserved at Brno. For illustrations, see Turner 1994, 340, 345.

16. Ward 1981, 56–7, reg no 1858, 8–21.1. The holograph of Mercator is not signed, but the handwriting is easily recognizable. It is now preserved in the archives of the Monastery of San Lorenzo El Escorial (Spain), MS &-III-8, fols 237–64. An associated holograph is signed by Mercator and dated 1551.

BIBLIOGRAPHY

Broecke, S Vanden 2001. 'Dee, Mercator, and Louvain instrument making: an uninscribed astrological disc by Gerard Mercator (1551)', *Annals of Science*, 58, 219–40

Cleempoel, K Van 2000. *A Catalogue Raisonné of Scientific Instruments Produced in Louvain between 1550 and 1600, with Introduction*, Louvain

Crossley, J (ed) 1851. 'The compendious Rehearsall of John Dee (Anno 1592)', in *Autobiographical Tracts of Dr. John Dee, Warden of the College of Manchester*, 5–7, Chetham Soc, 24, Manchester

Dekker, E 1999. 'The trepidation armillary sphere of Gualterus Arsenius', in *Globes at Greenwich* (E Dekker), 96–9, Oxford

Dunn, R 1994. 'The true place of astrology among the mathematical arts of late Tudor England', *Annals of Science*, 51, 151–63

Eade, J C 1984. *The Forgotten Sky: A Guide to Astrology in English Literature*, Oxford

Gabb, G H 1937. 'The astrological astrolabe of Queen Elizabeth', *Archaeologia*, 86, 101–3

Gunther, R T 1927. 'The Great Astrolabe and other scientific instruments of Humphrey Cole', *Archaeologia*, 76, 273–317

Gunther, R T 1937. 'The Astrolabe of Queen Elizabeth', *Archaeologia*, 86, 65–72

Shumaker, W (ed and transl) 1978. *John Dee on Astronomy*: Propaedeumata Aphoristica *(1558 and 1568)*, Berkeley

Turner, G L'E 1994. 'The three astrolabes of Gerard Mercator', *Annals of Science*, 51, 329–53

Turner, G L'E 2000. *Elizabethan Instrument Makers: The Origins of the London Trade in Precision Instrument Making*, Oxford

Turner, G L'E and Dekker, E 1993. 'An astrolabe attributed to Gerard Mercator, *c* 1570', *Annals of Science*, 50, 403–43

Ward, F A B 1981. *A Catalogue of European Scientific Instruments in the Department of Medieval and Later Antiquities of the British Museum*, London

X

An Astrolabe for Alessandro Farnese, Duke of Parma, by Erasmus Habermel

Erasmus Habermel (*c.* 1550–1606) held the post of instrument maker to the Holy Roman Emperor Rudolf II at his opulent court in Prague. Rudolf was a collector on a large scale, and attracted into his circle scholars, natural philosophers, artists, and craftsmen from all over Europe. Johannes Kepler compiled there his new and accurate astronomical tables, naming them after his patron. In the Age of Discovery, the instruments of science were in demand both by navigators and surveyors, and by those who, like the Emperor, were interested in the natural world, and the revolutionary new theories about the universe. Therefore it is not surprising that his court instrument maker was one of the finest craftsmen of the time.

Little is known about Habermel's life, but he was definitely working in Prague from 1585 until he died in 1606. The last two decades of the sixteenth century were a busy period for him and over 100 instruments from his workshop survive, the majority in museums. The present astrolabe can be regarded as his masterpiece. Nearly 15 inches in diameter, twelve-sided, with three plates, the instrument, made from gilded copper, is intricately inscribed and decorated and is signed *Erasmus Habermel fe:*, but not dated (Figs 1–3). However, a close date for the instrument can be arrived at from a study of the unusual and detailed perpetual calendar that is incorporated in the astrolabe. This is because a companion piece exists in the form of a calendar on a single plate, also engraved by Habermel, in a museum in Vienna. This plate has an inscription that dates it to 1587, and gives the name of its originator, a mathematician called Hermann Bulderus. Clearly, the plate came first, and was followed quite shortly by its twin, incorporated with this astrolabe. Therefore, the astrolabe must have been made after 1587.

More dating information comes from the coat of arms on the astrolabe, which is surrounded by the insignia of a famous Order (Fig. 5). The arms are those of Alessandro Farnese, Duke of Parma, who was awarded the Order of the Golden Fleece by King Philip II of Spain in 1585. Farnese was painted in 1590 by Jean Baptiste de Saive, wearing the chain of the Order, and the portrait is now in the national gallery at Parma. He died only two years later at the early age of forty-seven. So the astrolabe was either bought or presented to him between 1587 and

1592. Having all the signs of a princely gift, it may well have been commissioned by Philip of Spain to mark Farnese's success as governor of Flanders in the late 1580s.

The historical associations of the astrolabe are of great interest, but even more remarkable is the craftsmanship it embodies, and the amount of varied and detailed information inscribed on it. The back, in particular is a *tour de force*, with the fine engraving of the perpetual calendar, the vividly depicted Signs of the Zodiac, and the richly decorated shadow square. A detailed description reveals the full extent of the learning, art, and craft that went into its making.

Description of the Astrolabe

Throne The throne is attached to the top of the instrument by two screws, the heads of which are prominent. The front is decorated with strapwork and holds a magnetic compass (glass and needle missing). The cardinal points are engraved: *Septentrio / Oriens / Meridies / Occidens.* An arrow engraved 10 degrees East of North indicates the position of the magnetic pole at the period of manufacture (Fig. 4). The back of the throne is engraved with the coat of arms of Alessandro Farnese: tierced in pale, 1. per fess in chief Farnese, or, six fleurs-de-lis, azure, in base Austria impaling Burgundy-ancient, 2. gules, the Papal banner, the lance in pale or surmounted by the Papal keys in saltire, 3. per fes in chief Austria impaling Burgundy-ancient, in base Farnese. The crest is a unicorn's head; the whole surrounded by the chain of the Order of the Golden Fleece (Fig. 5). The shackle has two movements, and is surmounted by the ring.

Limb The limb is unusual in being twelve-sided, a feature found on one other Habermel astrolabe. It measures: outer diameter across the flats 370 mm, width 20–26 mm, thickness 8 mm. It is attached to the back plate by solder and twelve rivets. The face is divided into 360°, alternate degree divisions being shaded, labelled in the four quarters 10° to 90° (90° at top and bottom). The outer edge of the limb is divided in 24 hours, and is marked with the hours I to XII, and thence to XII, each hour subdivided to fifteen minutes, with alternate divisions shaded. At the top, on the inside, is a small aperture to locate the lugs on the plates to prevent rotation.

Mater The mater, or interior, of the astrolabe is engraved with a QVADRATVM NAVTICVM (Fig. 6). The four cardinal points are engraved in majuscules with elaborate swashes, a feature common with the instruments made during the sixteenth century in Louvain by Mercator, Arsenius, and others. Within the rectangular coordinate grid, whose sides are divided to 90° by single units, is the circular wind rose. This is framed by a circle with eight directions in

Italian: TRAMONTANA, Grecho, LEVANTE, Syrocho, MEZODI, Lebechio, PONENTE, Magistralis. The 32 pointers are inscribed with the name of the wind.

Back At the edge are divisions into degrees, as on the front. There follow the months in the year divided into 365 days. By each date is the day in the month, the Dominical Letter, and the martyr for the day. A complete list is given below. Next is the Zodiac circle, each Sign divided into 30°, marked every ten. The pictures of the constellations representing the Signs are engraved with considerable care (Fig. 8). The Zodiac, or Path of the Creatures, is the name for a celestial band divided into twelve sections whose course is traversed annually by the Sun through constellations bearing the names of living things. An apparent anomaly is Libra, but the scale pans were anciently the claws of the scorpion. Thought to have mystical and astrological associations, each ruled over one of the twelve months of the year. The beginning of Spring is denoted when the Sun enters the zodiacal sign of Aries, and on this astrolabe the date is 20.7 March, showing the equinox according to the reformed calendar of 1582.

Within the Zodiac are two mathematical features. In the lower half of the central portion is a shadow square divided in units and marked every ten units to 60 (the 45° point), a newer and more refined way of division. The same scale is also labelled in units five times larger, to 12 (the 45° point), which is the older method of division. The sides are marked in elaborate majuscules: VMBRA RECTA / VMBRA VERSA. These scales are employed for deducing the heights of buildings from shadow lengths. The space within the rectangle is embellished in a florid manner (Fig. 7).

The upper half contains a diagram for conversions between equal and unequal hours (also called planetary or temporary hours). The semicircle is divided in twelve hours by quarters (Horæ in Æquales); within this arc, the lines partition the day into twelve hours both before and after noon (HORÆ ÆQUALES) (Fig. 8).

Rete This is cut for 48 stars, and these are listed in the table below. The design is that familiar from the Louvain workshops with regard to the 'tulip' pattern at the centre (Fig. 9). The arrangement of the strapwork is, in fact, more reminiscent of the retes on the astrolabes of Gerard Mercator (Louvain, to 1552; Duisburg, after 1552) and Thomas Gemini (London, 1550s). The back of the Habermel rete shows no indication of laying-out marks. It is smoothly finished leaving a tarnished copper surface. The front is gilded, and no effort was made to treat the back, which is never seen when in use. There is an error in the division into degrees of the ecliptic circle. They are in Right Ascension and not in Mediation. This is proved by taking a line from the centre of the astrolabe to the divisions around the outer edge, when it will be found that the line cuts the ecliptic at the same degree position as on the edge. Astrolabe makers sometimes fall into this error, and it has been pointed out by Henri Michel that all Habermel's astrolabes are incorrect in this respect.

The star names engraved on the astrolabe are followed by their magnitude (brightness to the naked eye) from I (brightest) to 4.

1 *Sinister Cubitus Andro: 3*
2 *Venter Ceti 3*
3 *Extrema mandibula Ceti 3*
4 *Caput Medusæ 2*
5 *Dextrum latis Persei 2*
6 *Pectus Tauri 3*
7 *Oculus Tauri I*
8 *Hædi*
9 *Hircus I*
10 *Sinister pes Orionis I*
11 *Prior Cinguli Orionis 2*
12 *Dexter humeris Orionis I*
13 *Canis maior sirius I*
14 *Procÿon I*
15 *Extrema Nauis 3*
16 *Caput Hÿdræ*
17 *Lucida Hÿdræ 2*
18 *Cor Leonis I*
19 *Fundus Crateris 4*
20 *Cauda Leonis I*
21 *Rostrum Corui 3*
22 *Coma Berenices*
23 *Ala Dextra Corui 3*
24 *Preuindemiatrix 3*
25 *Spica Virginis I*
26 *Extrema Caudæ Vrsa maioris*
27 *Arcturus I*
28 *Sinister humeris Bootis 3*
29 *Lanx Borealis 2*
30 *Lucens in Corona 2*
31 *Media Frontis Scorpii 3*
32 *Dextrum Genu Ophiuchi 3*
33 *Caput Engonasi*
34 *Caput Ophiuchi 3*
35 *Lucida lyræ I*
36 *Aquila Vultur Volans 2*
37 *Boreale Cornu* [Sco] *3*
38 *Cauda Delphini 3*
39 *Cauda Cigni 2*
40 *Dexter hu: Cephi*
41 *Humerus sinister 3*
42 *Cauda* [Sco] *precedens*
43 *Dexter humerus* [Aqr] *3*
44 *Australis tibæ Dextræ* [Aqr] *3*
45 *Australis sequens in Flexu* [Aqr] *4*
46 *Vmbilcus Pegasi 2*
47 *Pectus Cassiopea 3*
48 *Extrema borea Cauda Ceti*

Plates There are three plates engraved on both sides for various latitudes and other features. One has the stereographic projections for latitudes 45° and 48° (Figs 10–11), another the same for 49° and 51°, while the third plate has a tablet of horizons on one side and a perpetual calendar on the other. The tablet allows the user to determine his latitude from the time of the Sun's rising or setting, or conversely. The calendar was devised by Hermann Bulderus of Prague, and an almanac in brass with the calendrical data engraved by Habermel was given in 1587 by Bulderus to his patron; it is now in Vienna. The perpetual calendar on the astrolabe is described fully below.

The stereographic projections are most expertly executed. Azimuths are cut every 2°, and altitudes every 1° (Fig. 10). The other dividing lines are the Tropic of Cancer, Equator, Houses of Heaven, and unequal hours. For each latitude, there is engraved a list of towns where that particular plate could be used. These locations, as inscribed, are:

a) Pro eleuatione poli 45 qualis est Geneuæ Veronæ Leoni Venetiæ Mantuæ Burdigalæ Rauennæ Belligradi Luodum

b) Pro eleuatione poli 48 qualis est Viennæ Linzæ Monachÿ Augustæ Ulmæ Salisburgæ Basilæ Parisÿ (Fig. 11)

c) Pro eleuatione poli 49 qualis est Noribergæ Strasburgi Trebonys Ratisbonæ Crumnouiæ Spiræ

d) Pro eleuatione poli 51 qualis est Vlatislaniæ Dresdæ Antuerpiæ Coloniæ Bruxellæ Lipsæ Neuburg Londini

Alidade The alidade (length 365 mm, width 12 mm, thickness 4 mm), used on the back of the astrolabe, is counter-changed, and is marked out with scales for use with the equal-unequal hour conversion diagram. On each arm is a scale of twelve hours, divided to 7.5 minute intervals. They are named: *Horæ Ortus* and *Horæ Occasus* (i.e., rising and setting hours). The notch and pinhole sighting vanes are hinged to lie flat when not in use.

Rule The rule (length 365 mm, width 12 mm, thickness 4 mm), which is used on the front of the instrument, is counter-changed at the middle. It is divided on one arm with a scale marked in degrees ranging from 75° North to 23½° South. The divisions on the bevelled edge are to one degree. North from the equatorial point (0°) is labelled: *Declinatio Septentrionalis*; the Southern part is labelled: *Declinatio Meridionalis*.

Pin The pin is integral with the centre of the alidade. It terminates in a thread, onto which screws a small disc (diameter 16 mm). This disc can be tightened by the insertion of a two-pronged key (not present) into the two tiny holes on a diameter. An offset hole is for a green silk cord (not modern) that is attached to the top of the throne by passing through a hole, with a bead (modern) on the opposite side to keep the place. Such a method for preventing the loss of this important item is not recorded elsewhere.

The Perpetual Calendar

The perpetual calendar was, and still is, a very popular reference work. It finds its most continuous use as a record of the feast days and saints' days of the Roman Church, and provides the days of Lent, Easter, and Advent which are not fixed, but change from year to year following the vestiges of a Lunar year. The Moon provided light for night-time journeys, so dates around when it was full were important to know, and conversely, the times of the New Moon were to be avoided. The dates of the full Moon during the year can be readily worked out given the age of the Moon on New Year's day, and this figure is called the Epact. Other information that could be provided in an Almanac, which is a written, or from the 1470s, a printed work, includes the time of Sunrise and Sunset, the length of the daylight through the year, the length of twilight. The eclipses of the Sun and Moon could be included in rather sophisticated almanacs. One such was that compiled in 1388 by Nicholas of Lynn, which he calculated to cover a period of 76 years specifically for the latitude and longitude of the City of Oxford.

The most renowned astronomer of the fifteenth century was Regiomontanus (1436–1476), who founded a printing works at Nuremberg expressly to publish almanacs that could be uniform, accurate, and cheap: his *Der deutscher Kalender* appeared in 1474. He, too, included a section on eclipses. Not only are these a wonder, but a night journey during an eclipse of the Moon is as undesirable as when the Moon is only slim.

The Julian year of 365¼ days was not sufficiently accurate to prevent the recurrence of a fixed astronomical event such as the Winter Solstice from slipping later and later by a fraction of time. Eventually the error built up to several days, which meant that Christmas was no longer close to the Solstice. The church called for reform of the calendar at the Council of Constance in 1414, but nothing came of it. Eventually the correction was made by a Papal Bull of Gregory XIII, whereby ten days were omitted from the calendar after 4 October 1582. Sunday 17 October was then regarded as the 18th after Pentecost, which would have agreed with an Easter Day on 25 April, and Advent on 28 November.

The back of the astrolabe has a table of 365 feast days and saints' days. In addition to the back, the circular plate provided with this Habermel astrolabe is a perpetual calendar that incorporates many of the features named above (Fig. 12). These features are:

a) Around the edge, are Easter calendars in both the Julian calendar and the Gregorian calendar. The first is labelled: STILO VETERE, and runs from 1550 to 1700, and listed are the year, Dominical Letter, and Easter Day. For example: 1599 G 8A (Dominical Letter G, and 8 April Easter). The second is labelled: STILO NOVT CALENDARII, and lists, as engraved: *Annus Domini, Dies Paschalis, Adventus Domini, Littera Dominicalis, Epactæ Lunaris*. It begins with 1582 and runs for over 300 year to 1900. The change to the new calendar in 1582 confused either the designer of the calendar or its engraver. The data for 1582 are: 18 April for Easter, 28 November Advent, C Dominical letter, and 26 the Epact. But the calendar change occurred in October, so Easter was actually under the old calendar on 15 April, whereas Advent was, in the new, on 28 November. That tables such as this are a test of the concentration of the engraver, who can only proceed slowly over weeks, is shown by a few other errors, some of little consequence as in the spelling of a name, others discovered and corrected. An example of an uncorrected error has just been given, and another is for the year 1649 in the Old Style, where Easter is 21 March, whereas it was actually celebrated on 25 March.

b) At the top are two panels to list in rows and columns the 12 hours of the night (TABVLA PLANETARVM NOCTIS) and day (TABVLA PLANETARVM DIEI) and the seven days of the week, placing in each square the symbol of the planet governing the hour. It is from the planets governing the first hour of each day that we derive our names for each day in the week, and their sequence.

c) Prominent in the middle of the plate is the table of eclipses of the Sun and Moon, entitled: TABVLA OBSCVRATIONIS SOLIS ET LVNAE. This runs from 1582 to 1620, and lists the year, month, day and time in hours and minutes of eclipses. Also given where possible are the beginnings and endings of the event and the duration. The last, right-hand, column gives the location in the Ecliptic for the eclipse, with the sign of the Zodiac, the degrees in the sign, and a symbol denoting the nodes of the Moon. These are either ascending or descending, and were known as the head and tail of the dragon. So, with a Lunar eclipse, it was said that the Moon had been swallowed by the dragon.

Such an elaborate perpetual calendar, which is a *tour de force*, could be expected to have some inspired mind behind it. And this is indeed so, because, incredibly, this astrolabe can be compared with a gilded brass plate in the Österreichisches Museum für angewandte Kunst, Vienna, that is the calendar alone, never having been part of an astrolabe. It must surely have been the precursor to the astrolabe plate, and the inventor is revealed. The perpetual calendar in Vienna has one side exactly the same as the astrolabe plate, while the reverse has the year calendar on Church events and a volvelle. It bears the coat of arms of Herzog Wilhelm von Rosenberg (died 31 August 1592) from Wittingau. The device is signed by the maker, bearing the inscription: *Geordnet und gerechnet durch Hermannum Bulderum von Ossenbrugh und durch Erasmum Habermel zugerichtet 1587*. Hermann Bulderus was a doctor and mathematician, whose studies in Prague (presumably at the Charles University) had been supported by von Rosenberg. A letter preserved in the archives at Wittingau, written by Bulderus on 15 October 1587 at Prague to the Herzog, offers the perpetual calendar as a gift to acknowledge the generosity he had received to enable him to study at Prague. In the letter he says how he had thought long to present calendrical information so that it might be read easily and rapidly, especially at a time of the reformed calendar. He says he had gone to the engraver daily for seventeen weeks so that nothing would be overlooked. This is indeed a rare glimpse of the time a highly skilled engraver would require to complete the task which, for the astrolabe, would involve only the calendar plate and the back of the astrolabe. The remainder would have taken nearly as much time again: let us say at least six months in all.

The Viennese plate tells us who designed the calendar and when the first one was made. The second, with the astrolabe, can hardly have been made before 1588. It is signed by Habermel, and was made for or presented to the Duke of Parma, who died of wounds in 1592. We may, therefore, accept *c.* 1590 as close to the astrolabe's actual date of manufacture.

The Church Calendar

IANVARIVS

1	A	NEWIAR
2	B	*Casparus*
3	C	*Melchior*
4	D	*Balthasar*
5	E	*Simeon bischoff*
6	F	H.3 KONIG
7	G	*Isidorus bischoff*
8	A	*Erhardus bischoff*
9	B	*Iulianus*
10	C	*Paulus einsidler*
11	D	*Hÿginus*
12	E	*Nicanor*
13	F	*Hilarius*
14	G	*Felix Priester*
15	A	*Maurus abt*
16	B	*Marcellus*
17	C	*Anthonius*
18	D	*Prisca iungf.*
19	E	*Martha*
20	F	*Fabian Sebastian*
21	G	*Agnes iungf:*
22	A	*Vincentius*
23	B*	*Emerentia*
24	C*	*Thimotheus eins:*
25	D	PAVLI BEKER
26	E	*Policarpus eins:*
27	F	*Iohannes Chrys:*
28	G	*Carolus magnus*
29	A	*Valerius*
30	B	*Adelgundis*
31	C	*Ignatius bischoff*

FEBRVARIVS

1	D	*Brigitta*
2	E	LIECHT MES
3	F	*Blasius*
4	G	*Veronica*
5	A	*Agatha*
6	B	*Dorathea*
7	C	*Theodosia*
8	D	*Helena*
9	E	*Apolonia*
10	F	*Scolastica iungf:*
11	G	*Eufusina*
12	A	*Ensalia*
13	B	*Agabus*
14	C	*Valentinus*
15	D	*Faustinus*
16	E	*Iuliana iungf:*
17	F	*Simeon bischoff*
18	G	*Constantia*
19	A	*Susanna*
20	B	*Eucharius*
21	C	*Hilarius*

* In 23, 24 January, the letters B, C have been stamped over A, B which have been rubbed down.

22 D PETER STVL
23 E *Tertullianus*
24 F MATTHIAS A:
25 G *Victor*
26 A *Alexander*
27 B *Nestor*
28 C *Romanus abt*

MARTIVS

1 D *Albinus*
2 E *Simplicus bapst*
3 F *Kunigund*
4 G *Hadrianus*
5 A *Fridericus*
6 B *Godfridus*
7 C *Thomas ag*
8 D *Perpetua Felicitas*
9 E *Cyrillus*
10 F *7 Bruder*
11 G *40000 Marterer*
12 A GREGORIVS
13 B *Rosina*
14 C *Zacharias*
15 D *Longinus*
16 E *Cÿriacus*
17 F *Gertrudis*
18 G *Anshelmus*
19 A *Ioseph*
20 B *Gordius*
21 C *Benedictus abt*
22 D *Pontanus***
23 E *Theodorus*
24 F *Agepitus*
25 G MARIE VERK
26 A *Castalius***
27 B *Rupertus*
28 C *Gedeon*
29 D *Quirinus*
30 E *Guido*
31 F *Abel*

APRILIS

1 G *Hugo bischoff*
2 A *Theodosius*
3 B *Inndcontius*
4 C *Ambrosius*
5 D *Ireneus*
6 E *Celestinus*
7 F *Thimotheus*
8 G *Calixtus*
9 A *Maria egipti*
10 B *Ezechiel*
11 C *Bohuslaus*
12 D *Iulius*
13 E *Euphemia*
14 F *Tibertius*
15 G *Olympias*
16 A *Calixtus*
17 B *Rudolphus*
18 C *Valerianus*
19 D *Hermogenes*
20 E *Florentius*
21 F *Sotherus*
22 G *Caius*
23 A GEORGIVS BI:
24 B *Adelbertus bischoff*
25 C MARCVS EVA
26 D *Cletussio*
27 E *Anastasius*
28 F *Vitalis Marterer*
29 G *Petrus*
30 A *Anacletus*

MAIVS

1 B PHILIPI IACO
2 C *Sigismundus*

** In 22 to 26 March, errors in placement have been rubbed down and the words re-engraved in the correct positions.

3 D +ERFIDVNG
4 E *Gothartus*
5 F *Florianus*
6 G *Iohan von der pfo:*
7 A *Damiscilla*
8 B *Stanislaus*
9 C *Iob*
10 D *Ephimachus*
11 E *Gangolphus*
12 F *Pangratius*
13 G *Seruatius*
14 A *Boniuatius*
15 B *Sophia*
16 C *Peregrinus*
17 D *Torpetus*
18 E *Patentia*
19 F *Sara*
20 G *Prudentia*
21 A *Constantinus*
22 B *Helena künigin*
23 C *Desiderius*
24 D *Dominicus*
25 E VRBANVS
26 F *Eleuterius*
27 G *Iohan bapst*
28 A *Wilhelmus*
29 B *Maximinus*
30 C *Felix bapst*
31 D *Petronella*

IVNIVS
1 E *Nicomedes*
2 F *Marcell*
3 G *Erasmus*
4 A *Mornica*
5 B *Bonifatius*
6 C *Phillipus diac*
7 D *Sigfridius*
8 E MEDARDVS
9 F *Primus felic:*
10 G *Onopheius*
11 A *Barnabas*
12 B *Basilia Ciri*
13 C *Anthonius bei:*
14 D *Basilius magnus*
15 E VITVS
16 F *Iustina*
17 G *Valcmarus*
18 A *Marcellian· ma:*
19 B *Geruasius Prat:*
20 C *Sylner bapst*
21 D *Albanus*
22 E *Vigilia*
23 F *Achatius*
24 G IOHANNE TAVF
25 A *Eulogius*
26 B *Ioan Paul*
27 C *7 Schlaffer*
28 D *Leo bapst*
29 E PETRI PAVLI
30 F PAVLI GEDECH

IVLIVS
1 G *Theobaldus*
2 A MARIA HEI:
3 B *Procopius*
4 C *Vlricus*
5 D *Demetrius*
6 E *Esaias*
7 F *Willibaldus*
8 G *Kilianus*
9 A *Cicillus*
10 B *7 Brueder*
11 C *Pins bapst*
12 D *Heinericus*
13 E MAGARETA
14 F *Bona ventura*
15 G APOSTEL TIEV
16 A *Euchstachius*
17 B *Alexus*
18 C *Arnolphus*
19 D *Ruffinus*

20	E	*Ioseph*
21	F	*Braxatis*
22	G	MARIA MAGDA
23	A	*Apollinaris*
24	B	*Cristina*
25	C	IACOB APOS
26	D	*Anna*
27	E	*Christophorus*
28	F	*Panthaleon*
29	G	*Martha*
30	A	*Abdon*
31	B	*Germanus*

AVGVSTVS

1	C	PETER KET
2	D	*Steffanus*
3	E	*Steffanus erfin:*
4	F	*Istinus*
5	G	*Oswald*
6	A	PERKLER CH
7	B	*Donatus*
8	C	*Cÿriacus*
9	D	*Romanus*
10	E	LAVRENTIVS
11	F	*Tiburtius*
12	G	*Clara iungf:*
13	A	*Hippolitus*
14	B	*Eusebius*
15	C	MARIE HIM
16	D	*Rochus*
17	E	*Veronica*
18	F	*Agapitus*
19	G	*Magnus marter:*
20	A	*Bernardus*
21	B	*Anastasius*
22	C	*Thimotheus*
23	D	*Zacheus*
24	E	BARTHOLOM:
25	F	*Ludouicus*
26	G	*Seuerinus*
27	A	*Ruffus*
28	B	*Augustinus*
29	C	IOHANNES Ē
30	D	*Felix*
31	E	*Paulinus*

SEPTEMBER

1	F	*Egidus*
2	G	*Absolon*
3	A	*Hermes*
4	B	*Moÿses*
5	C	*Victorinus*
6	D	*Magnus*
7	E	*Regina*
8	F	MARIE GEB:
9	G	*Gorgonius*
10	A	*Isaac*
11	B	*Prothus*
12	C	*Tobias*
13	D	*Maternus*
14	E	+ERHEBVNG
15	F	*Nicodemes*
16	G	*Ludomilla*
17	A	*Lambertus*
18	B	*Hildegardis*
19	C	*Ianuarius Ester*
20	D	*Fausta*
21	E	MATHEVS AP
22	F	*Mauritius*
23	G	*Tecla*
24	A	*Rupertus*
25	B	*Cleophas*
26	C	*Cÿprianus*
27	D	*Cosm: Dornian:*
28	E	WENCESLA
29	F	MICHÆLIS
30	G	*Hieronymus*

OCTOBER

1	A	*Remigus*
2	B	*Leodegarus*
3	C	*Simplicius*

4	D	*Franciscus*
5	E	*Aurelia Placid:*
6	F	*Fides*
7	G	*Marcus*
8	A	*Sergius*
9	B	*Dionisius*
10	C	*Gereon*
11	D	*Vedasius*
12	E	*Maximilianus*
13	F	*Colomanus*
14	G	*Calixtus bapst*
15	A	*Hedwig*
16	B	GALLVS
17	C	*Martha*
18	D	LVCAS EVA:
19	E	*Rosina*
20	F	*Cordula iungf*
21	G	*Vrsula*
22	A	*Seuerus*
23	B	*Seuerinus bisch*
24	C	*Cassiodorus*
25	D	*Crispinus*
26	E	*Amandus*
27	F	*Sabina*
28	G	SIMON IVD
29	A	*Narcissus*
30	B	*Felicianus*
31	C	*Wolfgangus*

NOVEMBER

1	D	ALLER HEIL:
2	E	*Aller Seelen*
3	F	*Theophilus*
4	G	*Ionas*
5	A	*Malachias*
6	B	*Leonhartus*
7	C	*Wilebrodus*
8	D	*Vier gekrönten*
9	E	*Theorodorus*
10	F	*Martinus bapst*
11	G	MARTINVS
12	A	*5 Brüeder*
13	B	*Briccius*
14	C	*Hubertus*
15	D	*Seraph*
16	E	*Othmarus*
17	F	*Hugo*
18	G	*Gelasius*
19	A	ELISABETH
20	B	*Pontianus*
21	C	MARIE OPF
22	D	*Cecilia*
23	E	*Klemens*
24	F	*Chrisogonus*
25	G	CATHARINA
26	A	*Virgilig bapst*
27	B	*Conradus*
28	C	*Gundherus*
29	D	*Saturninus*
30	E	ANDREAS

DECEMBER

1	F	*Candidus*
2	G	*Iouianus*
3	A	*Lucius*
4	B	BARBARA
5	C	*Sabba abt*
6	D	NICOLAVS
7	E	*Crispinus*
8	F	MARIA EMPF
9	G	*Ioachimus*
10	A	*Eulalia*
11	B	*Damasus*
12	C	*Iobst*
13	D	LVCIA *Ottilia*
14	E	*Nunsius*
15	F	*Valerianus*
16	G	*Lazarus*
17	A	*Adelheilis*
18	B	*Ignatius*
19	C	*Nemesius*
20	D	*Ammon*

21	E	THOMAS	26	C	STEFFAN
22	F	*Theodosius*	27	D	IOHANNES
23	G	*Victoria iungf:*	28	E	KINDLEINTAG
24	A	VIGILIA	29	F	*Athanasius*
25	B	CHRISTAG	30	G	*Dauid*
			31	A	*Sÿluester*

Alessandro Farnese, Duke of Parma

The coat of arms on the Habermel astrolabe is that of the Dukes of Parma, incorporating the impalement of Austria and Burgundy following the marriage of Duke Ottavio Farnese with Margherita, daughter of the Emperor Charles V. This identifies the coat of arms as that of Ottavio's son, Alessandro, who was a member of the Order of the Golden Fleece, the insignia of which surrounds the coat of arms.

Alessandro Farnese (1545-1592) Prince, and from 1586 Duke of Parma, was born in Rome, the son of Duke Ottavio and Margherita of Austria, natural daughter of the Emperor Charles V, and half-sister of King Philip II of Spain. Alessandro was brought up at the Spanish court, where he became a close friend of Don John of Austria, his uncle, and, like his mother, an illegitimate child of Charles V. In 1565, Alessandro married Maria of Portugal, and their son had a claim to the throne of Portugal; this, however was not pursued by his father, who was devoted to the house of Habsburg, and supported Philip II as rightful claimant of the Portuguese crown.

Prince Alessandro was trained as a soldier, and served with distinction at the Battle of Lepanto, 1571, under the command of Don John of Austria, at which the Turkish fleet was defeated and largely destroyed by a united naval force from Venice, Genoa, the Papal States, and Spain. The young Italian added to the legends surrounding this decisive battle by capturing the Turkish treasure galley at the head of a boarding party.

In the next few years, Prince Alessandro built up a considerable reputation as a military commander, serving in the Low Countries. In 1576, Flanders was in turmoil, and the Spanish troops were rioting because they had not been paid. After the soldiery sacked Antwerp, it was agreed to have them withdrawn, provided both sides agreed to a truce, but the truce was broken by the Calvinists. War followed, and the Spaniards, under the command of Alessandro Farnese, defeated the army of The Netherlands under William of Orange.

On the death of Don John in 1578, Prince Alessandro became governor of the Netherlands in his place, and proved as skilled a statesman as he was a soldier. He negotiated the Union of Arras with the southern provinces of the Low

Countries, resulting in the alliance of the Flemish nobility with Spain, and a lasting peace. He persuaded Philip II to provide regular pay for the Spanish army in the Low Countries, and succeeded in reconquering most of Flanders in the course of 1584, his victories culminating in the capture of Antwerp in August, after a long siege. This brought him into high favour with King Philip. He then moved north with his troops to try to deal with the more recalcitrant provinces of the northern Netherlands.

At this time, King Philip was planning the invasion of England. His instructions were that Prince Alessandro was to prepare his army to join the Spanish fleet that was to put in at one of the Dutch ports on its way to attack England. In the event, the Spanish army in the Netherlands was blockaded by a Dutch fleet, making any union of the forces impossible. The Armada was defeated in the Channel and driven into the North Sea by the English navy. In the attempt to return to Spain round the north of Scotland, more than half the remaining Spanish ships were destroyed by freak gales during the summer of 1588.

Alessandro Farnese fell from favour with King Philip because he had expressed doubts about the invasion of England, and he was recalled to Spain in the aftermath of the Armada disaster. He was, however, wounded in an action in France, and died of his wounds in 1592.

Note on the Order of the Golden Fleece

The coat of arms on the Habermel astrolabe is surrounded by the collar of the Order of the Golden Fleece. This Order was founded by Duke Philip the Good of Burgundy in 1430, and included the heads of most of the noble families of Flanders. The members of the Order held chapter meetings, and disciplined any of their number who dishonoured the Order. King Philip II took the Order into his personal control in 1568. He always wore its insignia himself, and bestowed it as a personal honour on members of the European nobility. Alessandro Farnese was made a member of the Order in 1585, following his success in pacifying Flanders. His portrait of 1590, about the time when the astrolabe was made, shows the Duke armed and wearing the chain of the Order, with a distant view of Antwerp in the background. This portrait by J.B. de Saive, is in the Galleria Nazionale, Parma.

Erasmus Habermel

Erasmus Habermel was without doubt one of the finest craftsmen of the sixteenth century making mathematical instruments, yet remarkably little is known of his career. The fact that the Emperor Rudolf II chose Prague instead of Vienna to be the seat of his court in 1583 would have helped trade were

Habermel already there, or he may have been attracted to Prague from elsewhere by the Emperor's interest in astronomy. Apart from his instruments, one of the very few written records is from 1593, with the account of payment made to him as "the astronomical and geometrical instrument maker" of the Emperor in Prague. This was a post he held until his death in 1606, and it is known that his workshop was visited by Tycho Brahe, who came to Prague from Denmark in June 1599, because Habermel made a 4½-foot sextant for him in 1600, which is preserved at the National Technical Museum in Prague. Unfortunately, Tycho died in October 1601, so he was not able to collaborate extensively with Johannes Kepler, the great mathematician, whom the Emperor had invited to Prague; he arrived at the beginning of 1600, and remained until 1612. His given task was to complete Tycho's planetary tables, and when these were eventually published in 1627, they were entitled *Tabulae Rudolphinae.*

There is recorded by Zinner and by Eckhardt an early example of Habermel's work at the National Museum of Antiquities, Edinburgh, bearing the date 1576. It is a compendium, was on loan from 1930 to 1961 to the Royal Scottish Museum, Edinburgh, and it is dated ***1596***. The earliest dated instruments in existence about which we may have confidence were made in 1585; four in that year, and nine in the following year. The remainder of that decade was a productive one. It is, in any case, reasonable to assume that Habermel would not have been chosen as instrument maker at the court of the Holy Roman Emperor unless he were well established. The question has always been, where did he learn his craft?

The suggestion has been made that he might have trained in Louvain, in the workshop of Arsenius, but this is unlikely. There is evidence from Habermel's astrolabes that his knowledge of astronomy was not learnt in the Low Countries. What is more plausible is that he may have spent some time in Augsburg, which was a great centre for the goldsmith's art, and the city of Christoph Schissler, another skilled and prolific instrument maker, but not of the highest quality.

Habermel's work is distinguished by the delicacy of the engraved lettering, suggesting that he may have learnt his skill in engraving copperplates for printing, and by the beauty of the decorative work, which he may have learnt from a goldsmith. Many of his instruments are made of gilded copper; he did not habitually use brass as did the Flanders instrument makers. Therefore one is led to the conclusion that Habermel was born either in or near Prague, and spent most of his life there, with an apprenticeship possibly elsewhere, perhaps in Vienna from whence he came in 1583 with the Emperor to Prague. He may have travelled in his youth to the Low Countries, or even to Italy. We know that he married in 1593, for the Emperor's marriage gift to him is recorded. Since he was in a very good way of business by then, he was probably in middle life, perhaps around forty. This gives a presumed birth date around 1550.

Well over 100 scientific instruments made by Habermel are known to exist, many, but not all, bearing his signature. Because of their beauty and intricacy, they were sought after by collectors during the nineteenth century, and subsequently the vast majority found their way into museums. A considerable number were made for noble or wealthy patrons, and bear coats of arms and/or inscriptions accordingly. A group of as many as twenty instruments bear the arms of Franciscus of Padua, a doctor and philosopher, who came from Friuli, and settled in Prague around 1590. He had a high reputation for his medical skill, and is reported to have successfully treated popes and princes; given the Emperor's poor health, his arrival in Prague is not surprising, nor is the fact that he was the subject of an engraved portrait in 1604 by the court painter, Agidius Sadeler. Other notable patrons included Prince Maximilian, the Emperor's brother, and Wilhelm von Rosenberg, to whom was presented the perpetual calendar devised by Bulderus that also appears on the present astrolabe.

We do not know whether this astrolabe, bearing the arms of the Duke of Parma, was a gift to him, or a commission from the Duke himself. It must have been made after 1587, on the evidence of the perpetual calendar, and before Farnese's early death in 1592.

The Historical Background

Politically, the end of the sixteenth century was a time of mounting conflict in Europe. The years of Rudolf II's reign as Holy Roman Emperor in Prague immediately preceded the beginning of the Thirty Years War. There was strife not only between Catholic and Protestant, but also between very different socio-economic systems: the old, feudal-style absolute monarchies of Austria and Spain, owing allegiance to Rome; and the new, bourgeois, commercial economies that were emerging in England and The Netherlands under the Protestant faith. It is interesting that the Habermel astrolabe was the work of a craftsman at a Habsburg court for an Italian/Spanish aristocrat, who made his reputation as governor of the Spanish Netherlands, the cockpit where Catholic power was challenged by Protestantism.

The political and religious unrest was fermented by a series of momentous events that had taken place over the period we now call the Renaissance. In the middle of the fifteenth century, the printing press was invented; learning was no longer the monopoly of the church, nor literacy the privilege of the very few. The questioning of long-held beliefs and theories, the thirst for new knowledge, were spread and encouraged through the printed word, and through the setting up of new universities.

At the end of the fifteenth century, the known world was hugely enlarged by the discovery of the American continent. The expansion of trade through deep-sea navigation provided an economic imperative, experienced by all the powers of Europe, to encourage what were known as "the mathematical arts", so that ships were not sacrificed to faulty charts and inadequate astronomical tables.

In 1543 was published the great work by the founder of modern astronomy, Nicholas Copernicus, *De revolutionibus orbium coelestium*, which placed the Sun, and not the Earth (as believed by the classical astronomers) at the centre of the universe. This new cosmology made possible the reform of the calendar, and the compilation of new and accurate astronomical tables by Johannes Kepler.

The new thinking and learning depended less on theory and more on accurate observation and measurement. There was a growing demand for instruments by the astronomer, the navigator, the land surveyor, and also from the universities for teaching. The skilled artisan was much in demand, and to meet the requirements of his customers, had to acquire some knowledge of the mathematical arts. The sixteenth century, therefore, saw the birth of the precision instrument-making trade in Europe, drawing its skill in engraving on metal from the printing industry. Of this trade, two of the most skilled and renowned early makers were Erasmus Habermel of Prague, and Gerard Mercator of Louvain and Duisburg.

Habermel found a patron in the Emperor Rudolf II, who made his court in Prague a centre for learning and the arts. Rudolf was deeply interested both in the new science – he invited Tycho Brahe and Kepler to his court – and in alchemy and the occult. He was also a collector, devoting much of his time, energy and wealth to amassing a huge "cabinet", which consisted of a systematic gathering of the wonders of the natural world and the finest human artefacts. This kind of patronage, as well as the requirements of the more humble users of instruments, helped to establish and maintain the instrument-making trade. Rudolf's patronage was certainly responsible for the richness and superb artistic quality of Habermel's instruments.

Note: This is the revised text of the catalogue entry provided for Christie's, to accompany the auction of the Habermel astrolabe on 11 October 1995 (single lot). I wish to thank Christie, Manson & Woods Ltd, for permission to reproduce the material, and Mr Jeremy Collins for providing the illustrations.

Literature

Braudel, Fernand, *The Mediterranean and the Mediterranean World in the Age of Philip II*, 2 vols (London, 1972, 1973).

Eckhardt, Wolfgang, "Erasmus Habermel: Zur Biographie des Instrumentenmachers Kaiser Rudolfs II.", and "Erasmus und Josua Habermel: Kunstgeschichte Anmerkungen zu den Werken der beiden Instrumentenmacher", *Jahrbuch der Hamburger Kunstsammlungen*, 21 (1976), 55–92; ibid., 22 (1977), 13–74.

Eisner, Sigmund, ed., *The Kalendarium of Nicholas of Lynn* (Chaucer Library, vol. 2) (London, 1980).

Evans, R.J.W., *Rudolf II and his World: A Study in Intellectual History 1576-1612* (Oxford, 1973).

Ginzel, F.K., *Handbuch der Mathematischen und Technischen Chronologie das Zeitrechnungswesen der Völker*, 3 vols (Leipzig, 1906–1914).

Lacarta, Manuel, *Biografia Historica: Felipe II* (Madrid: Silex, 1994).

Michel, Henri, *Traité de l'Astrolabe* (Paris, 1947).

Parma. *Der Glanz der Farnese: Kunst und Sammelleidenschaft in der Renaissance* (Munich: Haus der Kunst, 1995).

Prague. *Prag um 1600: Beiträge zur Kunst und Kulture am Hofe Rudolfs II.*, 2 vols (Freren: Luca Verlag, 1988).

Rohde, Alfred, *Die Geschichte der wissenschaftlichen Instrumente vom Beginn der Renaissance bis zum Ausgang des 18. Jahrhunderts* (Leipzig, 1923).

Šíma, Zdislav, "Prague Sextants of Tycho Brahe", *Annals of Science*, 50 (1993), 445–453.

Zinner, Ernst, *Der deutsche Kalender des Johannes Regiomontan, Nünberg, um 1474* (Leipzig, 1937).

Zinner, Ernst, *Deutsche und niederländische astronomische Instrumente des 11.–18. Jahrhunderts*, 2nd edn (Munich, 1979).

[See also chapter XII in this volume.]

Fig. 1 Front of the astrolabe

Fig. 2 Back of the astrolabe

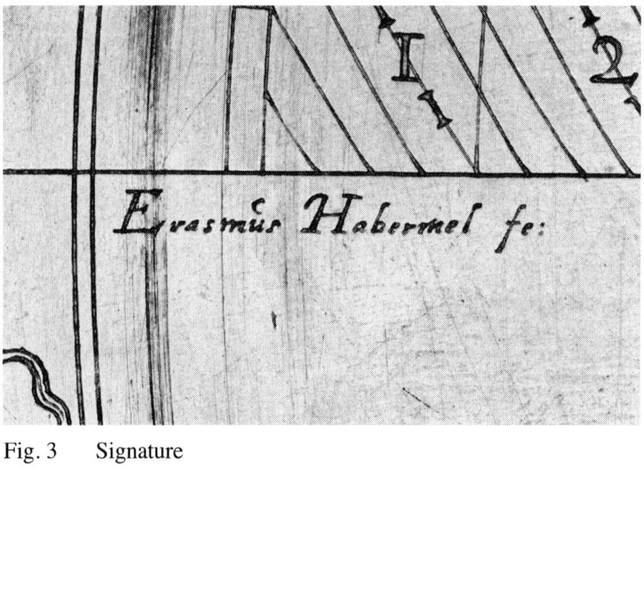

Fig. 3 Signature

Fig. 4 Front of throne showing compass.

Fig. 5 Back of throne showing coat of arms of the Duke of Parma.

Fig. 6 The mater showing the *quadratum nauticum*.

Fig. 7 Detail of the back showing the shadow square.

Fig. 8 The rete, cut for 48 stars.

Fig. 9 Detail of the back showing Signs of the Zodiac.

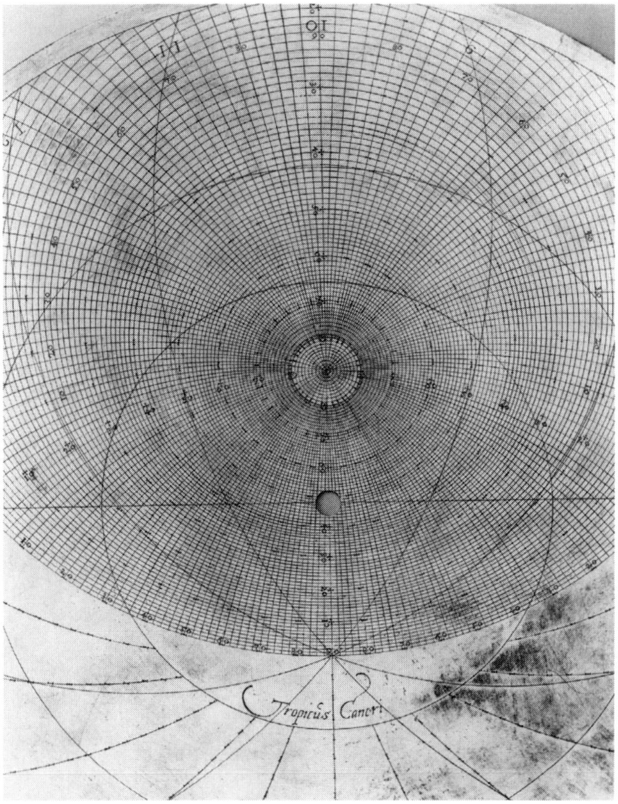

Fig. 10 Plate for 48°, detail around centre showing altitude to 1°, and azimuth to 2°.

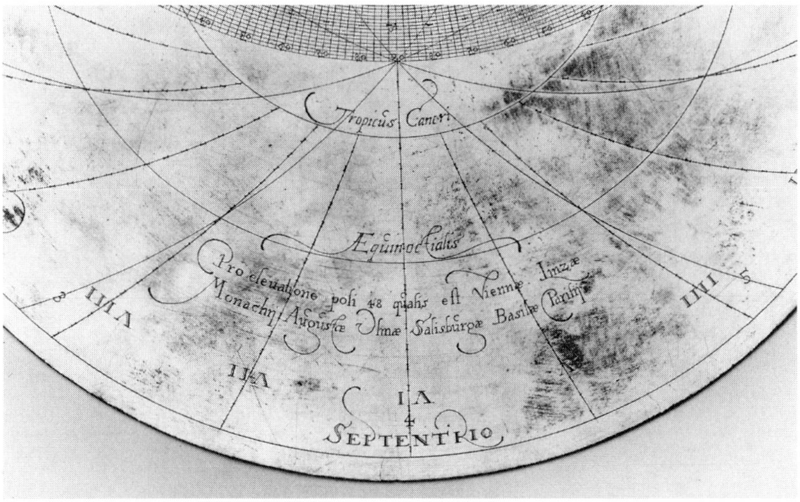

Fig. 11 Plate for 48°, detail, with names of cities on this latitude.

Fig. 12 Plate inscribed with the perpetual calendar.

XI

AN UNUSUAL ELIZABETHAN SILVER GLOBE BY CHARLES WHITWELL

By Elly Dekker and Gerard L'E. Turner, F.S.A.

In recent years a project has been carried out to catalogue the collection of globes and armillary spheres of the National Maritime Museum, Greenwich.[1] Among the treasures in this collection is a small silver globe of 63mm in diameter, which had not been studied in detail before. This small globe had been part of the Bonnefoy collection, and as such was exhibited at the Musée des Arts Décoratifs, Palais du Louvre, Paris, from March to April 1936. Afterwards, in May of that year, it was acquired by Sir James Caird, who presented it to the National Maritime Museum.

AN UNUSUAL PAIR OF GLOBES

The unusual construction of this silver globe immediately attracts the attention.[2] The single sphere is constructed out of two silver hemispheres which are joined at the equator by three bayonet catches (figs 1–2). The outer surface is engraved with a map of the world; at least 900 holes of different sizes have also been drilled into the surface from

Fig. 1. Silver globe (diameter 63mm), showing America not connected to Asia, loxodromes, and surface pierced with holes that represent the stars, which are to be viewed from inside the hemisphere. National Maritime Museum, Greenwich, GLB0025. *Photograph*: G.L'E. Turner

Fig. 2. Southern hemisphere of the globe showing the celestial coordinates, and the attempt to inscribe the constellation figure the River Eridanus. One of the three bayonet catches can be seen at the top of the picture. *Photograph*: G.L'E. Turner

the outside, representing the starry sky. Evidently this globe was intended to be used whole as a terrestrial sphere as well as in two parts in order to see, from the inside, the stars against a background light, exactly as seen from the Earth (fig. 3).

The terrestrial and celestial maps of this 'pair of globes' appear to have been copied from the well-known pair of globes made in Louvain by Gerard Mercator in 1541 and 1551, respectively. The overall outlines of land and sea depicted on the silver globe, as well as the sea monsters and most of the names engraved on it, are identical to those on Mercator's terrestrial globe. The inclusion of loxodromes is another well-known feature connected with that globe. The identification of the mapping of the celestial globe is less straightforward, because only one constellation figure, Eridanus, is drawn on the inside of the sphere. However, a number of peculiarities of Mercator's celestial globe, such as

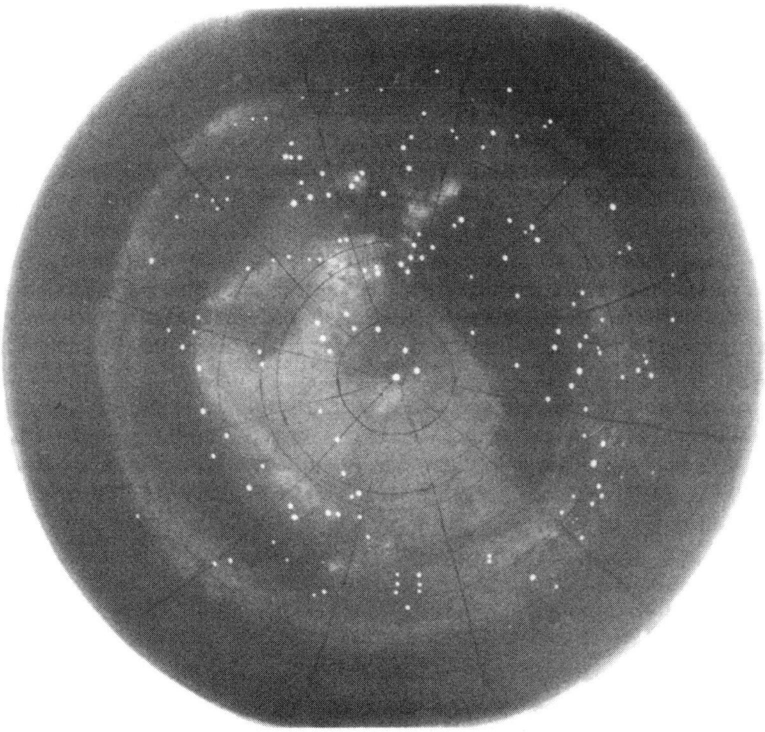

Fig. 3. Northern hemisphere of the globe viewed from inside against a bright light to show the holes that represent the star positions. Note the group at the bottom in the form of a U, which is the stellar magnitude table copied in form and location from Mercator's celestial globe of 1551. Because of the curved shape it is not possible to show on this photograph the star holes in a third of the globe's surface, which is why the photograph offers a dark outer ring. In fact, it is pierced with a few hundred holes. *Photograph*: G.L'E. Turner

the shape and place of the stellar magnitude table, are found in exactly the same manner on the silver globe.[3]

The use of the Mercator pair may indirectly have provided the idea for making this globe of unusual construction. A special feature of the terrestrial globe by Mercator is that, in addition to the outlines of land and sea, some fifty stars are marked on it.[4] This idea was clearly taken over by the maker of this silver globe, but instead of marking the stars on the surface he drilled holes in the terrestrial sphere. By adding a mechanism to take the sphere apart an unusual concave celestial globe was created. The normal, convex celestial globe shows the stars from the viewpoint of an imaginary observer outside the sphere, hence not as they are seen by an observer on Earth. The stellar configurations and their mutual arrangement on a normal, convex globe are the mirror images of those on a concave globe. For example, the zodiacal signs on a normal globe run from left to right, whereas in the sky we see them in the opposite direction. Globe makers like Isaac Habrecht II (1621) and Vincenzo Coronelli (1688) in the seventeenth century tried to remedy this shortcoming of normal, convex celestial globes by designing printed gores to be pasted on the inside of a hollow sphere.[5] Concave globes of considerable size as employed by these makers cannot have been very manageable, and this explains why no examples of such hollow celestial globes have survived.[6] The design of the pocket globe invented by Joseph Moxon (1654), on the other hand, lends itself well to the making of such concave celestial globes.[7] However, the then existing tradition in globe making apparently initially prevailed in the sense that the celestial gores pasted *inside* the cases of Moxon's pocket globes, and also those of other early English makers such as Charles Price and John Senex, show the sky in 'globe view', that is as it is seen by an outside observer.[8] Later makers of pocket globes, in particular Abraham van Ceulen (1697), Johann Baptist Homann (1715), and Richard Cushee (1731), put this right.[9] The solution offered by the maker of the present globe was only possible because it is of metal and because it is small, but this makes it no less original.

The question now arises, when was this unusually constructed silver globe pair made, and by whom? The use of the Mercator pair as the model provides a first clue to the date. This pair was the most popular pair of globes in the second half of the sixteenth century. It is not amazing to find that it was used as the model for the silver globe provided that the latter was made before or about 1600.

THE MAKER

Since the globe is not signed only the engraving can help to establish the maker. On first inspection it is obvious that the small, unsigned and undated, silver globe was made in England and most probably during the reign of Queen Elizabeth. Therefore only a handful of men skilled in the craft of mathematical instrument making could have produced it. Those to be considered are: Humfrey Cole (*c.* 1530–91); Augustine Ryther (fl. 1576–95); James Kynvyn (fl. 1570–1610); Charles Whitwell (*c.* 1576–1611).[10] Whitwell's apprentice, Elias Allen ended his apprenticeship in 1612, and can be eliminated on date and on the individual characteristics of his engraving hand. All these craftsmen used the same style of engraving for their letters and numbers, one that is distinctively English, effectively from London. Their style was that introduced from the Louvain workshops by Thomas Gemini (*c.* 1500–62), who had come to London by 1540 as an engraver, publisher, and instrument maker. His hand is close to that of Gerard Mercator (1512–94). There is so far no evidence that Humfrey Cole was apprenticed to Thomas Gemini, but he must have been aware of his products, which include copper-plate book illustrations, astrolabes, and quadrants. One may regard Gemini as having given a fresh impetus to the London trade in precision instrument making, which developed quickly and became supreme in Europe during the seventeenth century.

As mentioned above, the silver globe is derived from the 1541 terrestrial globe and the 1551 celestial globe of Gerard Mercator. Thus the initial date is 1551. The last date is likely to be 1592, when Emery Molyneux (died 1598/9) introduced his new pair of 24 inch globes to Queen Elizabeth's Court. After this one may presume that the Mercator globes of forty and fifty years earlier would no longer be acceptable as a model. Of the possible engravers in this period, it is Cole, Ryther, Kynvyn, and Whitwell who have to be considered, and three of them eliminated.

One of us (G.L'E.T.) has examined twenty-six mathematical instruments by Cole, all but one of which are signed, and all but three dated. Of his other products, a map and book illustrations are known. He may be said to be heavy in the handling of the burin, cutting deeply into the brass, with stocky lettering. When allowed plenty of space, as in his magnificent 2 foot astrolabe dated 1575, his lettering even rivals that of Mercator. One of the conclusions that can be drawn from an examination of Cole's lettering is that he developed and practised a strong personal style, which remained consistent throughout his working life. Cole used standard letter-forms from which he rarely deviated. Nearly all of these are illustrated in Mercator's *Literarum latinarum* (Louvain 1540),[11] and the influence of Mercator on Cole is thereby established. Cole's first known piece is a pocket compendium dated 1568, and the last piece known is another signed compendium, dated 1590. Cole died in 1591, and was said to be an old man. Humfrey Cole is ruled out as the maker of the silver globe through fine points in his engraving technique and in the details of his style. And there are stronger contenders to be dealt with.

Augustine Ryther is known for the series of maps concerning the defeat of the Spanish Armada in 1588, for two English county maps engraved for Saxton, which are dated 1576, and for two planispheres (1590) and a sea chart (1592) to the designs of Thomas Hood. Of two instruments known bearing Ryther's signature, one is a theodolite at Florence, and the other is a pocket compendium at the British Museum.[12] He engraved his upper case letters with unusually wide risers, pronounced serifs at each end of his S, a square E, and extended descenders to his *x* and *y*. Ryther's numbers are similar to Cole's, but he had greater difficulty in forming perfectly round zeros, his 2 has a 'fishtail' serif to the bar, and his 4 is elongated. In all these elements his hand differs from his apprentice, Charles Whitwell.

James Kynvyn can, like Cole, be omitted from further consideration on similar grounds. A skilled engraver, his nine known instruments show that he habitually made simple errors. For example, he has engraved 15 instead of 105, and used a carat mark for the small zero he inserted. Other errors he would ignore, or rub out and over-engrave. The globe shows none of these characteristics, nor his versions of some numbers. His symbols for the signs of the zodiac are clearly different from those used on the globe.

Charles Whitwell must have been born in about 1568 since he was apprenticed on 15 December 1582 to Augustine Ryther of Leadenhall in the City of London. Whitwell took his Freedom on 10 November 1590. The earliest instrument signed and dated by Whitwell is his pocket vertical sundial (1593) made for Nathaniell Torporley (1564–1632), whose patron was the ninth Earl of Northumberland (fig. 4).[13] So far, twenty-eight instruments by Whitwell have been identified; the latest is a compendium dated 1610. He died before September 1611. Apart from his mathematical instruments, Whitwell is known for a number of maps, including Asia (1595), Jerusalem (1595), Kent (1596), and Surrey (*c.* 1604). He was responsible for the engraved illustrations of instruments in William Barlowe, *The Navigators Supply* (London, 1597), and Thomas Hood, *The Making and vse of the Geometricall Instrument, called a Sector* (London, 1598). Particularly important is the group of ten Whitwell instruments taken to Florence in 1606 by Sir Robert Dudley.[14] Several were commissioned by Dudley, two of which are dated 1595. It is clear that by 1595 Whitwell, then in his middle twenties, was executing major commissions. Some were grand, such as the delicate nautical hemisphere to the design of Michael Coignet of Antwerp, and the large astrolabe to

(a)

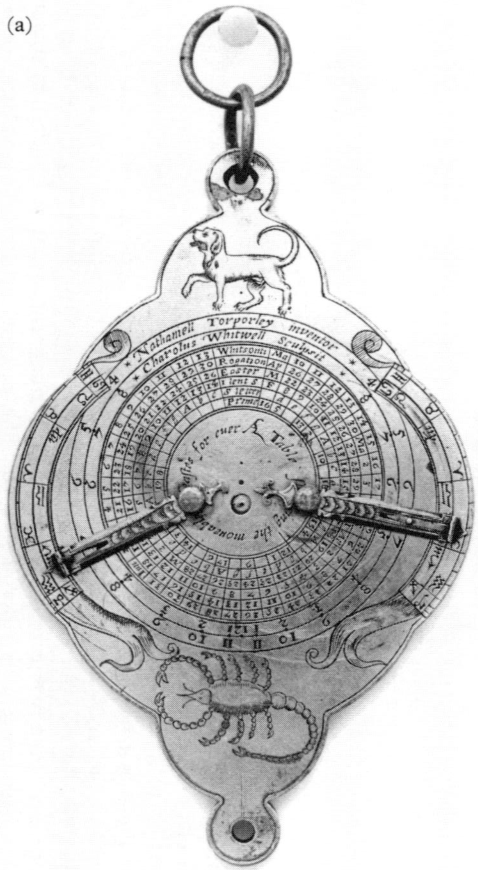

(b)

Fig. 4a, b. Vertical dial (diameter 70mm), signed: *Charolus Whitwell Sculpsit*; calendar begins at 1593. In 4b the alidade has been removed for clarity. Note the zodiac symbols. Museum of the History of Science, Oxford, LE 65. *Photograph*: G.L'E. Turner

the design of John Blagrave of Reading. There were also simple commissions, such as the labelling of the declination scale and the numbering of hours on the equinoctial line of a horary quadrant dated 1399.[15]

Whitwell's hand shows readily distinguishable features. For example: the two strokes in an italic *p* are not connected at the top; the capital E is not square as the lower foot protrudes to the right; the 6 and 9 are small open circles with a long straight riser or descender; the 7 has a fishtail serif on the top bar made by two nicks of the burin; the top of the risers in the italic *h*, *f*, and *l* end in a graceful curve in some of his signatures. Clear and emphatic styling characteristics are to be found in the symbols of the signs of the zodiac. The examination of many instruments from different locations and periods shows that engravers acquire their own particular set of shapes for these things. The equi-curved wings of Aries, the slanting Libra, long straight strokes like a pair of his sixes in Cancer, the long stroke to the left in Capricorn, the back-to-back Cs joined by a short bar in Pisces, all add up to point firmly to Whitwell as the maker of the silver globe. One receives the impression that it may be an early piece, and with the ingenuity in the novel feature of viewing the stars, one could be excused for thinking the silver globe might have been Whitwell's masterpiece.

THE EARLIEST EXTANT GLOBE MADE IN ENGLAND?

The trade in mathematical instruments in the City of London developed during the reign of Queen Elizabeth I. Towards the end of the sixteenth century the most precise engraver of instruments was Charles Whitwell, and it is not surprising that he should be asked to make a small silver globe. There was a lively interest in globe making in London in the last decades of the sixteenth century encouraged by the circumnavigations of Francis Drake (1577–80) and Thomas Cavendish (1585–8), the voyages of exploration of Martin Frobisher (1576–8) to North America, and the attempts to find the North-West Passage by John Davis (1585–7).[16] In 1592 the first printed globes were published in London by Emery Molyneux, the copperplates for which were engraved by the Flemish refugee, Jodocus Hondius Snr, who lived in London from 1583–93.[17] The publication of this pair was well advertised; Queen Elizabeth received a pair in 1592 as a gift from William Sanderson, the merchant who had financed the production of these globes. As the late Helen Wallis has pointed out, the publication of the Molyneux pair of globes was a public event. The terrestrial globe was even alluded to in Shakespeare's *The Comedy of Errors*, Act III, scene ii, when he describes the kitchen maid: 'No longer from head to foot than from hip to hip; she is spherical, like a globe; I could find out countries in her.'[18] Understandably, the new pair of globes of Molyneux was superior to that of Mercator. It appears, therefore, that the use of a Mercator pair by a London maker can be easily understood if this took place before 1592, because until that date Mercator's globes were the chief globes in use in England. This situation changed drastically in 1592,[19] and because the new Molyneux globes were so well advertised, it is hard to believe that a London engraver and instrument maker involved in making navigational instruments would have remained unaware of them. This would point to a date for our silver globe of before 1592, and would make this unusual globe by Charles Whitwell the earliest still extant globe manufactured in London.

ACKNOWLEDGEMENT

We are grateful to the Director of the National Maritime Museum, Greenwich, for allowing the globe to be shown to the meeting on 7 March 1996.

NOTES

1. The project was made possible by the Dr Sackler Fellowship, awarded to Elly Dekker during the academic years 1993–5.
2. Inventory no. GLB0025. A complete description of the globe will be published in Dekker forthcoming.
3. On both silver globe and Mercator's, longitude 100° to 105°, latitude 48° to 55°.
4. The idea of plotting stars on a terrestrial globe was taken over from the terrestrial globe made by Gemma Frisius in co-operation with Gaspar van der Heyden and Gerard Mercator in about 1536 (see van der Krogt 1993, 55). The stars and their names on the terrestrial globes of Gemma Frisius and Gerard Mercator have been listed in Elly Dekker and Peter van der Krogt 1994.
5. The concave globes by Habrecht are mentioned in the cartouche of his terrestrial globe of 1621, a copy of which is in the National Maritime Museum, GLB0111. For the concave globes of Coronelli, see the 'introduction' to Wallis 1969.
6. Copies of the gores of Coronelli's concave globe have survived. In mounted form they are seen only on the outside of a sphere, not on the inside as they should.
7. Dekker and van der Krogt 1993, 111–12.
8. *Ibid.*, 109, pl. 31.
9. *Ibid.*; on Van Ceulen, 55, fig. 24; Homann, 82, pl. 20; Cushee, 112, fig. 53.
10. On Cole, see Turner 1983, 98–9; on Ryther, see Brown 1979, 24, 58–60 and Hind 1952, 138–49; on Kynvyn, see Bruce 1866, 348; on Whitwell, Turner 1983, 98–9 and Brown 1979, 24.
11. Printed in facsimile as part 2 of Osley 1969, 121–75.
12. Theodolite dated 1590 in the Museo di Storia della Scienza, inv. no. 240; compendium dated 1588 in the British Museum, (Registration no. 66 2–21 1). This was the subject of the paper by John Bruce (1866).
13. Vertical sundial and perpetual calendar, signed: *Charolus Whitwell Sculpsit*; calendar begins at 1593. Also inscribed: *Nathaniell Torporley, inventor.* Museum of the History of Science, Oxford, inv. no. LE 65. Torporley was a mathematician from Christ Church, Oxford. On known instruments, Whitwell writes his first name as Charles (once), Charolus (three times), Carolus (once), and with just C. (five times). On a map he tried to remove the intrusive *h* in Charolus.
14. Leader 1895; Righini Bonelli 1968, 27–8; Miniati 1991, 21–2, 32–4.
15. Turner 1985. The Whitwell quadrant at Florence, dated 1595, is illustrated on pl. 99b.
16. Waters 1958.
17. The globes by Emery Molyneux have been discussed in a series of papers by Wallis: 1951, 1955; with Crinò 1987, 1989. See also van der Krogt 1993, 107–9, 138.
18. Wallis 1962; Wallis and Crinò 1989, 94.
19. On the Continent the globes by Mercator were replaced in about 1584 by those made in Amsterdam by Jacob Floris van Langren; see van der Krogt 1993, 107–12. These globes were known in England, if only because the map of the celestial globe of Emery Molyneux was copied from Van Langren's celestial globe of 1589 (see the description by E. Dekker in Bott 1992, no. 1.27). However, there is no indication that the Van Langren globes were widely known in England at the time. The celestial maps published in London in 1590 by Thomas Hood and engraved by A. Ryther, are copied from the celestial globe of Gerard Mercator; see Warner 1979, 127; and the reproduction of these maps in Stott 1991, 54–7.

BIBLIOGRAPHY

Bott, G. 1992. *Focus Behaim Globus*, vol. 2, Nuremberg

Brown, J. 1979. *Mathematical Instrument-Makers in the Grocers' Company* 1688–1800, London

Bruce, J. 1866. 'Description of a Pocket-dial made in 1593 for Robert Devereux, Earl of Essex', *Archaeologia*, 40, 343–56

Dekker, E. forthcoming. *The Globes and Armillary Spheres of the National Maritime Museum*

Dekker, E. and van der Krogt, P. 1993. *Globes from the Western World*, London

Dekker, E. and van der Krogt, P. 1994. 'De Globes', in *Gerardus Mercator Rupelmundanus* (ed. M. Watelet), 243–67, Antwerp

Hind, A.H. 1952. *Engraving in England in the Sixteenth and Seventeenth Centuries*, part i, Cambridge

Leader, J.T. 1895. *Life of Sir Robert Dudley*, Florence (reprinted Amsterdam, 1977)

Miniati, M. 1991. *Museo di Storia della Scienza: Catalogo*, Florence

Osley, A.S. 1969. *Mercator: A Monograph on the Lettering of Maps etc. in the* 16th Century Netherlands, with a facsimile and translation of his Treatise on the Italic Hand and a Translation of Ghim's Vita mercatoris, London

Righini Bonelli, M.L. 1968. *Il Museo di Storia della Scienza a Firenze*, Milan

Stott, C. 1991. *Celestial Charts: Antique Maps of the Heavens*, London

Turner, G. L'E. 1983. 'Mathematical Instrument-Making in London in the sixteenth Century', in *English Map-Making* 1500–1650 (ed. S. Tyacke), 93–106, London

Turner, G. L'E. 1985. 'Charles Whitwell's Addition, *c.*1595, to a Fourteenth-Century Quadrant', *Antiq. J.*, 65, 445–54, 476, pls 99–101

van der Krogt, P. 1993. *Globi Neerlandici: The Production of Globes in the Low Countries*, Utrecht

Wallis, H. 1951. 'The First English Globe: A recent Discovery', *Geogr. J.*, 117, 275–90

Wallis, H. 1955. 'Further Light on the Molyneux Globes', *Geogr. J.*, 121, 311

Wallis, H. 1962. 'Globes in England up to 1660', *Geogr. Mag.*, 35, 267–79.

Wallis, H. 1969. *V. Coronelli Libro dei Globi* 1693 (1701), facsimile edition, Amsterdam

Wallis, H. and Crinò, A.M. 1987. 'New Researches on the Molyneux Globes', *Der Globusfreund*, 35–7, 11–20

Wallis, H. and Crinò, A.M. 1989. 'Emery Molyneux, Jodocus Hondius and the first English globes', in *Theatrum Orbis Liborum: Liber amicorum Presented to Nico Israel on the Occasion of his Seventeenth Birthday* (eds T.C. van Uchelen, K. van der Horst and G. Schilder), 94–104, Utrecht

Warner, D.J. 1979. *The Sky Explored: Celestial Cartography* 1500–1800, New York and Amsterdam

Waters, D.W. 1958. *The Art of Navigation in England in Elizabethan and Early Stuart Times*, London

XII

Zinner's Ghosts and a Curious Date: 1576

G.L'E. Turner,
Imperial College, London
and
A.D. Morrison-Low,
National Museums of Scotland, Edinburgh

In the continuing study of astronomical and mathematical instruments, especially those of the Renaissance, it is important to establish the location and duration of workshops. This brings together examples scattered over many museums so that the extent and nature of the trade centres can be known.[1] Part of the study is to examine critically the literature on the whereabouts of objects, their makers, dates, and provenance. The task is made more difficult when the secondary sources were written before World War II, because of losses and dispersals. The account now given serves as an introduction to some of the problems, and provides the present locations for instruments once on loan to the Royal Scottish Museum, Edinburgh, which were dispersed at auction. The entry point is through an instrument signed by Erasmus Habermel at Prague that is dated 1596, but which has consistently been given the wrong date of 1576 (Figs 1–3). It is essential to clear this date from the records because it affects the perception of the place and the time span of Habermel's career.

Erasmus Habermel produced fine instruments of which over 100 are extant. There may be more, but no studies that involve detailed examination have been

1. The identification of twenty-six instruments, of which only four are signed, scattered around the world shows that a master craftsman was working in Florence during the later sixteenth century: G.L'E. Turner, "The Florentine Workshop of Giovan Battista Giusti, 1556–*c.*1575", *Nuncius: Annali di Storia della Scienza*, 10, fasc. 1 (1995), 132–172, plus 16 plates. The identification of three unsigned and undated astrolabes proved that instruments actually made by Gerard Mercator still exist: G.L'E. Turner and Elly Dekker, "An Astrolabe attributed to Gerard Mercator *c.*1570", *Annals of Science*, 50 (1993), 403–443; G.L'E. Turner, "The Three Astrolabes of Gerard Mercator", *Annals of Science*, 51 (1994), 329–353; a corpus of eleven German astrolabes of the mid-fifteenth century were identified as the product of one or two workshops, with Vienna as the possible location for one: D.A. King and G.L'E. Turner, "The Astrolabe presented by Regiomontanus to Cardinal Bessarion in 1462", *Nuncius: Annali di Storia della Scienza*, 9, fasc. 1 (1994), 165–206, plus 16 plates. Vienna as a craft centre for scientific instruments has received scarcely any attention.

performed. Ernst Zinner, in his great work published in 1956,[2] listing all instruments by Germanic craftsmen, began his section on Habermel with:

> 1576. Messingene Büchse in Gestalt Bucheinbandes mit Papiereinlage. Im Deckel verschiedene Scheiben, aber Keine Sonnenuhr. Bezeichnet "Erasmus Habermel Pragae 1576". Edinburgh National Museum of Antiquities.

W. Eckhardt, in his biography and listing of Habermel's products published in 1977,[3] began with this same instrument::

> 1 Einband eines Notizbuches und Kompendium: Lt. Zinner, S. 331: Edinburgh, National Museum of Antiquities, dort nicht nachweisbar[25]. Angeblich bez.: "Erasmus Habermel Pragae 1576". – Messing.

The National Museum of Antiquities of Scotland, Edinburgh, was based on the museum of the Society of Antiquaries of Scotland (founded in 1780), and was amalgamated with the former Royal Scottish Museum in 1985 to form the National Museums of Scotland, well after Zinner's death.[4] Ernst Zinner (1886–1970) was a working astronomer who became interested in the history of his subject, which he pursued with exhaustive thoroughness; he produced 394 publications. He visited the Royal Scottish Museum in June 1931, but somehow all the instruments were published by Zinner as being in the Museum of Antiquities, thus creating a *ghost location.* Furthermore, the date on the instrument, which

2. Ernst Zinner, *Deutsche und niederländische astronomische Instrumente des 11.–18. Jahrhunderts* (Munich, 1956; 2nd enlarged edn 1967; reprinted 1972 and 1979), quotation on p. 331. The Habermel entries in the first and second editions have been compared, and they are identical.

3. Wolfgang Eckhardt, "Erasmus Habermel: Zur Biographie des Instrumentenmachers Kaiser Rudolfs II", and "Erasmus und Josua Habermel: Kunstgeschichte Anmerkungen zu den Werken der beiden Instrumentenmacher", *Jahrbuch der Hamburger Kunstsammlungen*, 21 (1976), 55–92, ibid., 22 (1977), 13–74, quotation on p. 23. His note 25 refers to p. 60 of his first article where he explains: "Aber sind denn die von Zinner verzeichneten frühen, gleichermaßen signierten wie datierten und lokalisierten Instrumente – Bucheinband von 1576, ein Wegmesser und eine Bussole von 1580 – wirklich eigenhändige Arbeiten des Instrumentenmachers, ja, existieren sie überhaupt alle? Der angeblich im National Museum of Antiquities of Scotland in Edinburgh verwahrte Bucheinband hat sich weder dort noch im Royal Scottish Museum, weder in der National Library of Scotland noch in der Edinburgh University Library nachweisen lassen." So Eckhardt's No. 1 could not be traced in Edinburgh in 1976.

4. Diedrich Wattenberg, "Ernst Zinner: 2.2.1886 – 30.8.1970", *Astronomische Nachrichten*, 293 (1971), 79–80; *Journal of the History of Astronomy*, 2 (1971), 132. He was Director of the Remeis-Sternwarte in Bamberg from 1926 to 1953.

actually is **1596**, was incorrectly recorded by Zinner himself, and so the *ghost date* of 1576 was created, along with a *ghost signature.*

Zinner came to London for the Second International Congress of the History of Science and Technology, based at the Science Museum from 29 June to 3 July 1931.[5] As part of the programme, visits were made to The British Museum, Greenwich, and Oxford.[6] Evidence from his diary shows that Zinner went first to Edinburgh, and visited the Royal Scottish Museum on 20 June, 1931.[7] A copy of a letter to him from the Royal Scottish Museum dated 29 June 1931, says that a photograph of a Chinese planisphere was enclosed.[8] There is no letter from Zinner requesting the photograph; most likely he ordered it in person. Zinner's letter of 17 July 1931 acknowledges the photograph, and says that he thinks a sundial in the Findlay collection in case 106 was made by Erasmus Habermel, a famous maker in Prague in 1580.[9] We know from the Zinner archive that he had also seen the signed and dated Habermel compendium in case 103; but had written down the date as 1576: the "76" is underlined with a red pencil.[10]

The compendium that begins Zinner's extensive published list of Habermel instruments was, in fact, part of the collection of Sir John Ritchie Findlay (1866–1930) that was placed on loan at the Royal Scottish Museum after his death.[11] The collection was withdrawn in 1961, and sold at two auction sales by Sotheby's on 4 December 1961 and 26 February, 1962. In the later sale the compendium was lot 232: "An important writing tablet with gilt-metal covers... inscription and date

5. Zinner says that he joined the Congress, and also looked for astronomical manuscripts and instruments of German origin in English observatories, museums and libraries: E. Zinner, *Die Remeis-Sternwarte zu Bamberg 1889–1939* (Veröffentlichungen der Remeis-Sternwarte zu Bamberg Band IV) (Bamberg, 1939), p. 28.

6. *Isis*, 16 (1931), 126–129; an outline of the programme. For an account of the proceedings, see Gary Werskey, *The Visible College: A Collective Biography of British Scientists and Socialists of the 1930s* (London, 1988), 138–149.

7. The papers of Ernst Zinner were left by him to the Institut für Geschichte der Naturwissenschaften in Frankfurt am Main. (His books were sold to the San Diego State College Library, California.) The papers are named Sammlung Zinner. The Edinburgh visit is in S Zi 25, I: Reisen 1894–1943. "14. Juni – 15. Juli 1931: Reise nach England. E[rnst] + S[usi] Zinner"; "20.VI.1931... Dann bis abends 9 Uhr in R. Scottish Museum... Sehr schön die wissenschaftlichen Instrumente der Findlay-Sammlung im Erdgeschoß und einige Instrument im 2. Stock."

8. NMS Technology Correspondence: Outgoing letter 29 June 1931.

9. NMS Technology Correspondence: Incoming letter 17 July 1931.

10. Zinner archive (note 7), S Zi 2, III. A ringbinder marked "E–Ha" contains notes on makers whose names begin with these letters. In Zinner's hand: "Bucheinband...Edinburgh, Pragae 1576. Erasmus Habermel"; the "76" is underlined in red pencil. In a different hand: "Edinburgh, Nation. Mus. of Antiquities Schrank 103. Büchse in Gestalt eines Buchdeckels mit Papiereinlage. Verschiedene Scheiben, aber keine Sonnenuhr. Messing, vergoldet. Pragae 1576."

11. See A.D. Morrison-Low, "Sold at Sotheby's: Sir John Findlay's Cabinet and the Scottish Antiquarian Tradition", *Journal of the History of Collections*, 7 (1995), 197–209.

Pragae Fecit Erasmus Habermel 1596, 3½*in.*" It was purchased by Mannheimer of Switzerland for £570.[12] From the photographs in the auction catalogue, it can be seen that Zinner, and hence Eckhardt, had jumbled the signature, and it should read as here, but with a lower case *f* in *fecit.*

The compendium appears again, this time at Christie's, who sold it on 30 March 1977 as lot 105: "An important German gilt-metal compendium and aide-memoire, *dated* 1596. Signature: Prague fecit Erasmus Habermel 1596". The description is more extensive than before, and is reasonable. All four sides are illustrated, and there is a frontispiece to the catalogue in colour.

Like a returning comet, another sighting occurred in November 1977, this time in a catalogue from the Parisian antiquarian bookseller, Alain Brieux.[13] The compendium is illustrated in a colour photograph with a description, which is a version in French of the Christie text. Again, the transcription of the signature is slightly in error: *Pragae, Fecit Erasmus Habermel 1596.*[14]

Eckhardt has picked up both the Christie and Brieux references under his entry no. 31; however, he ends: "Nicht bei Zinner". Eckhardt's entry under No. 31 (not realizing this was the same instrument as his No. 1) illustrates yet again top and bottom and gives the correct signature: *Pragae fecit Erasmus Habermel 1596.* All three sale catalogues show the instrument, and it is clearly the same one because of little diagnostic features.

The first of Eckhardt's entries, the Findlay compendium wrongly dated 1576, can now be removed from his list, because he unknowingly included it as his No. 31, derived from the Christie sale of 1977 and the Brieux catalogue of 1977. Eckhardt's Nos 2 to 5 and 8 are fakes, and 6 and 7 are lost.[15] At present, the earliest signed and dated pieces that may be accepted with some confidence are some or all of those dated 1585, listed as Nos. 9 to 12 in Eckhardt. It is important to remove the wrong dates from the beginning of Habermel's career, because the conjecture can now be made that Habermel followed the Emperor Rudolf II from Vienna in 1583 when the Court moved to Prague.[16] It is the false dating that precluded this up till now.

---

12. Sebastian Whitestone, "Edgar Mannheimer" [obituary], *Independent*, 12 January 1994.

13. Alain Brieux, *Histoire des Sciences: Livres, instruments, autographes* (Paris, Alain Brieux, November 1977), 133, no. 9497.

14. The compendium is now at the Time Museum, Rockford, Illinois (no. 1439).

15. Eckhardt (note 3), 23–24. Nos 2–5 are described as "plumpe Fälschung", 6 and 7 are lost (if they ever truly existed), and 8 is given reasons for being false. No. 8 is in the Whipple Museum, Cambridge, see D.J. Bryden, *Catalogue 6: Sundials and Related Instruments* (Cambridge, 1988), no. 382 in the section of forgeries, where it is said to have been made *c.* 1920. For 2 to 4 and 8, Eckhardt has relied on D.J. Price, "Fake Antique Scientific Instruments", *Actes du 8ème Congrès international d'Histoire des Sciences* (Florence, 1957), I, 380–394.

16. G.L'E. Turner, "An Astrolabe for Alessandro Farnese, Duke of Parma, by Erasmus Habermel", *Catalogue* (London: Christie, Manson & Woods, 11 October 1995), 8–30. [See Essay X in

Some authors have sought to find locations where Habermel might have learnt his craft. Henri Michel began his sketch: "n. Regensburg vers 1538; d. Prague, 1606". He continued:[17]

> Né probalement en Allemagne du Sud, nous présumons qu'il a travaillé pour Arsenius à Louvain, entre les années 1558 et 1576. Après la Furie Espagnole, ou après la Peste de Louvain, il disparaît quelque temps.

These conjectural dates could follow from Michel's entry for Gautier Arsenius of Louvain, whose working dates are given as 1557 to 1575. Then comes the extraordinary statement: "ses [Arsenius] astrolabes présentent une telle analogie avec les travaux d'Erasme Habermel, qu'il est permis de supposer que ce dernier fut au nombre des ouvriers d'Arsenius". Michel had already in 1935 published a list of Habermel instruments known to him.[18] There are 45, of which 22 are signed and 5 dated (1585–1588). Museums in Oxford and Dublin provide examples, but Edinburgh is not mentioned. On 1 August 1937, Michel wrote to the Royal Scottish Museum saying he was listing Habermel's instruments, and that he had heard from Zinner: "that some instruments by Habermel might be presently in your Museum, and I take the liberty to ask you if this information is right". The Keeper of the Technological Department replied on 9 August 1937: "I regret, however, that in our collection, which contains many 16th century instruments, we have no signed instruments by this celebrated maker".[19] Yet the Findlay Collection was present with a signed Habermel. It seems, therefore, that only Zinner, and not the Keeper or Michel, was aware of the Habermel in the museum at Edinburgh.

Ernst Zinner's position as Director of the Remeis-Sternwarte in Bamberg enabled him to publish his historical research in the institute's scientific Reports. He wrote extensively in the three published during the 1930s: No. 26 (1932), No. 27 (1934), No. 28 (1939), which were also issued as monographs.[20] The last contains an index to all three. No. 26 is no help to us; presumably it was written before the London Congress. No. 27 gives the evidence for the visit to Britain; instruments are located in: the "National Museum of Antiquities in Edingburgh

this volume.]

17. Henri Michel, *Traité de l'Astrolabe* (Paris, 1947), 171.

18. Henri Michel, "L'Art des instruments de mathématiques en Belgique au XVIe siècle", *Bulletin de la Société Royale d'Archéologie de Bruxelles*, no. 2 (March–April 1935), 26–34.

19. NMS Technology Correspondence: Incoming letter 1 August 1937; Outgoing letter 9 August 1937.

20. Ernst Zinner, *Untersuchungen zur Geschichte der Sternkunde* (26. Bericht der Naturforschenden Gesellschaft in Bamberg) (Bamberg, 1932); idem, *Die fränkische Sternkunde in 11. bis 16. Jahrhundert*, op. cit. (27: 1934); idem, *Die ältesten Räderuhren und modernen Sonnenuhren*, op. cit. (28: 1939).

[*sic*]", the "Royal Observatory in Edinburgh", and the "Evans Collection Oxford", and several other museums. Naturally, he brings in the Emperor's craftsman:

> Hier müßte auch Erasmus Habermel († 1616), der sehr geschickte astronomische und geometrische Instrumentenmacher Kaiser Rudolfs, genannt werden, da er angeblich aus Nürnberg stammt [ref. to Rohde, p. 18]. Jedoch ist die Nachricht zu unbestimmt. Eher ist anzunehmen, daß er in Regensburg geboren ist, weil er dort die Feinmechanik erlernte und im Jahre 1576 ein Josua Habermel [*sic*] dort eine Nachtuhr mit waagrechter und senkrechter Bewegung herstellte [ref. Rohde, p. 67]. Erasmus Habermel selbst, ... arbeitete von 1576 bis 1600.[21]

It is true that for both Zinner and Eckhardt, Habermel's last *dated* instrument is 1600. But this is only Eckhardt's No. 37, and his next section is of signed but *undated* instruments, which runs up to No. 133.

It would seem that Zinner was relying on Rohde for more than a note on Habermel, for it is possible that his own encyclopedic compilation of 1956 was inspired by the book published in 1923.[22] Rohde wrote in an extraordinarily muddled fashion: "Erasmus Habermel wirkt in Süddeutschland und Prag" (p. 12); "Vermutlich in Süddeutschland (oder Italien?) ist eine Sonnenuhr von Erasmus Habermel entstanden (Abb. 17), die 1585 datiert und 'Erasmus Habermel Exculptor' bezeichnet ist." (p. 17); "Der Kompassmacher Erasmus Habermel stammt vermutlich aus Nürnberg, wo er das Uhrmacherhandwerk gelernt hatte. Später lernte er in Regensburg die Feinmechanik." (p. 18) In the footnote Rohde acknowledges archivists in Prague, Vienna, and Nuremberg, none of whom have found a birth date for Habermel. Indexing a man named Josua Habermehl [*sic*], Rohde has a footnote to link goldsmith Josua Habermehl and goldsmith Hans Gabriel Habermehl, both of Regensburg (p. 116). On p. 67, fig. 90, Rohde illustrates what he calls a "Universalinstrument", then in Dresden, which is apparently signed and dated: "Josua Habermehl Me Fecit 1576 in Civitate Ratis Bonae" (p. 64).[23] There is sufficient here to make one suspicious, and keen to examine the object in detail, but that is now impossible. Be that as it may, the key point for this story is the repetition of the date 1576. It has thus become easy

---

21. Ibid. (27: 1934), 59–60.

22. Alfred Rohde, *Die Geschichte der wissenschaftlichen Instrumente vom Beginn der Renaissance bis zum Ausgang des 18. Jahrhunderts* (Leipzig, 1923). References to Habermel on pp. 7, 12, 17–18, 64, 66–67, 116.

23. No. 182 and fig. 33 in Eckhardt (note 3) where the signature is rather different: IOSVA HABERMEL ME FECIT 1576 IN CIVITATE RATIS BONAE. Note the absence of the penultimate 'h' in the name. This theodolite was lost in World War II.

enough for some to assume that Erasmus was part of the Regensburg group of goldsmiths, that he was born in Regensburg, and that he was manufacturing instruments in 1576. What we do not know is whether Zinner misrecorded 1596 as 1576 through a simple human error, or whether he had 1576 set in his mind from reading Rohde.

It is certain, however, that Zinner kept to this date through and through. The Henri Michel correspondence at Brussels reveals that during 1948 Michel and Zinner were writing to each other. On 31 October 1948, Zinner mentions that E. Habermel was in Prague from 1576, and Michel replies on 8 November: "Habermel: Vous n'indiquez sa présence à Prague en 1576. La date la plus ancienne où j'ai vu sa signature datée de Prague est 1587....Avez-vous une autre indication sure?". Zinner answers on 29 November: "E. Habermel hat leider nur zelten Jahresstahl und ort angegeben – Meine Aufzeichnungen ergebens, 1576 Prague, 1580 Prague, 1587 Pragae... Das älteste stuck befindet sich in Edinburg Nat. Museum of Antiquities Schrank 103."[24]

Both the late C.H. Josten and Mr F.R. Maddison in their respective parts of the *Catalogue* to the Billmeir Collection,[25] wrote: "Erasmus Habermel was called to Prague in 1576" (No. 19); "Habermel's dated instruments cover the period from 1576 to 1600" (No. 198). The late F.A.B. Ward, taking Zinner as his authority, says: "[Habermel's] signed instruments show that he was already at work in Prague from 1576".[26] Such is the legacy of a simple error in recording. Dear reader, please take care!

The opportunity has now been taken by one of us (A.D.M.-L.) to list all those instruments published by Zinner in 1956 as being located in Edinburgh; most of them were dispersed in the Findlay sale. The Royal Scottish Museum is referred to as the National Museum of Antiquities in five different ways.

---

24. The private correspondence of Henri Michel is held by Mr Jan de Graeve, 5 Avenue de Meysse, 1020 Brussels, Belgium; information provided February 1996. Printed material, a large group of offprints, some correspondence, and photographic archives are in the Centre National d'Histoire des Sciences, Bibliothèque Royale Albert 1er, Brussels; see Roger Calcoen, *Inventaire du Fonds Henri Michel* (Brussels, 1991).

25. [C.H. Josten], *A Catalogue of Scientific Instruments from the 13th to the 19th Centuries from the Collection of J.A. Billmeir C.B.E.* (London, 1954; 2nd edn 1956), 14, no.19; F.R. Maddison, *A Supplement to a Catalogue of Scientific Instruments in the Collection of J.A. Billmeir, Esq., C.B.E.* (London, 1957), 53, no.198. Josten was writing before the publication of Zinner's great work (note 2), and he probably referred to Rohde, cited in the Bibliographical Note published in *A Supplement*, p. 11.

26. F.A.B. Ward, *A Catalogue of European Scientific Instruments in the Department of Medieval and Later Antiquities of the British Museum* (London, 1981), 145.

Acknowledgements

We are grateful for the assistance received from Mr Angus Macdonald, Librarian, Royal Observatory, Edinburgh, Mr A.V. Simcock, Librarian, Museum of the History of Science, Oxford, Ms Patricia Attwood, The Time Museum, Rockford, Illinois; Mme Dominique Brieux provided the photograph. Dr Silke Ackermann kindly looked through the Zinner archive at Frankfurt for confirmatory papers, and Mr Jan de Graeve examined the Michel archive in Brussels: we are indebted to both.

Concordance to Table

In all, Zinner recorded 55 instruments in Edinburgh, all but one at the incorrect address of the National Museum of Antiquities of Scotland, variously described: this should have read 'Royal Scottish Museum'. The single exception, an ivory diptych dial by Hans Tucher, remains at the Royal Observatory, Edinburgh. All the Edinburgh-located instruments are recorded, under the names under which they appear in Ernst Zinner, *Deutsche und niederländische astronomische Instrumente des 11.-18. Jahrhunderts* 2nd edition (Munich, 1967, reprinted 1972), together with their lot number (where appropriate) in the two Sotheby's sales of the Findlay Collection, 4 December 1961 and 26 February 1962. Present locations are given, where known:

Adler = Adler Planetarium, Chicago
Fort Caroline = Fort Caroline National Memorial, Timucuan Ecological and Historic Preserve, Jacksonville, Florida
Harvard = Harvard University, Collection of Historical Scientific Instruments
Leiden = Museum Boerhaave
MHSO = Museum of the History of Science, Oxford
NMAH = National Museum of American History, Smithsonian Institution, Washington D.C.
NMI = Egestorff Collection, National Museum of Ireland
RSM = Royal Scottish Museum
ROE = Royal Observatory, Edinburgh
Time Museum = Time Museum, Rockford, Illinois
† represents an unknown location.

The Scientific Instruments in Edinburgh Listed by Zinner

Zinner Page no.	*Signature on instrument in Zinner*	*Findlay sale lot Number*	*Current Location*
109	unsigned	Lot 123	MHSO, inv. no: 61-149
241	*BP*	n/a	†
247	*David Beringer, 1736* [2 examples]	part Lot 93	Harvard, inv. no: CHSI WP 7418
		part Lot 94	†
	David Beringer, no date [2 examples]	Lot 105	†
		n/a	RSM, inv. no: NMS T.1924.157 [1]
308	*Joan Engelbrecht Beraunensis 1801*	Lot 153	RSM, inv. no: NMS T.1962.12
	Joan Engelbrecht Beraunensis in Bohemia	Lot 224	†
319	*Johannes Gebhart Norimb, Anno LVII*	Lot 200	†
325	*L. Gräsl*	Lot 187	†
331	*Erasmus Habermel Pragae 1576*	Lot 232	Time Museum, inv. no: 1439 [2]
368	unsigned Hartman-type astrolabe	Lot 261	†
385	*J. N. Hölderich*	Lot 67	†
409	*J. G. Kleininger fecit*	part Lot 93	Harvard, inv. no: CHSI WP 7421
418	*Johann Krabble 1582*	Lot 262	Fort Caroline
420	*Krigener Marienburg*	Lot 189	†

Zinner Page no.	*Signature on instrument in Zinner*	*Findlay sale lot Number*	*Current Location*
434	*KM*	not identified	†
440	*Johann Martin*	Lot 77 or 188 or 190	†
447	*Lienhart Miller 1605*	Lot 203	†
	Lienhart Miller 1618	n/a	RSM, inv. no: NMS T.1924.10.[3]
449	*Leonhart Miller, 1640* (mit Lilie als Meistermarke)	Lot 204	†
450	(Lilie als Meistermarke), no date	not identified	†
	Lienhart Miller	not identified	†
	Leonhart Miller	? Lot 205	†
451	Nikolaus Miller, 1651	Lot 201 or 202	†
455	*LTM*	Lot 71	†
474	*Marcus Purman Monachi Faciebat Elevatio Poli 48 Grad Anno 1599*	Lot 124	MHSO, inv. no: 61-148
488	*Paulus Reinman Norenbergae faciebat [1603]*	Lot 206	†
489	Paulus Reinman Norimbergae faciebat	Lot 207	†
	[twice]	Lot 208	NMI
494	*J. C. Rosenthal* [1717]	Lot 79	†
497	*LS [1518]*	Lot 210	NHAH, inv. no: 320148
499	*Michael Scheffelt [1705]*	Lot 246	†

Zinner Page no.	*Signature on instrument in Zinner*	*Findlay sale lot Number*	*Current Location*
512	*Christophorus Schissler 1571 Augusta faciebat*	Lot 233	NMAH, inv. no: 320151
514	[Christophorus Schissler 1580] SCALA HOROLOGII SOLIS	Lot 234	† [4]
517	Ohne Jahr und Hersteller, wohl aus Schißlers Werkstatt. [3 examples]	not identified	†
		not identified	†
		not identified	†
522	See page 525	n/a	n/a [5]
525	*15 VS 84*	Lot 98	'Berlin Slg. Thorban' [6]
530	*Johann Schrettegger in Augsburg*	Lot 75 or 183	†
560	*Hans Ducher*	Lot 81	†
562	*Hans Tucher*	n/a	ROE [7]
563	[Hans Tucher] Ohne Jahr und Hersteller. [2 examples]	Lot 82	†
		Lot 83	RSM, inv. no. NMS.T.1961.70 [8]
567	[Thomas Tucher] Ohne Jahr und Hersteller. [3 examples]	Lot 82	†
		Lot 84	†
		not identified	†
572	*And Vogl*	Lot 184	† [9]
592	*Johann Willebrand in Augspur 48*	n/a	RSM, inv. no: NMS T.1924.7 [10]
597	*George Zorn, 1627*	Lot 132	'Berlin Slg. Thorban' [11]

Zinner Page no.	*Signature on instrument in Zinner*	*Findlay sale lot Number*	*Current Location*
679	Cor, Drebel, 1597 [addition p.298]	Lot 95	†
681	*T. E. Nholderich in Augsperg* [correction p.385]	Lot 67	†
683	M – Mettelin [addition p.445].	Lot 109	†
684	*Petrus ab aggere (Van Rampaart), 1558* [addition p. 463]	Lot 121	Adler, inv. no: DPW-40 [12]
686	*Henricus Sneewins* [addition p. 537]	Lot 37	Leiden, inv. no: 3091 [13]
687	*Hans Tucher, 1612* [correction p. 560]	Lot 83	RSM, inv. no: NMS T.1961.70
	F I V 1766 [addition p. 568]	Lot 115	Harvard, inv. no: CHSI WP 7345
	Caspar Vopel astronomus profes. Faciebat Colonie 1541 [addition p. 579]	Lot 106	NMAH, inv. no: 319950

Notes to table:

1. Presented to the Royal Scottish Museum by Sir John Findlay, Bt., in 1924.
2. Subsequently sold at Christie's, 30 March 1977, Lot 105; offered for sale by Librairie Alain Brieux, November 1977. Sold by the Adler at Sotheby's Olympia, 30 October 2002, Lot 7; catalogue with six colour plates.
3. Presented to the Royal Scottish Museum by Sir John Findlay, Bt., in 1924.
4. The Findlay Collection instrument had been acquired from the 1923 Rosenheim sale, and previously from the 1911 Roussel sale; recorded twice by Zinner.
5. There was no instrument with Alexius Schniep's signature or initials in the Findlay Collection, nor in the collections of either the Royal Scottish Museum or the National Museum of

Antiquities of Scotland.
6. Purchased by a Mr Thorban at the Findlay sale.
7. Acquired with the Crawford Collection in 1890s.
8. Stamped with the crowned serpent flanked by the initials H.T. and dated 1612.
9. Lots 68 and 185 were also signed 'And. Vogler', but are not noted by Zinner: the present whereabouts of Lot 68 is unknown; Lot 185 was subsequently acquired by the National Museum of American History, Smithsonian Institution, Washington D.C., inventory number: 320150.
10. Presented to the Royal Scottish Museum by Sir John Findlay, Bt. However, Findlay sale Lot 73, also signed by Johannes Willebrand, is not mentioned by Zinner: whereabouts unknown.
11. Purchased by a Mr Thorban at the Findlay sale, and subsequently acquired by the National Museum of American History, Smithsonian Institution, Washington D.C., inventory number: 319948.
12. Published in Enrico Morpurgo, *Orologi Preziosi dal XVI al XIX secolo* (Milan, n.d.), 48-49, as part of the Zoccai Collection. Sold at Sotheby's 13 March 1967, Lot 45. Now in the Adler Planetarium.
13. Sold again through Sotheby's, with Findlay Collection provenance, 31 March 1978, Lot 17 and acquired by the Museum Boerhaave, Leiden.

Fig. 1 Astronomical compendium in gilt-metal, signed *Pragae fecit Erasmus Habermel 1596*, outer faces, with signature.

Fig. 2 Astronomical compendium in gilt-metal, signed *Pragae fecit Erasmus Habermel 1596*, inner faces.

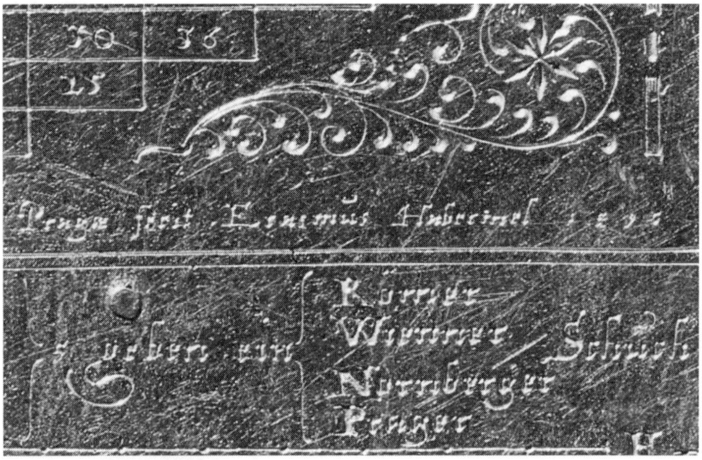

Fig. 3 Detail of signature.

NAME INDEX